METHODS IN MOLECULAR BIOLOGY

Series Editor
John M. Walker
School of Life and Medical Sciences
University of Hertfordshire
Hatfield, Hertfordshire, AL10 9AB, UK

For further volumes:
http://www.springer.com/series/7651

Yeast Functional Genomics

Methods and Protocols

Edited by

Frédéric Devaux

Laboratoire de biologie computationnelle et quantitative, Sorbonne Universités, Paris, France

 Humana Press

Editor
Frédéric Devaux
Laboratoire de biologie computationnelle et quantitative
Sorbonne Universités
Paris, France

ISSN 1064-3745   ISSN 1940-6029   (electronic)
Methods in Molecular Biology
ISBN 978-1-4939-3078-4   ISBN 978-1-4939-3079-1   (eBook)
DOI 10.1007/978-1-4939-3079-1

Library of Congress Control Number: 2015951231

Springer New York Heidelberg Dordrecht London

Printed on acid-free paper

Humana Press is a brand of Springer
Springer Science+Business Media LLC New York is part of Springer Science+Business Media (www.springer.com)

Preface

This volume of the "Methods in Molecular Biology" series aims at reflecting the state of the art of yeast functional genomics. Since the publication of its genome sequence in 1996, yeast functional genomics has been at the forefront of technological advances and never stopped evolving. Ten years ago, 90 % of the publications in this field were made of microarray-based transcriptome and chromatin immunoprecipitation analyses, and the reader will find in this volume the most recent protocols for these "classics" which are still widely used and up to date. Since then, yeast functional genomics have diversified in many ways.

First, the emergence of high-throughput sequencing technologies considerably enlarged our capacity to investigate yeast transcriptomes and genomes. Hence most of the chapters of this volume present protocols based on new generation sequencing technologies.

Second, all aspects of gene expression regulation, from nuclear architecture to translational rates and metabolite steady states, can now be studied at a genome-wide scale. This volume provides a panel of protocols for the study of DNA-DNA contact maps, replication profiles, transcription rates, RNA secondary structures, protein-RNA interactions, ribosome profiling, and quantitative proteomes and metabolomes.

Third, the availability of genome sequences for tens of yeast species and hundreds of strains in some species allowed for yeast comparative functional genomics and yeast populations genomics and opened the way to a common use of the natural or laboratory-generated genetic polymorphism to identify functional relationships between genes and gene-phenotype interactions in a powerful and comprehensive way. This volume includes protocols for yeast comparative transcriptomics, yeast high-throughput genetic screens, yeast QTL mapping, and yeast experimental evolution. Moreover, several protocols presented here were optimized for other species than *S. cerevisiae*.

Finally, the accumulation of these genome-wide data of various natures pushed forward the development of bioinformatics tools and methods to make available, represent, and analyze the properties of large yeast cellular networks. Most of the protocols presented in this volume emphasized both "wet lab" and in silico analyses aspects. Moreover, two chapters were specifically dedicated to the integration of high-throughput data in evolutionary models and to data mining of global regulatory networks, respectively.

Obviously, the field is so diverse that this book could not be comprehensive. For instance, just the different methods nowadays available for yeast quantitative proteomics would have filled the whole volume. Our goal was rather to make this issue of Methods in Molecular Biology as representative of its time and as useful to a broad audience as possible. Did we achieve this goal? I believe the answer is yes but actually, this is to the reader to tell. So…

Have a nice reading!

Paris, France   *Frédéric Devaux*

Contents

Contributors

NICOLAS AGIER • *Biologie Computationnelle et Quantitative, Sorbonne Universités, UPMC Univ. Paris 06, CNRS, UMR 7238, Paris, France; Biologie Computationnelle et Quantitative, CNRS-Université Pierre et Marie Curie, UMR 7238, Paris, France*

LISBETH-CAROLINA AGUILAR • *Institut de recherches cliniques de Montréal, Montréal, QC, Canada*

GWENAEL BADIS • *Institut Pasteur, Génétique des Interactions Macromoléculaires, Centre National de la Recherche Scientifique, Paris, France*

AGNÈS BAUDIN-BAILLIEU • *Institute for Integrative Biology of the Cell (I2BC), UMR 9198 CEA, CNRS, Université Paris Sud, Orsay, France*

CORINNE BLUGEON • *Plateforme Génomique, Ecole Normale Supérieure, Institut de Biologie de l'ENS, IBENS, Paris, France; Inserm, U1024, Paris, France; CNRS, UMR 8197, Paris, France*

GERALDINE BUTLER • *School of Biomolecular and Biomedical Science, Conway Institute, University College Dublin, Belfield, Dublin, Ireland*

JEAN-MICHEL CAMADRO • *Mass Spectrometry Laboratory, Institut Jacques Monod, UMR 7592, CNRS—Univ Paris Diderot, Sorbonne Paris Cité, Paris, France; Mitochondria, Metals and Oxidative Stress group, Institut Jacques Monod, UMR 7592, CNRS—Univ Paris Diderot, Sorbonne Paris Cité, Paris, France*

HOWARD Y. CHANG • *Howard Hughes Medical Institute and Program in Epithelial Biology, Stanford University School of Medicine, Stanford, CA, USA*

AXEL COURNAC • *Institut Pasteur, Department Genomes and Genetics, Groupe Régulation Spatiale des Génomes, Paris, France; CNRS, UMR 3525, Paris, France*

HUGO DEVILLERS • *INRA, UMR 1319 Micalis, Jouy-en-Josas, France; AgroParisTech, UMR Micalis, Jouy-en-Josas, France*

MAITREYA J. DUNHAM • *Department of Genome Sciences, University of Washington, Seattle, WA, USA*

SCOTT G. FILLER • *Division of Infectious Diseases, Los Angeles Biomedical Research Institute at Harbor-UCLA Medical Center, Torrance, CA, USA*

GILLES FISCHER • *Biologie Computationnelle et Quantitative, Sorbonne Universités, UPMC Univ Paris 06, CNRS, UMR 7238, Paris, France; Biologie Computationnelle et Quantitative, CNRS-Université Pierre et Marie Curie, UMR 7238, Paris, France*

MALLORY A. FREEBERG • *Life Sciences Institute, University of Michigan, Ann Arbor, MI, USA; Department of Computational Medicine and Bioinformatics, University of Michigan, Ann Arbor, MI, USA*

CAMILLE GARCIA • *Mass Spectrometry Laboratory, Institut Jacques Monod, UMR 7592, CNRS—Univ Paris Diderot, Sorbonne Paris Cité, Paris Cedex 13, France*

STEPHEN HAMMEL • *School of Biomolecular and Biomedical Science, Conway Institute, University College Dublin, Belfield, Dublin, Ireland*

TING HAN • *Department of Biochemistry, UT Southwestern Medical Center, Dallas, TX, USA*

ISABELLE HATIN • *Institute for Integrative Biology of the Cell (I2BC), UMR 9198 CEA, CNRS, Université Paris Sud, Orsay, France*

AARON D. HERNDAY • *Department of Molecular and Cell Biology, School of Natural Sciences, University of California, Merced, Merced, CA, USA*

JING HOU • *Department of Genetics, Genomics and Microbiology, CNRS, UMR7156, Université de Strasbourg, Strasbourg, France*

ANTONIO JORDÁN-PLA • *Departamento de Bioquímica y Biología Molecular and ERI Biotecmed, Facultad de Biológicas, Universitat de València, València, Spain; Department of Molecular Biosciences, The Wenner-Gren Institute, Stockholm University, Stockholm, Sweden*

JOHN K. KIM • *Life Sciences Institute, University of Michigan, Ann Arbor, MI, USA; Department of Biology, Johns Hopkins University, Baltimore, MD, USA*

CHRISTIAN KLOSE • *Lipotype GmbH, Dresden, Germany*

SARA A. KNAACK • *Wisconsin Institute for Discovery, University of Wisconsin at Madison, Madison, WI, USA*

PISIWAT KONGSOMBOONVECH • *Department of Molecular and Cell Biology, School of Natural Sciences, University of California, Merced, Merced, CA, USA*

ROMAIN KOSZUL • *Institut Pasteur, Department Genomes and Genetics, Groupe Régulation Spatiale des Génomes, Paris, France; CNRS, UMR 3525, Paris, France*

RACHEL LEGENDRE • *Institute for Integrative Biology of the Cell (I2BC), UMR 9198 CEA, CNRS, Université Paris Sud, Orsay, France*

THIBAUT LÉGER • *Mass Spectrometry Laboratory, Institut Jacques Monod, UMR7592, CNRS—Univ Paris Diderot, Sorbonne Paris Cité, Paris Cedex 13, France*

GAËLLE LELANDAIS • *Institut Jacques Monod, CNRS UMR 7592, University of Paris Diderot, Paris, France*

MATTHEW B. LOHSE • *Department of Microbiology and Immunology, University of California, San Francisco, San Francisco, CA, USA*

MARIA MADRIGAL • *Department of Molecular and Cell Biology, School of Natural Sciences, University of California, Merced, Merced, CA, USA*

CHRISTOPHE MALABAT • *Génétique des Interactions Macromoléculaires (UMR3525-CNRS), Institut Pasteur, Paris, France*

MARTIAL MARBOUTY • *Institut Pasteur, Department Genomes and Genetics, Groupe Régulation Spatiale des Génomes, Paris, France; CNRS, UMR 3525, Paris, France*

JAWAD MERHEJ • *Laboratoire de Biologie Computationnelle et Quantitative, Sorbonne Universités, UPMC University of Paris 06, UMR 7238, Paris, France; Laboratoire de Biologie Computationnelle et Quantitative, CNRS, UMR 7238, Paris, France*

ANA MIGUEL • *Departamento de Bioquímica y Biología Molecular and ERI Biotecmed, Facultad de Biológicas, Universitat de València, València, Spain*

AARON P. MITCHELL • *Department of Biological Sciences, Carnegie Mellon University, Pittsburgh, PA, USA*

PEDRO T. MONTEIRO • *INESC-ID, Instituto Superior Técnico, Universidade de Lisboa, Lisbon, Portugal*

NICOLAS MORIN • *INRA, UMR1319 Micalis, Jouy-en-Josas, France; AgroParisTech, UMR Micalis, Jouy-en-Josas, France*

JULIEN MOZZICONACCI • *LPTMC, Université Pierre et Marie Curie, Paris, France*

OLIVIER NAMY • *Institute for Integrative Biology of the Cell (I2BC), UMR 9198 CEA, CNRS, Université Paris Sud, Orsay, France*

CÉCILE NEUVÉGLISE • *INRA, UMR1319 Micalis, Jouy-en-Josas, France; AgroParisTech, UMR Micalis, Jouy-en-Josas, France*

CLARISSA J. NOBILE • *Department of Molecular and Cell Biology, School of Natural Sciences, University of California, Merced, Merced, CA, USA*

Chapter 1

Using RNA-seq for Analysis of Differential Gene Expression in Fungal Species

Can Wang, Markus S. Schröder, Stephen Hammel, and Geraldine Butler

Abstract

The ability to extract, identify and annotate large amounts of biological data is a key feature of the "omics" era, and has led to an explosion in the amount of data available. One pivotal advance is the use of Next-Generation Sequencing (NGS) techniques such as RNA-sequencing (RNA-seq). RNA-seq uses data from millions of small mRNA transcripts or "reads" which are aligned to a reference genome. Comparative transcriptomics analyses using RNA-seq can provide the researcher with a comprehensive view of the cells' response to a given environment or stimulus.

Here, we describe the NGS techniques (based on Illumina technology) that are routinely used for comparative transcriptome analysis of fungal species. We describe the entire process from isolation of RNA to computational identification of differentially expressed genes. We provide instructions to allow the beginner to implement packages in R such as Bioconductor. The methods described are not limited to yeast, and can also be applied to other eukaryotic organisms.

Key words *Candida*, Next-generation sequencing, Illumina, Bioconductor

1 Introduction

Transcriptome analysis using RNA-seq quantifies transcription levels across the entire genome [1]. Transcript numbers are measured by sequencing; the number of reads obtained from a specific gene in a test condition compared to a control is used to measure changes in gene expression [2, 3]. Studying the transcriptome in different yeast species has helped elucidate the mechanisms behind key cellular processes and pathways [4–7]. As Next-Generation Sequencing (NGS) technologies drop in price, RNA-seq has become widely used as a method for analyzing gene expression under an array of conditions, and holds many advantages over other similar analytical techniques such as microarrays [2, 8]. Although most of the techniques required can be carried out "in-house," there are many private companies that now provide NGS services. They will sequence user-provided libraries, but will also

Frédéric Devaux (ed.), *Yeast Functional Genomics: Methods and Protocols*, Methods in Molecular Biology, vol. 1361, DOI 10.1007/978-1-4939-3079-1_1, © Springer Science+Business Media New York 2016

isolate poly(A) RNA from total RNA preparations and construct libraries, at additional cost. This makes RNA-seq accessible to almost any laboratory.

In this chapter, we describe how to carry out RNA-seq analysis from RNA isolation to computational analysis. Labs without access to next-generation sequencing technologies can use commercial companies for the sequencing steps, and move straight to the computational analysis section, where the interpretation of results is discussed. We describe a series of tools implemented in the R statistical language [9]. A wide variety of bioinformatics tasks and collections of R packages, such as Bioconductor [10] or CRAN [11] make it possible to utilize R for almost any task associated with analyzing and visualizing sequencing data. We have provided a set of instructions that make it possible for even the beginner to implement tools such as DeSeq2 [12] on a laptop or personal computer, to analyze changes in gene expression.

2   Materials

2.1 Specialized Equipment

1. NanoDrop spectrophotometer.
2. Agilent 2100 Bioanalyzer.
3. Qubit Fluorometer.
4. Dark Reader-Blue Light Transilluminator.
5. Bead beater.
6. Next-Generation DNA Sequencer (e.g., Illumina platforms Genome Analyzer IIx, HiSeq 2500, or MiSeq), or commercial sequencing services.
7. PC or laptop with Linux or Mac OS X as operating system and Internet access.

2.2 Kits

1. Yeast RNA Extraction Kit (e.g., Ribopure, Ambion).
2. RNA 6000 Nano Kit (Agilent).
3. High-sensitivity DNA Kit (Agilent).
4. Zinc RNA Fragmentation Kit.
5. Gel Excision Tips (e.g., GeneCatcher).
6. PCR Purification Kits.
7. Qubit dsDNA High Sensitivity Assay Kit (or equivalent).
8. Quick Ligation Kit.

2.3 Buffers and Reagents

1. 1× Binding Buffer: 20 mM Tris–HCl pH 7.5, 1.0 M LiCl, 2 mM EDTA.
2. 1× Washing Buffer: 10 mM Tris–HCl pH 7.5: 0.15 M LiCl: 1 mM EDTA.

3. Tris–NaCl Buffer (50 μl 1 M Tris–HCl pH 7.5, 10 μl 5 M NaCl, and 940 μl Nuclease-free water).

4. 10 mM Tris–HCl, pH 8.5.

5. RNAlater.

6. Dynabeads Oligo (dT)$_{25}$ (e.g., Dynal from Ambion).

7. dNTP mix (10 mM dATP, dTTP, dCTP, and dGTP).

8. UTP mix (10 mM dATP, dCTP, dGTP, 20 mM dUTP).

9. Reverse transcriptase with buffers (e.g., Superscript III).

10. RNaseOUT.

11. DNA Polymerase I.

12. Klenow DNA Polymerase I.

13. Klenow Fragment (3′ → 5′ exo-).

14. T4 DNA polymerase.

15. T4 DNA Ligase.

16. T4 polynucleotide kinase.

17. High Fidelity PCR polymerase.

18. Ribonuclease H.

19. Uracil DNA glycosylase.

20. G-50 column (e.g., Illustra Microspin).

3   Methods

3.1   Preparation of RNA

3.1.1   Cell Growth

1. Inoculate overnight cultures in 5 ml growth medium incubated at a relevant temperature (often 30 °C, shaking at 200 rpm).

2. Sub-culture to an A_{600nm} of 0.2 in 50 ml growth medium and grow to mid-log phase (incubation time depends on growth rates of different species being studied). At this point, an additional treatment can be used, for example treatment with a drug, or a change in temperature or oxygen concentration.

3. Following treatment, retrieve cells from 25 ml culture either by centrifugation for 5 min at 4 °C at 3160×g, or to avoid stress [13] by collection on a filter (0.45 μm nitrocellulose membrane filter) using a vacuum source.

4. Resuspend the cell pellet in 100 μl RNAlater stabilization solution and store at −80 °C until required. The RNAlater solution inactivates any RNases and prevents any changes in expression of the RNA.

3.1.2   RNA Isolation, Yield, and Quality

1. Treat all lab surfaces and pipettes with 70 % ethanol and an RNase decontamination solution (e.g., RNaseZap) to remove any unwanted RNases.

2. Thaw cells on ice.

3. Extract RNA using a commercial kit, following the manufacturer's instructions. We use Ribopure Yeast RNA Extraction Kit from Ambion (*see* **Note 1**).

4. Determine RNA concentrations below 50 ng/µl by measuring absorbance with a Qubit fluorometer. Use a NanoDrop to identify contaminants [14]. A reading at 260 nm is used to determine concentration (A_{260} of 1 = 40 µg/ml). Proteins absorb at 280 nm. The $A_{260/280}$ ratio therefore provides a measurement of the purity of the RNA; the ratio should lie between 1.8 and 2.2. Ethylenediaminetetraacetic acid (EDTA), carbohydrates, and phenol all have absorbance near 230 nm. The $A_{260/230}$ ratio is therefore used as a secondary measure of nucleic acid purity. Expected $A_{260/230}$ values lie in the range of 2.0–2.2. If the ratio is appreciably lower than this, contaminants are probably present and the sample should not be used.

5. Measure RNA quality with a fluorometric based analytical systems, e.g., Bioanalyzer from Agilent Technologies, following the manufacturer's instructions. Analysis on a Bioanalyzer generates a graphical visualization in the form of an electropherogram of ribosomal peaks (28S and 18S), peak ratio, RNA concentration, a calculated RIN (RNA Integrity Number) value, and a gel-like image of the RNA sample. The RIN value is a measurement of the overall integrity of a given RNA sample that is not affected by sample concentration but by the overall RNA content and background degradation. RIN values >6 are considered to be of acceptable quality. The quantitative range for the RNA 6000 Nano Kit is 5–500 ng/µl.

3.2 Library Generation

The library protocol described here was developed for sequencing on an Illumina Genome Analyzer IIx. Most analysis is now carried out with the more recent HiSeq and MiSeq systems from the same company. The protocol described here may be adapted for use with the newer platforms (HiSeq/MiSeq/NextSeq) with some minor updates (*see* Subheading Adapter Synthesis). The steps required for library generation are shown in Fig. 1. This protocol generates strand-specific information by incorporating dUTP during the synthesis of the second strand cDNA synthesis [15–17]. This is subsequently removed by digestion with uracil DNA glycosylase (UDG). There are several variations of the dUTP method, including combining with Illumina TruSeq kits [15, 18]. Other methods for generating strand-specific data are described by Levin et al. [16].

3.2.1 Purification of Poly(A) RNA (Fig. 1a)

1. Dilute 10 µg total RNA in 50 µl using nuclease-free water.

2. Incubate at 65 °C for 5 min to disrupt RNA secondary structures, and then place on ice.

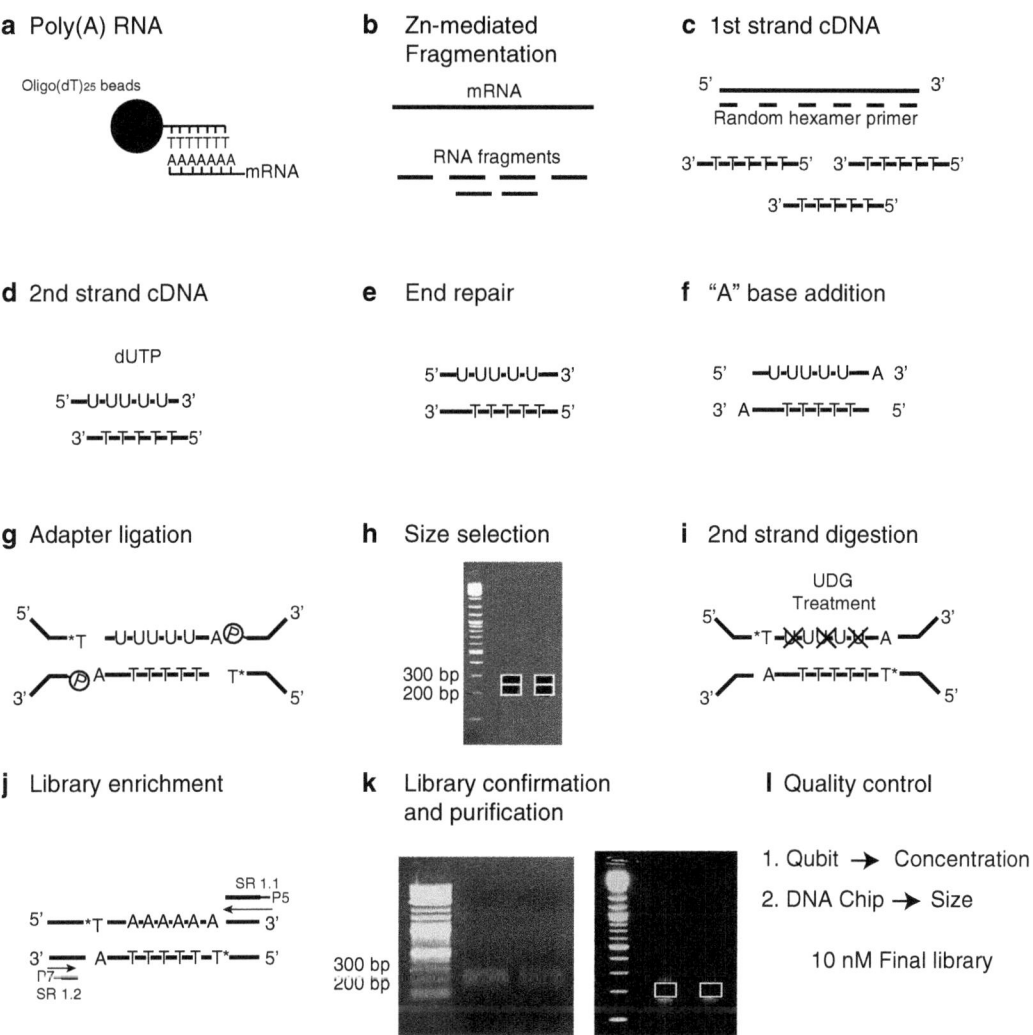

Fig. 1 The workflow for constructing strand-specific libraries from total RNA. Each step is described in detail in the text. (**a**) Poly(A) RNA is selected by binding to oligo(dT)$_{25}$ Dynabeads. (**b**) mRNA is fragmented using Zn-mediated fragmentation. (**c**) First strand cDNA is synthesized using random hexamer primers. (**d**) Second strand cDNA is synthesized incorporating U instead of T (**e**) Ends of the cDNA fragments are repaired. (**f**) "A" bases are added to the 3′ ends of the cDNA fragments (**g**) Y-shaped iAdapters anneal to the cDNA fragments by overlapping "T" and "A" bases (**h**) cDNAs ranging from size 200 to 250 bp and 250 to 300 bp are isolated from the gel (shown with *arrows*). (**i**) The *bottom strand* is copied by priming from SR1.2, and the library is amplified using primers SR1.2 and SR1.1 (which adds the P5 sequence, *see* Fig. 2 for alternatives). (**k**) Amplification of the library is confirmed by electrophoresis on a 2.5 % agarose gel, and the library (ranging from 200 to 250 bp) is purified from a gel. (**l**) Library concentration and size are estimated and are diluted to 10 nM for sequencing

3. Wash 100 μl Dynabeads Oligo (dT)$_{25}$ with 100 μl 1× Binding Buffer twice using a magnetic rack and resuspend the beads in 50 μl 1× Binding Buffer.

4. Mix 50 μl heated RNA from **step 2** and 50 μl washed beads from **step 3** and rotate the mixture for 5 min at room temperature. Recover the beads using a magnetic rack and wash twice with 100 μl 1× Washing Buffer.

5. Elute the mRNA in 20 μl Tris–HCl (10 mM, pH 7.5) by heating at 80 °C for exactly 2 min.

6. Wash the beads twice with 1× Washing Buffer.

7. Add 80 μl 1× Binding Buffer to the beads and the 20 μl mRNA from **step 5**, and repeat the poly(A) selection.

8. Elute the poly(A) RNA in 10 μl 10 mM Tris–HCl (pH 7.5) by heating at 80 °C for exactly 2 min.

9. Recover the RNA from the beads immediately using a magnetic stand, and transfer 9 μl to thin wall PCR tubes. Store the mRNA at –80 °C.

3.2.2 Zinc-Mediated Fragmentation of mRNA (Fig. 1b)

1. Add 1 μl 10× Fragmentation Buffer (from kit) to 9 μl purified mRNA (poly(A) RNA) in a PCR tube.

2. Incubate the mixture at 70 °C in a thermocycler for 5 min.

3. Add 1 μl Stop Buffer (from kit) and incubate briefly on ice.

4. Add 1 μl 3 M sodium acetate (pH 5.2), 2 μl 5 μg/μl glycogen, and 30 μl 100 % ethanol and precipitate the mRNA at –80 °C for ≥30 min followed by centrifugation at 17,000×g at ≤4 °C for 25 min.

5. Remove the supernatant carefully and wash the pellet with 700 μl 80 % ethanol.

6. Air-dry the pellet and resuspend it in 10.5 μl Nuclease-free water.

3.2.3 First Strand cDNA Synthesis (Fig. 1c)

1. Add 1 μl of random hexamer primer (3 μg/μl, Invitrogen) to 10.5 μl fragmented mRNA.

2. Incubate the mixture at 65 °C for 5 min and then place on ice.

3. Add 4 μl 5× First Strand Buffer (supplied with reverse transcriptase), 2 μl DTT (100 mM), 1 μl 10 mM dNTP mix, and 0.5 μl RNase OUT (40 units/μl), incubate at 25 °C for 2 min and then add 1 μl of Reverse Transcriptase (Superscript III is recommended) to each sample.

4. Incubate the mixture at 25 °C for 10 min, 42 °C for 50 min and then 70 °C for 15 min.

5. Store the first strand cDNA on ice.

6. Remove dNTPs and hexamers by centrifugation through a G-50 spin column. Centrifuge the G-50 column at $2000 \times g$ for 1 min. Add the first strand cDNA sample carefully to the top and center of the resin and collect by centrifuging for 2 min at $2000 \times g$.

7. Immediately carry out second strand cDNA synthesis.

3.2.4   Second Strand cDNA Synthesis with dUTP (Fig. 1d)

1. Incubate all reagents on ice for 5 min prior to use.

2. Add 1.3 µl 5× First Strand Buffer, 20 µl 5× Second Strand Buffer (supplied with reverse transcriptase), 3 µl dUTP mix (10 mM dATP, dCTP, dGTP, 20 mM dUTP), 1 µl DTT (100 mM), 5 µl *E. coli* DNA Polymerase I (10 units/µl), and 1 µl Ribonuclease H (2 units/µl) to the first strand samples.

3. Add Nuclease-free water to bring the volume to 100 µl.

4. Incubate at 16 °C for 2.5 h, and purify the cDNA in 30 µl 10 mM Tris–HCl, pH 8.5 or equivalent solution supplied with PCR purification kit as per manufacturer's guidelines, and store at −80 °C.

3.2.5   End Repair (Fig. 1e) and Addition of a Single "A" Base (Fig. 1f)

Treating the cDNA fragments with a combination of T4 DNA polymerase and *E. coli* DNA polymerase I Klenow fragments removes 3′ overhangs via the 3′–5′ exonuclease activity, while the polymerase fills in any 5′ overhangs. Both these steps are necessary to facilitate ligation of sequencing adaptors. A single adenosine base is added to the 3′-end of the cDNA fragments to facilitate ligation to the sequencing adapter.

1. Add 45 µl Nuclease-free water, 10 µl T4 DNA Ligase buffer with 10 mM ATP, 4 µl dNTP mix (10 mM), 5 µl T4 DNA polymerase (3 units/µl), 1 µl Klenow DNA Polymerase (5 units/µl), and 5 µl T4 Polynucleotide Kinase (10 units/µl) to 30 µl cDNA. Incubate at 20 °C for 30 min, and purify by elution with a PCR purification kit as per manufacturer's guidelines. This is a safe stopping point, and samples may be stored at −80 °C.

2. To add an A base to the 3′ end, add 5 µl Klenow buffer, 10 µl dATP (1 mM), and 3 µl Klenow Exo Fragment (5 units/µl , 3′ → 5′ exonuclease) to end repaired cDNA, in a total volume of 50 µl.

3. Incubate the reaction at 37 °C for 30 min and purify by elution using a PCR purification kit as per manufacturer's guidelines. Samples with an A overhang should not be stored for long periods as they are unstable.

3.2.6   Adapter Synthesis

Several libraries can be pooled together and sequenced on the same run. This is achieved by ligating specific adapters containing different barcode (or index) sequences, to the DNA fragments.

The barcodes are used to separate library specific data after sequencing [19]. We originally used home-made single read (SR) Y-shaped adapters with short six nucleotide barcodes, designed by Dr. Amanada Lohan UCD, and based on the 2008 Illumina customer letter and Craig et al. [19, 20] (Fig. 2a).

These adapters are made from two single stranded oligonucleotides (Oligo-1 and Oligo-2) with both complementary regions and noncomplementary regions that when annealed create a Y-shaped adapter bound together at the hinge (complementary) region (Fig. 2a). The top oligonucleotide (Oligo-1) contains a T overhang with a phosphorothioate linkage required for stabilization and resistance to nuclease digestion, that is designed to ligate to the A overhang added to the insert DNA during library generation. Oligo-2 is phosphorylated at the 5' end during synthesis, and is complementary to the P7 region, which anneals to sequences on the flowcell. An equivalent P5 region is added at the opposite end during library amplification (**step 10**, Fig 1j). The 6 nucleotides barcode (index) is added to both Oligo-1 and Oligo-2. Table 1 shows six barcode indexes, allowing six libraries to be combined in a single lane (multiplexing). The choice of barcode depends on the number of libraries combined (Table 2). It is now possible to design and synthesize long adapter sequences, using updated recommendations from Illumina [21] that remove the necessity to add the P5 region during library amplification (Fig. 2b, [21]) (*see* **Note 2**).

1. Synthesize the oligonucleotides commercially, and using HPLC purification. To construct the Y-shaped adapters, resuspend lyophilized oligonucleotides in 10 mM Tris–HCl at 100 pmol/μl and anneal them together to form a forked adapter (iAdapter), making sure that each oligo contains the same barcode/index sequence.

2. Add 20 μl of the relevant indexed Oligo-1, 20 μl indexed Oligo-2 and 10 μl Tris–NaCl Buffer in 0.2 ml PCR tube.

Fig. 2 (continued) of the barcode index (using index sequence of iSR-6 (Table 1) as an example). The inserted cDNA is shown in *italics*. The P7 sequence is highlighted in *dark grey*. Before library amplification the first (*top*) strand (contain U residues) is degraded, and the *bottom strand* is copied using SR1.2 as a primer. Subsequent amplification with primers SR1.1 and SR1.2 adds the P5 sequence (*light grey*). (**b**) Library generation using updated Illumina recommendations [21]. Two oligonucleotides (the universal adapter and indexed adapter) are synthesized for each library, and a Y-shaped adapter is generated as in 2A. The cDNA is ligated at the *arrow*. The universal adapter sequence contains the P5 sequence and the indexed adapter contains the P7 sequence. The P7 sequence also contains an index/barcode (In). The libraries are amplified with primers 1 and 2. The number of multiplexed samples can be increased by also including indexes in the universal primer (dual indexing, not shown). For single reads a sequencing primer for the P7 end is used, for paired-end reads, primers from both P7 and P5 ends are used. Advice on designing and synthesizing longer adapters is available from refs. [15, 45]. The *asterisk* indicates a phosphorothioate linkage, and the *P* indicates a phosphorylated nucleotide

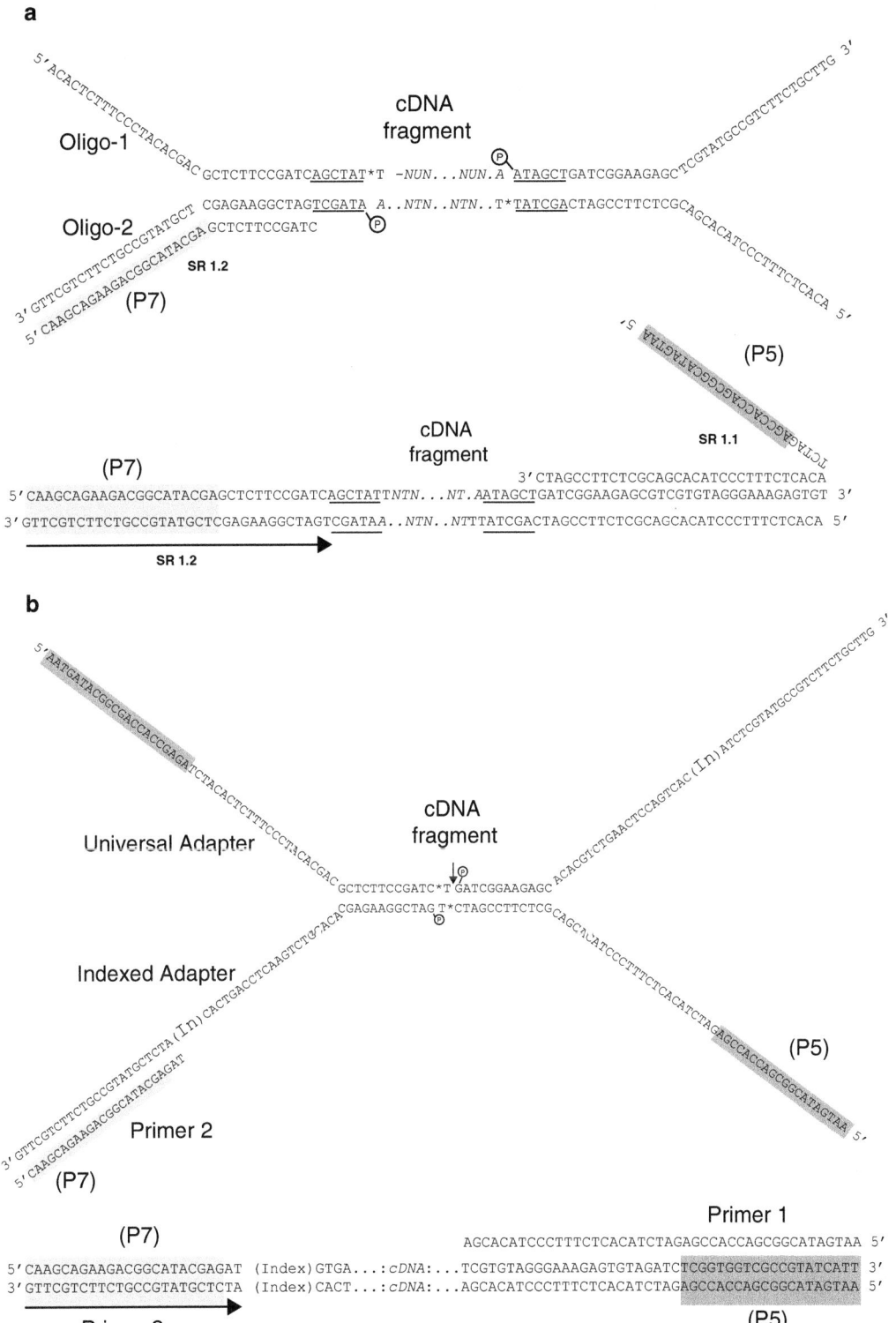

Fig. 2 Adapters used for library generation for Illumina sequencing. (**a**) iAdapters generated using the original legacy SR (single read) adapters described in the Illumina customers letter prior to 2009. Two oligonucleotides, Oligo-1 and Oligo-2 are synthesized for each adapter. The *underlined* six nucleotide sequences show the location

Table 1
Barcode sequences

Adapter ID[a]	Index/barcode[b]
iSR-6	AGCTAT
iSR-10	CGATCT
iSR-20	GATCGT
iSR-11	GCTAGT
iSR-13	TAGCTT
iSR-16	TCGATT

[a]SR indexed adapters are designed by adding a barcode of six nucleotides, which have to maintain color balance for each base. A/C bases are identified by the red laser and G/T bases by the green laser on a Genome Analyzer IIx. We show six adapters designed according to Illumina indexing guidelines [21]. Many more are possible (for example, *see* ref. [45])

Table 2
Pooling strategies

Number of libraries	Best combinations
2 in the lane	(iSR 6, iSR 20) or (iSR 10, iSR13)
3 in the lane	(iSR 10, iSR 11, iSR13) or (iSR 6, iSR16, iSR 20)
5 in the lane	(iSR 6, iSR 10, iSR 13, iSR 16, iSR 20)
6 in the lane	(iSR 6, iSR 10, iSR 11, iSR 13, iSR 16, iSR 20)

3. Incubate at 97 °C for 2 min, followed by a stepdown of −1 °C/min for 72 cycles, and finally at 25 °C for 5 min.

4. Store the 40 μM iAdapter master stock at −20 °C. For most RNA-seq libraries a 15 μM working stock is used. However when <1 μg starting RNA is available further dilution may be required.

3.2.7 Adapter Ligation (Fig. 1g)

1. Add 25 μl 2× Quick DNA ligase buffer, 1 μl iAdapter mix (15 μM), and 2 μl Quick T4 DNA ligase (NEB) to 22 μl end-repaired cDNA with an A overhang (from the end repair step).

2. Incubate at 20 °C for 15 min and purify by elution in 10 μl 10 mM Tris–HCl, pH 8.5 using a PCR purification kit.

3. Store at −80 °C.

3.2.8 Gel Purification (Fig. 1h)

1. Separate samples on a 2.5 % agarose gel by prepared using ultrapure TAE buffer and high-resolution agarose containing Ethidium Bromide at a final concentration of 1 μg/μl (*see* **Note 3**). Use a suitable DNA ladder (*see* **Note 4**).

2. Add 10 μl adapter-ligated cDNA to 6 μl gel loading dye (such as Orange G from Promega) and add to the same volume of DNA ladder.

3. Load the wells of the gel with this solution very slowly to prevent overflow and spilling.

4. Electrophorese at 80 V for 3 h until sufficient separation of the 100 and 200 bp bands of the DNA ladder has occurred.

5. Visualize the DNA on a Dark Reader Transilluminator, which operates at a wavelength that does not damage the sample, unlike normal UV transilluminators.

6. Excise regions corresponding to 200–250 bp and to 250–300 bp (for backup) using Gel Excision Tips (for example from GeneCatcher).

7. Elute the adaptor-ligated cDNA in 30 μl of 10 mM Tris–HCl, pH 8.5 by using a Gel Extraction kit. Store the samples at –80 °C.

3.2.9 Second Strand Digestion (Fig. 1i)

1. Aliquot 26 μl gel purified adapter-ligated material (200–250 bp) to sterile 200 μl PCR tubes.

2. Add 3 μl 10× uracil DNA glycosylase (UDG) buffer and 1 μl uracil DNA glycosylase (1 unit/μl).

3. Incubate in a thermal cycler at 37 °C for 20 min.

4. Terminate the reaction by heating at 94 °C for 10 min and 4 °C for 5 min. Store the digested DNA samples at –80 °C.

3.2.10 Amplification of Adapter Ligated DNA Templates (Fig. 1j)

The adaptor-ligated cDNA samples are amplified by PCR to ensure there is a sufficient quantity for sequencing.

1. Add 10 μl 5× buffer (provided with enzyme), 0.8 μl PCR primer SR 1.1 (Fig. 2a), 0.8 μl PCR primer SR 1.2 (Fig. 2a) or other suitable primers (Fig. 2b), 0.8 μl dNTP mix (25 mM), and 0.8 μl High Fidelity polymerase (e.g., cloned Phusion polymerase from NEB) to 20 μl digested DNA and bring the total volume to 50 μl with Nuclease-free water. If a different polymerase is used, ensure that it is not inhibited by dUTP.

2. Amplify at 98 °C for 30 s, followed by 12–14 cycles of 98 °C for 10 s, 65 °C for 30 s, 72 °C for 30 s, and then 72 °C for 5 min. In the method shown in Fig. 1j and Fig. 2a the P5 sequence, which is required to hybridize to the sequence on the flow cell, is added to the end of the adapter by amplification with oligonucleotide SR 1.1. The second oligonucleotide SR 1.2 contains the P7 sequence. Adapters shown in Fig. 2b can be amplified using the oligonucleotide primers shown.

3. Visualize library quality using 1.5 μl of the amplified DNA reaction on a 1 % agarose gel (Fig. 1k). The adapter/dimers should be 100–120 nucleotides long.

4. Purify the remaining 48.5 μl amplified cDNA by elution in 10 μl 10 mM Tris–HCl, pH 8.5 with a PCR purification kit if a product is visible.

5. Separate the DNA library samples on a high-resolution grade 2.5 % agarose gel. Visualize using a Dark Reader Transilluminator and excise fragments in the range of 200–250 bp, as described in the gel purification step (see **Note 3**). Store the samples at −80 °C.

3.2.11 Quality Control and Purification of Final Library (Fig. 1l)

1. Quantify the amplified library samples using a Qubit Fluorometer and the Qubit High-sensitivity dsDNA assay, as per manufacturer's guidelines. 10 nM library dilution is typical for starting point dilution for cluster generation.

2. Check the quality of cDNA library using a High-sensitivity DNA chip assay on a Bioanalyzer, as per manufacturer's instructions.

3.2.12 Pooling Strategy

1. Normalize cDNA libraries to 10 nM based on Qubit and DNA chip values. Ensure that the libraries contain a single peak of approximately 200 nucleotides, with little or no evidence of adapter dimers.

2. Dilute the library with 10 mM Tris–HCl, pH 8.5 with 0.1 % Tween 20 recommended for stability of library. Add 10 μl of each adapter-ligated library per lane. It is recommended that only certain barcodes (indexes) are combined together (Table 2). A minimum of 10 μl (one library) is required for clustering process.

3.2.13 Cluster Generation and Sequencing

We carry out cluster generation and sequencing using an in-house Genome Analyzer IIx platform (see **Note 5**). The multi-indexed library mix is loaded on the Illumina 8 channel flowcell. For the first step cluster generation, hundreds of millions of templates are hybridized to a lawn of oligo nucleotides immobilized on the flow cell surface. Immobilized DNA template copies are amplified by isothermal bridge amplification. The process is repeated on each template by cycles of isothermal denaturation and amplification to create millions of individual copies. Each cluster of dsDNA bridges is denatured and reverse strand is removed by specific base cleavage, leaving the forward DNA strand. After strand blocking on the flow-cell surface, the sequencing primer is hybridized to the complementary sequence on the adapter on unbound ends of the templates in the clusters and each cycle of sequencing identifies a single base.

3.3 Prerequisites to Computational Analysis

A substantial part of RNA-seq experiments consists of computational processing and analysis of the data. These analyses range from filtering the raw reads obtained from the sequencing machine, to differential gene expression analysis and biological interpretation of the results.

Table 3
Sequencing Read Archive accession numbers for samples used as an example

SRA accession number	Description
SRR1278968	C. parapsilosis planktonic replicate 1
SRR1278969	C. parapsilosis planktonic replicate 2
SRR1278970	C. parapsilosis planktonic replicate 3
SRR1278971	C. parapsilosis biofilm replicate 1
SRR1278972	C. parapsilosis biofilm replicate 2
SRR1278973	C. parapsilosis biofilm replicate 3

In the following sections we describe how to download, process, and analyze RNA-seq data using Mac OS X or a Linux distribution (such as Ubuntu) as the operating system. A server or computer cluster (e.g., Amazon EC2) can also be used.

To illustrate the use of the software we use a subset of recently published data from an experiment investigating the differences between the transcriptome of *Candida parapsilosis* grown as biofilms and under planktonic growth conditions (Table 3) [22]. This is strand-specific transcriptional profiling data obtained from a commercial company (BGI, Hong Kong) using an Illumina HiSeq 2000 with paired end reads of 90 bases. We describe in some detail how to visualize the results using a combination of R [9] and Bioconductor [10].

3.3.1 Download Data Set

1. Download the dataset from the Gene Expression Omnibus (GEO [23]) under GEO accession number GSE57451. It includes wild type *C. parapsilosis* cultures grown under planktonic and biofilm conditions. The individual reads are stored in the Sequence Read Archive (SRA [24]) under accession number SRP041812. Download the data using your favorite tool, or use the Unix command "wget" to import the files to your server or hard drive. The total size is 10 GB.

2. Alternatively, all required data and output files generated by working through the exercises in the computational analysis section can be downloaded from http://www.cgob.ie/supp_data. The directory structure follows the Unix setup described under "Common Unix Setup", which makes it easy to verify that your locally generated results are correct.

3.3.2 Common Unix Setup

1. Use a structured directory system for all sequencing related software and data to help to keep track of files and installed software. The setup described here is applicable for both Mac OS X and Linux operating systems. In Mac OS, use the

"Terminal" application to execute commands and manipulate files on the hard drive. For the purpose of the RNA-seq workflows below, create a general folder called "ngs" and subfolders for "data" and "applications". If you are using a server and do not have root access, the folder "~/local" and "~/local/bin" should be created as well. The "~/" represents the home directory and "~/local/bin" is the general location where executables and file links (Unix command "ln –s") to applications are stored. Use the following commands create the directories:

mkdir ~/ngs

mkdir ~/ngs/data

mkdir ~/ngs/applications

mkdir ~/local

mkdir ~/local/bin

2. Extend the PATH variable to include the "~/local/bin" folder. This enables the execution of the installed software anywhere on the system by typing the name of the software, rather than the full path to the directory where the software is installed. Use the following commands to extend and view the PATH variable:

export PATH=$HOME/local/bin:$PATH

echo $PATH

This change to the PATH variable is temporary and will be lost after logging out of the current session. To permanently extend the PATH variable add the command "export PATH=$HOME/local/bin:$PATH" to the end of either the ".profile" or ".bashrc" file in the home directory using for example emacs or vim.

3. Most software can be executed directly after unpacking downloaded archives. More advanced users, or users working with operating systems other than Linux or Mac OS X can build software from source (*see* **Note 6**). Use the following commands:

   – Downloading files to a server/local hard drive from the command line: wget http://some.web.address/file.tar.gz.

   – Unpacking an archive: tar xvfz file.tar.gz.

   – Creating a folder: mkdir name-of-folder.

   – Configuration of the installation script: ./configure --prefix=$HOME/local/bin.

   – Building the software: make; make test; make install.

   – Linking the new software to a folder that is included in the $PATH variable: ln -s $PWD/new-software-executable ~/local/bin/

If a program depends on external tools that need to be installed, the README or INSTALL files from the downloaded archive provide further details.

3.3.3 Installing Software All software required for the analysis workflow under "Data processing" are listed in Table 4. After downloading the individual files, execute the commands below inside the Terminal application in Mac OS X or Linux.

1. SRA Toolkit is available as compiled binaries. Download the archive into "~/ngs/applications", unpack, and link the executable to "~/local/bin". For the Ubuntu SRA Toolkit version 2.3.5, use the commands:

    tar xvfz sratoolkit.2.3.5-2-ubuntu64.tar.gz

    cd sratoolkit.2.3.5-2-ubuntu64/bin

    ln -s $PWD/fastq-dump ~/local/bin/

2. SAMtools (version 1.0 and above) is available as compiled binaries for Linux and Mac OS X that include SAMtools, BCFTools and HTSlib. The current version of SAMtools (v1.1) is not yet compatible with TopHat and we therefore recommend using SAMtools v0.1.19. This issue might be fixed with a TopHat version above 2.0.12. Download SAMtools to the applications folder and execute the following commands (for SAMtools version 0.1.19):

    tar xvfz samtools-0.1.19.tar.bz2

    cd samtools-0.1.19/

    make

    ln -s $PWD/samtools ~/local/bin

3. FastQC is based on Java and platform independent. Java is installed by default on current Linux and Mac OSX operating systems. To test which version of Java is installed execute:

    java -version

    Install the Mac OS X version of FastQC by copying the FastQC bundle into the applications folder. Download the Linux version into the applications folder and execute (for FastQC version 0.11.2):

    unzip fastqc_v0.11.2.zip

    cd FastQC

    chmod 755 fastqc

    ln -s $PWD/fastqc ~/local/bin

    The "chmod" command changes the "fastqc" file permission to make it executable.

Table 4
Overview of file formats, software, and resources used for processing and analyzing RNA-seq data

Format	Description
SRA	Used by the Sequence Read Archive to store and provide sequencing data
FASTQ	For storing sequencing data and corresponding quality scores
GTF/GFF	General Feature Format/Gene Transfer Format. Standardized formats for storing gene information
SAM	Sequence Alignment Map. Tab-delimited file format for storing alignment information from sequencing reads
BAM	Binary version of SAM format with a significantly smaller file size
BED	Format to store specific meta-data for regions of the genome. Used by the UCSC Genome Browser
BEDGRAPH	Based on the BED format. Stores scored data for specified genomic regions
BIGWIG	Very large collections of the BEDGRAPH format can be transformed into the binary BIGWIG format

Software/resources	Reference/website	Description
SRA/SRA Toolkit	[24] ncbi.nlm.nih.gov/Traces/sra	Storage for raw sequencing reads Collection of scripts for handling SRA files
SAMtools	[37] www.htslib.org/	Collection of scripts to handle SAM files
Skewer	[30] sourceforge.net/projects/skewer	Tool for quality trimming and filtering of sequencing reads
FastQC	[29] bioinformatics.babraham.ac.uk/ projects/fastqc	Generates quality reports for sequencing files
Bowtie2	[36] bowtie-bio.sourceforge.net/ bowtie2	Tool for aligning sequencing reads to a reference genome
TopHat	[31] ccb.jhu.edu/software/tophat	Splice-aware aligner for sequencing reads to a reference genome
HTSeq	[38] www-huber.embl.de/users/ anders/HTSeq	Python based tools to analyze sequencing data. The script htseq-count calculates read counts per gene
R	[9] www.r-project.org	Statistical language used for a variety of computational biology tasks
Bioconductor	[10] www.bioconductor.org	Large collection of R packages for biological data
CRAN	[11] cran.r-project.org	Large collection of R packages
CGD	[40] www.candidagenome.org	Extensive resource for *Candida* species
IGV	[25] www.broadinstitute.org/igv	Integrative Genomics Viewer for displaying sequencing data
Cytoscape	[64] www.cytoscape.org	Open source platform for visualizing and analyzing network data
Python	[65] www.python.org	Programming language commonly used for Bioinformatics tasks

4. Skewer is available as compiled binaries for Mac OS X and Linux. Download the respective binary and execute the following commands:

mkdir skewer

mv skewer-0.1.118-linux-x86_64 skewer/

cd skewer

chmod +x skewer-0.1.118-linux-x86_64

ln -s $PWD/skewer-0.1.118-linux-x86_64 ~/local/bin/skewer

5. For both MacOS and Linux, download and unpack the Bowtie2 archive and link "bowtie2" and the genome indexer "bowtie2-build" to "~/local/bin" with the commands (for version 2.2.3):

unzip bowtie2-2.2.3-linux-x86_64.zip

cd bowtie2-2.2.3

ln -s $PWD/bowtie2 ~/local/bin

ln -s $PWD/bowtie2-build ~/local/bin

6. Installation of TopHat is very similar to bowtie2 with the following commands (for version 2.0.12):

tar xvfz tophat-2.0.12.Linux_x86_64.tar.gz

cd tophat-2.0.12.Linux_x86_64

ln -s $PWD/tophat ~/local/bin

7. Install Python (version number above 2.5 and below 3.0) before installing HTSeq. Most servers and computer clusters will have one or several versions of Python already installed. Check the version of Python using:

python --version

If Python is not installed, or installed with a wrong version number, use the Unix tool "apt-get" to install Python. The user needs to have "sudo" rights for the following command to install Python 2.7, and the two additional packages that HTSeq requires, numpy and matplotlib:

sudo apt-get install build-essential python2.7-dev python-numpy python-matplotlib

If no "sudo" rights are available and Python or the numpy or matplotlib packages are not installed, contact the systems administrator to install them. To verify that numpy and matplotlib are installed, execute the following commands (the "python" command will enter the Python command line):

python

import numpy

import matplotlib

If no error messages are displayed the packages are installed and ready to use (*see* **Note 7**).

8. To install HTSeq download and unpack the HTSeq archive and install it using Python with the following commands (for version 0.6.1):

tar xvfz HTSeq-0.6.1.tar.gz

cd HTSeq-0.6.1

python setup.py install –user

ln –s $PWD/build/scripts-2.6/htseq-count ~/local/bin

9. For Mac OS X, an installation package for R is provided. Download the newest version, double-click the downloaded file and follow the instructions of the Mac OS X installer. On Linux systems execute the command:

sudo apt-get install r-base r-base-dev

If no "sudo" rights are available, download the source package of R, unpack the archive and build R with the following commands:

tar xvfz r-base_3.1.1.orig.tar.gz

cd R-3.1.1

./configure --prefix=$HOME/local/bin

make

ln –s $PWD/bin/R ~/local/bin

10. The Integrative Genomics Viewer [25] can be downloaded and used locally or launched directly from a web browser with varying amounts of allocated memory.

11. The Bioconductor package DESeq2 is required for the differential expression analysis described under "Generating HTML reports". The Bioconductor package ReportingTools is used to create HTML reports from DESeq2 results. Start R and execute the following commands:

source("http://bioconductor.org/biocLite.R")

biocLite("DESeq2")

biocLite("ReportingTools")

3.3.4 File Formats

Several different file formats are required. The user should become familiar with the various types listed in Table 4 and described below.

SRA: File format used by the Sequence Read Archive [24] to store and provide sequencing data. SRA format files can be converted into several commonly used formats using SRA Toolkit. SRA files in the sequence read archive format (file ending ".sra") can be transformed into the FASTQ format using "fastq-dump" from SRA-tools (*see* **Note 8**).

FASTQ: A text based format for storing sequencing data and quality scores. Each entry (read) in the FASTQ format consists of four lines. These represent (1) sequence identifier and description,

(2) the sequence, (3) an optional line that starts with a "+" and most commonly includes the sequence identifier and description again and (4) the Phred scale that is used to measure the base quality. Phred scores indicate the probability of incorrect base calls and the Phred scale is based on ASCII characters. For current Illumina sequencing data the ASCII encoded scores have an offset of 64 and raw base qualities normally range from character @ (quality 0) to i (quality >40). A shortened example of one FASTQ file entry is shown below:

1. @SRR1278968.1.1   FCC1WYWACXX:1:1101:1238:2126 length=90

2. TGGGNCTGTACGTGGTTCTTCAATTGCTTGTTTGTT CAATGGTAAATTCG[…]

3. +SRR1278968.1.1   FCC1WYWACXX:1:1101:1238:2126 length=90

4. ___cBQ\accgg^ee[ddeegghfff gghbe_cegffaa_c^_aeedc`[…]

GTF/GFF: The General Feature Format (GFF) and Gene Transfer Format (GTF) are two very similar formats used to store feature (gene) information. These include the genomic locations of exons, Coding Sequences (CDS), transcripts, 3′ and 5′ UnTranslated Regions (UTRs), tRNA, etc. The GFF file for an organism is used to assign features to sequencing reads that are mapped to the genome. An example of a line from a *C. parapsilosis* GFF file is shown below:

Cp_c1 . exon 94585 95295 . - . gene_id "CPAR2_100565_ exon"; transcript_id "CPAR2_100565_mRNA"

It stores the chromosome name (Cp_c1), feature type (exon), start and stop positions, strand (-), as well as additional attributes that are used for feature annotation, for example the transcript name (CPAR2_100565_mRNA). Single dots "." in this example indicate missing/empty information (*see* **Note 9**).

SAM: The Sequence Alignment/Map (SAM) format is a tab-delimited file to store alignment information for sequencing reads. There are eleven mandatory columns for each entry, which include information such as the sequence identifier, a bitwise FLAG that provides a summary of the read alignment, mapping position and quality score of the mapping. Additional columns can contain more specific information, such as the number of times the sequence mapped to the genome (NH), comments (CO), or mate pair information if the sequence data is generated from paired-end reads (MC, MQ).

An example of a line from a SAM file is shown below:

HWI-D00382:125:C48G6ACXX:8:1101:1134:59125   137 Cp_c8 2005578 50 101M * 0 0 AGCTGGTATCTTGTTG ACCCCAACTTTTGTCAAGTTGATTGCTTGGTACGATA

ACGAATACGGTTACTCCACCAGAGTTGTTGATTT
GTTGGAAAAATTTG CCCFFFDEHHHGHIGHHGGIID:
CGHEHHGHFGEH>HEHIGIIIHEGHIG=FHGIIG=
ACGHEHAAH;C==BBDE(.(6>A?B@;A@CACAA3(:<?B@C4
AS:i:0 XN:i:0 XM:i:0 XO:i:0 XG:i:0 NM:i:0 MD:Z:101 YT:
Z:UU XS:A:+ NH:i:1

From the beginning it lists the sequence ID, FLAG, chromosome name, leftmost mapping position, mapping quality,
CIGAR string for the alignment (101 matching bases),
sequence ID of mate or read pair ("*" means information
unavailable), position of mate, observed template length, raw
sequence, and Phred-scaled base quality. Information about
the optional fields, such as AS or XN, are available from
SAMtools' GitHub repository [26].

BAM: Smaller binary version of SAM format, can be viewed
using the "samtools view" command. BAM files are commonly
used to display alignment data in genome browsers because of
their smaller size compared to the non-binary SAM format.

BED: Mostly used for displaying genomic data in a genome
browser. Three fields specify the chromosome, start and end
position. Nine additional fields can be used to provide more
specific values for the genomic location, i.e., name, score,
strand, thickStart, thickEnd, itemRgb, blockCount, block-
Sizes, and blockStarts. An example of a bed file generate from
TopHat showing deletions found in RNA-seq data compared
to the reference genome is below:

track name=deletions description="TopHat deletions"

Cp_c1 1474 1475 - 1

Cp_c1 1509 1511 - 2

Cp_c1 1771 1772 - 1

BEDGRAPH: More specific format for displaying continuous-
valued data in a genome browser. Based on the BED and WIG
formats, the BEDGRAPH format can be used to display con-
tinuous-numeric values for genomic regions, for example transcriptome data. An example of a BEDGRAPH file is below.
The columns represent chromosome name, start and stop
position, and the numeric value, e.g., a user-defined score or
coverage information.

Cp_c1 665 756 -2

Cp_c1 1039 1042 -1

Cp_c1 1042 1067 -2

5. BIGWIG: For very large collections of data the BEDGRAPH
files can be converted into the binary BIGWIG format to save
disk space.

3.4 Data Processing

3.4.1 Quality Control and Trimming of Raw Data

After successfully installing the required software listed in Table 3 on Linux or Mac OS X, all further commands can be executed in the Terminal. Throughout the workflow below, we will provide example commands for sample SRR1278968 (Table 3).

1. To successfully execute downstream analyses these commands need to be executed separately for all six samples downloaded from SRA (Table 3). This data is already stored in separate files for each sample. If multiple samples are sequenced in the same lane on a sequencing machine, e.g., an Illumina HiSeq 2500, the raw sequencing reads must be separated using the bar-code/indexing information, which is usually the first six bases of each read. The fastx_barcode_splitter tool from the FastX-Toolkit can be used to achieve this [27, 28].

   The data in Table 3 was obtained from strand-specific 90 base paired-end sequencing. For each sample there are two different files, one containing the first read of the pair and one containing the second read. The standard file naming convention for paired-end reads ends is "_1" and "_2. Generate the files by providing fastq-dump with the option "--split-3" as shown here:

   fastq-dump --split-3 SRR1278968.sra

2. To confirm that the conversion from SRA to FASTQ was successful, use the Unix commands "head" or "less" to briefly examine the generated files. Each sample should have two additional files with the endings "_1.fastq" and "_2.fastq". The first read in the FASTQ file looks like this:

   @SRR1278968.1   FCC1WYWACXX:1:1101:1238:2126 length=90

   TGGGNCTGTACGTGGTTCTTCAATTGCTTGTTTGT TCAATGGTAAATTCGAGTCATCATGATGTGTTGGAGT TTGATTGGTGATTGTTTG

   +SRR1278968.1 FCC1WYWACXX:1:1101:1238:2126 length=90

   ___cBQ\accgg^ee[ddeegghfff`gghbe_cegffaa_c^_ aeedc`Xe^aeebfa]beg\beb\bZc_bcgR\`^`V^R^__]]bB

3. Once all the files are generated, check the overall quality of the data using FastQC, a java based tool that creates extensive summary reports. Use the following commands:

   mkdir qc

   fastqc SRR1278968_1.fastq -o qc 1>qc/SRR1278968_1.log 2>qc/SRR1278968_1.err

   fastqc SRR1278968_2.fastq -o qc 1>qc/SRR1278968_2.log 2>qc/SRR1278968_2.err

The "mkdir" command creates the "qc" folder where the results will be stored. For each file, FastQC generates a fastqc_report.html file, which can be displayed in any available browser. The quality report provides a basic summary of the sequencing reads in each file, overall per base and per sequence quality scores of a random subsample of all reads and several statistics to assess the quality of the sequencing run and the sequenced material. FastQC also reports basic sequence analysis results, such as over-represented sequences and relative kmer enrichments.

To assess of the quality of the sequencing data, it is helpful to look at the per base sequence quality boxplot. Figure 3 shows plots from two different fastq files, one from the sample data. The data in Fig. 3a is of a high quality. The boxplot shows that the quality scores from the base calling rarely fall below 30 for all 90 bases in the reads. Applying quality filtering on this sample will result in discarding a very small number of reads.

Figure 3b shows an example of a sample with reads of mixed quality. The black boxes for the first 80 bases are close to or above a base quality of 30, which indicates that the majority of reads has a high quality. However, a large number of reads in the lower 25th percentile fall below an acceptable base quality threshold (e.g., [15]) and have to be trimmed or removed.

The quality of reads from a sequencing experiment can vary significantly, the reasons vary from poor quality (degraded) or contaminated starting material to mistakes during the sequencing run itself (e.g., temporary shortage of solutions in the sequencing machine, or bubbles in the flowcell) [29].

4. After inspecting the FastQC report, trim the raw reads using Skewer [30] (*see* **Note 10**). Trimming sequencing data is an important step to ensure that only high quality data is analyzed and the results are not influence by poor quality reads. Skewer was developed primarily to improve adapter trimming of next-generation sequencing data, but it is also one of the fastest tools to remove poor quality bases from paired-end RNA-seq reads. It can utilize multiple processors to further speed up the quality trimming [30].

To run Skewer a few options must be specified. These include "-m pe" for paired-end trimming. Additionally recommended thresholds for trimming are a minimum read length of

Fig. 3 (continued) and below each represent 25 %. *Light, medium*, and *dark grey* background colors indicate poor, medium, and good per base quality, respectively. Quality scores are encoded in Illumina 1.5 format (**a**) and >1.3 format (**b**). Expected quality for raw sequencing data in both formats ranges from 0 to 40, with the exception that in Illumina format version 1.5 and above the quality score 2 represents the Read Segment Quality Control Indicator and 0 and 1 are unused

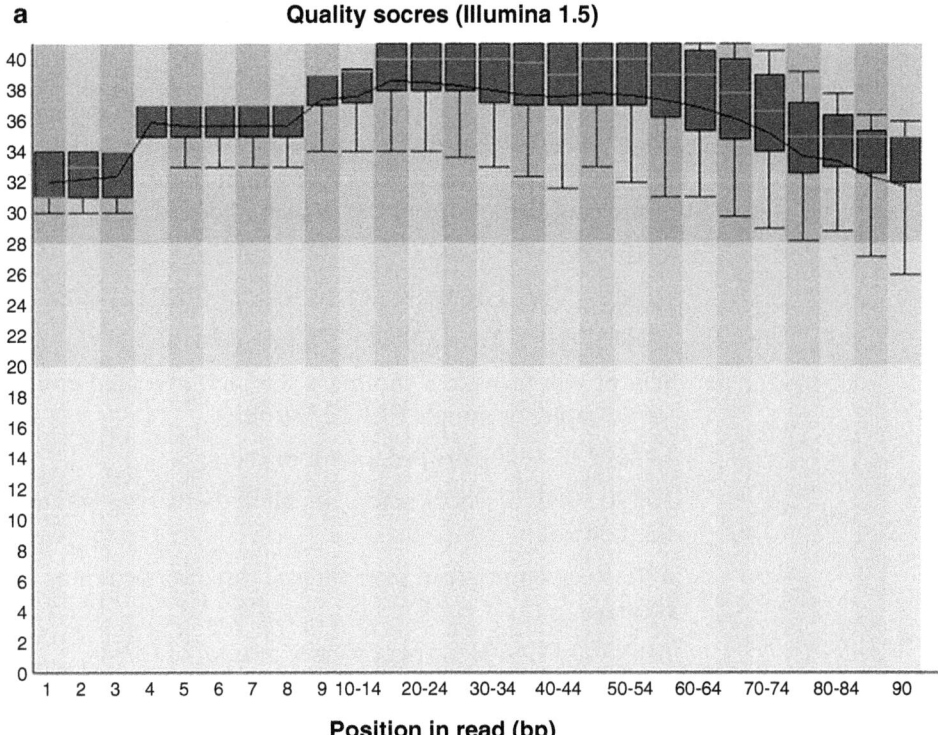

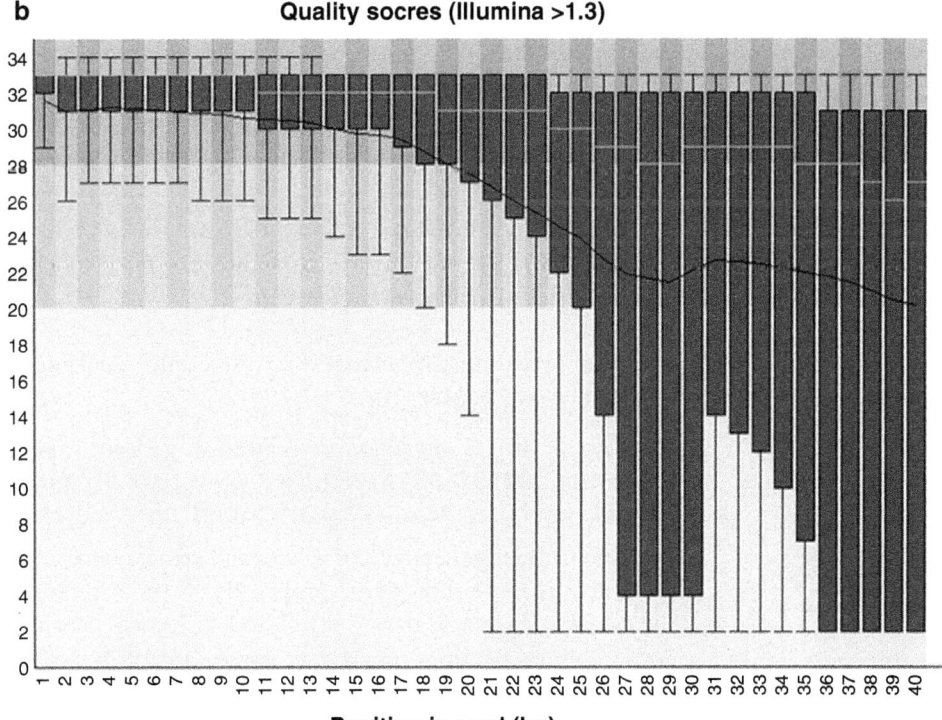

Fig. 3 FastQC per base sequence quality boxplot example for high quality (**a**, sample SRR1278968) and low quality data (**b**, plot adapted from [29]). Each *black box* represents 50 % of the reads and the *black lines* above

36 bases after trimming "-l 36"; a quality threshold for removing bases of the 3′ end with scores lower than 15 "-q 15"; a threshold of 15 for the mean quality of a read "-Q 15". In the example below the output directory for the trimmed data is "trim" and "-t 4" specifies that four cores should be used when executing Skewer. The two fastq files for our sample are listed at the end of the command line, with "_1.fastq" before "_2.fastq".

mkdir trim

skewer -m pe -l 36 -q 15 -Q 15 -o trim/SRR1278968 -t 4

SRR1278968_1.fastq SRR1278968_2.fastq

Skewer will provide a summary for each executed command, for example for sample SRR1278968:

13722223 read pairs processed; of these:

26390 (0.19 %) short read pairs filtered out after trimming by size control

0 (0.00 %) empty read pairs filtered out after trimming by size control

13695833 (99.81 %) read pairs available; of these:

2902491 (21.19 %) trimmed read pairs available after processing

10793342 (78.81 %) untrimmed read pairs available after processing

Only 0.19 % of the reads were discarded after trimming, since their length was shorter than 36 bases. From the other reads, 21.19 % were trimmed by a varying number of bases from the 3′ end because the base quality fell below the specified threshold of 15 ("-q 15").

5. After trimming the data, run FastQC again, this time using the output FASTQ files from the trim folder. This ensues all the data is of high-quality (not shown).

3.4.2 Mapping Reads to the Genome

All RNA-seq reads must be mapped to a reference genome, using an aligner such as TopHat [31].

1. Download the *C. parapsilosis* reference genome and gene annotation [22, 32–35] from http://www.cgob.ie/supp_data. The files are called "cpar.fa" and "cpar.gff", respectively.

2. Rename the files generated by Skewer and create an index for the reference genome. For paired-end reads, TopHat requires that the FASTQ files end in "_1.fastq" and "_2.fastq", for the first and second mate respectively. For single-end reads no naming convention or order of samples exists. Renaming the trimmed FASTQ files that currently end with "pair1.fastq" and "pair2.fastq" is easily achieved using the Unix rename command:

rename 's/pair/pair_/' *pair*

3. To create the reference genome index use bowtie2-build [36] with the FASTA genome file and output folder "cpar-index":
mkdir cpar-index

bowtie2-build cpar.fsa cpar-index/cpar

4. Execute TopHat with the following command:
tophat -p 12 -o SRR1278968 -G cpar.gff -g 1 --b2-very-sensitive

--library-type fr-firststrand cpar-index/cpar

trim/SRR1278968-pair_1.fastq   trim/SRR1278968-pair_2.fastq

The option "-p" sets the number of processing cores TopHat utilizes, "-o" sets the output folder, "-G" is optional and provides genome annotation and "-g 1" sets the maximum amount of times a read can map to the genome before it is reported as ambiguously mapped.

The preset "--b2-very-sensitive" is specified, which includes a number of settings (-D 20 -R 3 -N 0 -L 20 -i S,1,0.50). The D and R options specify "effort" options of TopHat. The higher these numbers are, the higher the amount of attempts TopHat will execute to realign reads or extend existing alignments. The N, L and i options fine-tune how TopHat tries to align the reads. The number of mismatches that are allowed during seed alignment (N), the length of the seed substring (L) and the function for the interval between substrings (i). Further information on TopHat can be found in the online manual (follow link in Table 4).

The very-sensitive option is used to increase the probability of mapping reads, as well as the length of the alignment. If the full read does not map to the reference genome, it is cut into smaller pieces (seeds) that TopHat tries to realign.

To specify that the RNA-seq data is strand-specific, set the library-type option for TopHat ("--library-type fr-firststrand") (see Note 11).

The *C. parapsilosis* genome, like many other eukaryotes, includes several introns [34, 35]. For this reason, a splice-aware aligner is used to map reads to the reference genome. TopHat has this ability, as do other tools (see Note 12).

Aligning the reads generated several files. The most important one is "accepted_hits.bam", which includes all reads that were successfully mapped to the genome, as well as the mapping quality and the mapping position. The "unmapped.bam" file lists all reads that were not successfully mapped to the genome. The files "deletions.bed" and "insertions.bed" show positions where reads were successfully mapped to the genome, but compared to the reference genome bases were either missing in the read (included in "deletions.bed") or additional bases were present in the read ("insertions.bed"). The file "junctions.bed" lists all positions in the genome where reads would span a region. This includes regions where reads span an intron. Visualizing the junctions in a genome browser can help identify different isoforms of a gene.

5. To check the data for properly aligned mate pairs, generate a summary of the alignment file from TopHat using the flagstat script from SAMtools [37] with the following command:

samtools flagstat accepted_hits.bam

The summary lists the total number of reads with a detailed breakdown of the paired reads. An example output is shown below. Here, 97.25 % of aligned reads are properly paired and only a very minor subset of reads without a mate or with a mate mapping to a different chromosome are present.

25680571 + 0 in total (QC-passed reads + QC-failed reads)

0 + 0 duplicates

25680571 + 0 mapped (100.00%:-nan%)

25680571 + 0 paired in sequencing

12793758 + 0 read1

12886813 + 0 read2

24974160 + 0 properly paired (97.25%:-nan%)

25135208 + 0 with itself and mate mapped

545363 + 0 singletons (2.12%:-nan%)

41182 + 0 with mate mapped to a different chr

41182 + 0 with mate mapped to a different chr (mapQ>=5)

6. When the reads are mapped to the genome, prepare the aligned ".bam" files for further analysis and viewing in a genome browser. Visualizing the reads at this point in the analysis is a good way to identify any potential problems, such as incorrect mapping of paired mates, or incorrect orientation of strand-specific data.

First index the "accepted_hits.bam" file. This is essential for genome browsers to display reads efficiently. The following SAMtools command generates a ".bai" file, which contains the bam index.

samtools index accepted_hits.bam accepted_hits.bam.bai

3.4.3 Visualizing the Mapped Reads in a Genome Browser

1. Install the IGV browser as described in "Installing Software" (*see* **Note 13**).

2. Load the reference genome sequence and the annotation. This is achieved from a single dialog box under "Genomes -> Create .genome File...".

3. Enter a unique identifier for the genome, and select the fasta file that was used earlier for building the Bowtie2 genome index as the "FASTA file". Alternatively, select the genome annotation GFF file that was used with TopHat as the "Gene file".

4. To load BAM files into IGV, select the reference genome from the drop-down menu at the top left corner of IGV.

5. Select a BAM file generated by an aligner from the local file system, a server or URL. The accompanying BAM index file must be in the same directory as the BAM file itself.

6. Load the GFF file to display the annotation. Initially, no sequencing reads are visible. This is because the default view in IGV is to show the entire chromosome, and displaying all reads mapped to the chromosome requires too much memory. Sequencing reads will be displayed if the visible region of the chromosome is set to below 100 kb in length. A snapshot of IGV with a BAM file and genome annotation loaded is shown in Fig. 4.

7. The reads can be displayed in three different ways: collapsed, squished and expanded. This is selected by right-click on the "accepted_hits.bam" label on the left side of the track. It also can be helpful to visualize paired-end RNA-seq data. To display read pairs as connected reads, select "View as pairs" again by right-click on the "accepted_hits.bam" label. By default IGV will display read pairs in different colors since the reads have different directions. To adjust the coloring schema to

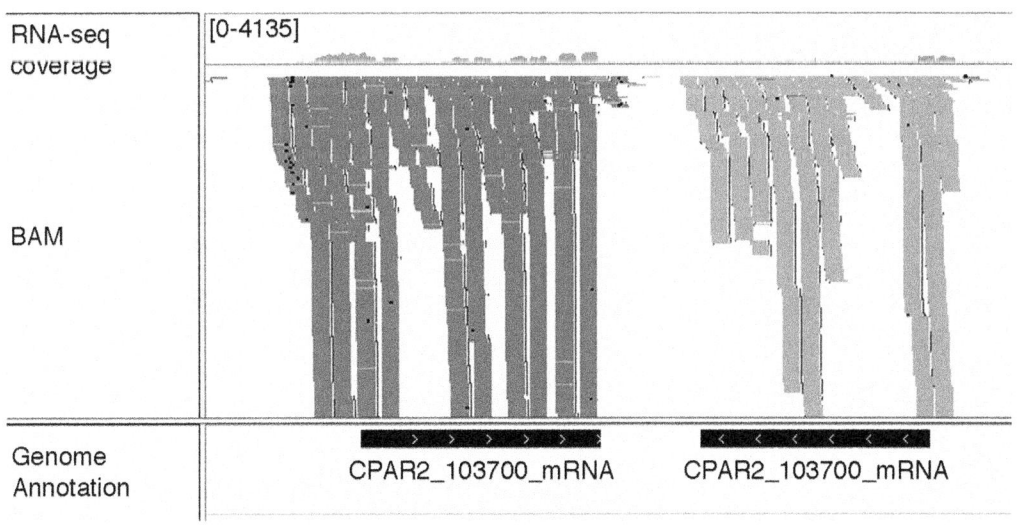

Fig. 4 Snapshot of IGV showing strand-specific *C. parapsilosis* RNA-seq data [22]. Included are RNA-seq coverage (*top track*), BAM file (*middle track*) and genome annotation (*bottom track*). Reads on the forward strand are *dark grey* and *light grey* on the reverse strand. *Arrows* inside the annotation track indicate the direction transcription. The data range of the RNA-seq coverage is indicated in *square brackets* [0-4135]. BAM file reads are displayed using the "squished" visualization option and colored using the "first-of-pair strand" option

show transcriptional orientation, right-click into the "accepted_hits.bam" label again and choose "Color alignments by" -> "first-of-pair-strand".

3.4.4 Counting Transcripts

To measure transcripts, the number of mapped reads for each gene must be counted. The Python script htseq-count from HTSeq [38] does exactly this. However, in order to run htseq-count, mapped reads must first be sorted in the bam file according to their location on the genome, and the file converted into the non-binary and significantly larger SAM format.

1. Sort the reads by location (option "-n") and convert the sorted BAM file to the SAM format using the following two SAMtool commands. A descriptive header for the SAM file is included with the option "-h".

   samtools sort -n accepted_hits.bam accepted_hits.sorted

   samtools view -h -o accepted_hits.sorted.sam accepted_hits.sorted.bam

2. To run htseq-count, specify the following command for each sample:

   htseq-count -m union -s reverse -t exon -i transcript_id

   -o accepted_hits.sorted.sam.htseq accepted_hits.sorted.sam

   ../cpar.gff 1>accepted_hits.sorted.sam.htseq.count

   2>accepted_hits.sorted.sam.htseq.count.log

   The option "–m union" specifies how the HTSeq algorithm assigns a read to a gene (also referred to as "feature"). The union option is recommended in most cases [38]. The strand direction (-s), the feature type (-t, third column in the GFF file), and the attribute for that htseq-count should report the read counts (-i), e.g., for each exon, or for each transcript, should also be specified.

   After successfully executing htseq-count the count data will be stored in "accepted_hits.sorted.sam.htseq.count". This is a tab-delimited file with transcript names in the first column and read counts in the second:

   CPAR2_100010_mRNA 271

   CPAR2_100020_mRNA 454

   CPAR2_100030_mRNA 3277

   In the following section we will explain how to identify differentially expressed genes from read count data and subsequently uncover the biological differences between different conditions.

3.4.5 Differential Gene Expression Analysis

Identify differentially expressed genes from read count data using the Bioconductor package DESeq2, which assumes that the read count data follows a negative binomial distribution [12]. (*see* **Note 14**).

Run Packages in R from the R command line or from the graphical user interface. (*see* **Note 15**).

1. Once R and DESeq2 are installed (*see* "Installing Software" for instructions) open R and load DESeq2 and ReportingTools [39] into the R environment using the following commands:

   library("DESeq2")

   library("ReportingTools")

2. Set the working directory to the analysis folder, for example:

   setwd("~/ngs/data/")

3. The count data are stored inside each TopHat output folder as specified for the TopHat command in "Mapping reads to the genome". To read these into R, use the function read.table():

   samples <- c("SRR1278968","SRR1278969","SRR1278970",

   "SRR1278971","SRR1278972","SRR1278973")

   cDataAll <- NULL

   for(i in 1:length(samples)){

   file <- read.table(

   sprintf("%s/accepted_hits.sorted.sam.htseq.count",

   samples[i]))

   cDataAll <- cbind(cDataAll, file[,2])

   }

   rownames(cDataAll) <- file[,1]

   colnames(cDataAll) <- samples

4. The count data from all six samples is now stored in the "cDataAll" variable. The example data (Table 3) includes measurements from three planktonic and three biofilm samples. To specify the conditions for each sample create the variable "groups" with "P" for planktonic and "B" for biofilm (*see* **Note 16**):

   groups <- factor(x=c(rep("P", 3), rep("B", 3)), levels=c("P", "B"))

5. The htseq-count data from Subheading "Counting transcripts" includes additional rows that indicate how many reads could not be associated with a unique feature, or where the alignment quality was too low. Exclude these five rows from the analysis using the match() function:

   cData <- cDataAll[-match(x=c("__no_feature", "__ambiguous", "__too_low_aQual", "__not_aligned", "__alignment_not_unique"), table=rownames(cDataAll)),]

6. DESeq2 uses raw count data as input, because it normalizes the data internally. To get a compact overview plot of all count

data, use Tags Per Million (TPM) normalization and the density function:

```
tpm <- t(t(cData)/colSums(cData))*1e6
inlog <- log(tpm)
colLabel <- c(rep("#E41A1C", 3), rep("#377EB8", 3))
colTy <- c(rep(1:3, 3), rep(1:3, 3))
plot(density(inlog[,1]), ylim=c(0,0.4), main="Density plot of
counts per gene", lty=colTy[1], xlab="Log of TPM per gene",
ylab="Density", col=colLabel[1])
for(i in 2:ncol(tpm)){
lines(density(inlog[,i]), lty=colTy[i], col=colLabel[i])
}
legend("topright", legend=colnames(tpm), lty=colTy, col=
colLabel)
```

This generates a density plot that shows the distribution of the log transformed TPM count data for each sample. It should follow a negative binomial distribution and with maximum log(TPM) between 4 and 5. All samples should have a very similar distribution. If this is not the case for any one sample, it is an early indicator that the transcriptome is very different to other samples. This could be due to a number of reasons. If no quality issues were detected in the raw sequencing reads and the mapping frequency to the reference genome was above 95 %, it may indicate that there is a problem with the biological sample.

7. The Principal Coordinate Analysis (PCoA) plot can also be used to characterize how similar samples are. PCoA returns coordinates that represent the dissimilarities between samples as distances. Use the normal plot() function to create the PCoA plot:

```
d <- dist(t(tpm))
fit=cmdscale(d, eig=TRUE, k=2)
x=fit$points[,1]
y=fit$points[,2]
plot(x, y, type="p", pch=20)
text(x, y, labels=row.names(t(tpm)), cex=1, adj=c(-0.25,-0.25))
```

Figure 5 shows the PCoA plot using data from Table 3. Control and test samples should cluster separately and samples within each group should cluster together. The x-axis in Fig. 5 (PCoA dimension 1) clearly separates the control and test samples and the y-axis (PCoA dimension 2) indicates small differences within each group, but with a 10× smaller scale.

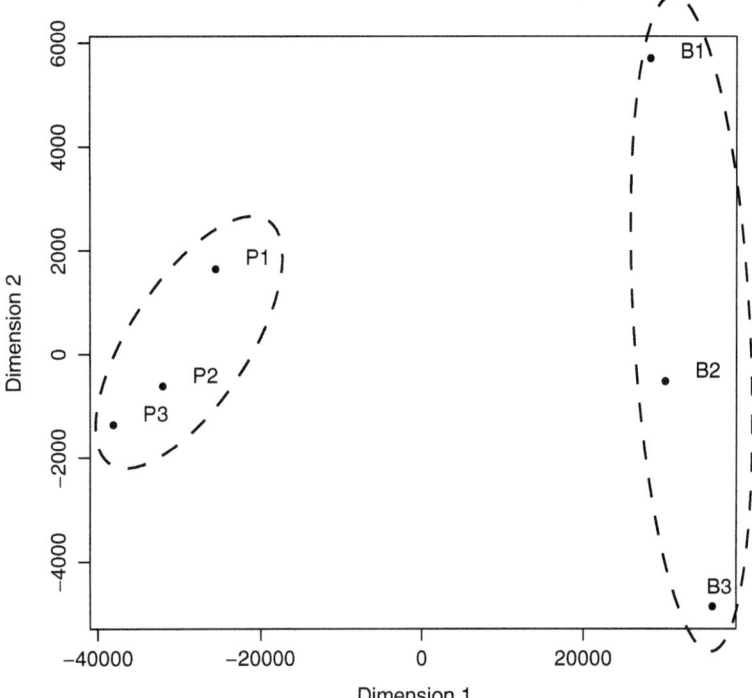

Fig. 5 PCoA plot showing transcriptional profiles of *C. parapsilosis* cells grown in planktonic conditions (P1-P3, SRA IDs: SRR1287968, SRR1278969, SRR1278970) and biofilm conditions (B1-B3, SRA IDs: SRR1278971, SRR1278972, SRR1278973) [22]. The two groups are visually separated by Dimension 1. There is minor variation among the biological replicates, which is indicated by the ten-fold smaller scale for Dimension 2 compared to Dimension 1

8. When you are confident that the data does not have any bias or other confounding factors, carry out differential expression analysis. To keep structure in the analysis directory, create the folder DESeq2 and set it as the working directory.

if (file.exists("DESeq2")){

setwd("DESeq2")

} else {

dir.create("DESeq2")

setwd("DESeq2")

}

9. Use the following script to run DESeq2:

colData <- DataFrame(condition=groups)

dds <- DESeqDataSetFromMatrix(cData, colData, formula (~condition))

dds <- DESeq(dds)

res <- results(dds, cooksCutoff=FALSE)

The Data.Frame "colData" contains the group factor that DESeq2 uses to separate the samples. The DESeqDataSet FromMatrix() function takes the count data, sample groups, and the formula for comparing the samples as input and creates a DESeqDataSet object, which can be used to run DESeq2. The function DESeq() executes DESeq and writes all results into the DESeqDataSet object. Use the results() function to create a Data.Frame containing the results, such as gene name, log2 fold change, and adjusted *p*-value.

10. To write the results to a file that can be opened with Excel use the write.csv() function. In addition the result Data.Frame can be ranked by the log2 fold change to enable easier analysis of the data.

    res <- res[order(res$log2FoldChange, decreasing=TRUE),]

    write.csv(as.data.frame(res), file="p-vs-b_results.csv")

    Below are the first rows of the results from the CSV file.

geneID	baseMean	log2FoldChange	lfcSE	stat	pvalue	padj
CPAR2_203270_ mRNA	5478.47	11.33	0.42	26.44	3.91e–154	7.59e–152
CPAR2_807700_ mRNA	20856.65	9.68	0.17	54.60	0	0

11. To extract the numbers of significantly upregulated and down-regulated genes you must specify the results thresholds. We recommend using a log2 fold change greater than 1 for genes with increased expression or lower than –1 for genes with decreased expression. Set the significant adjusted p-value (padj) to less than 0.01.

    up <- rownames(res[!is.na(res$padj) & res$padj <= 0.01 & res$log2FoldChange >= 1,])

    down <- rownames(res[!is.na(res$padj) & res$padj <= 0.01 & res$log2FoldChange <= -1,])

    sprintf("%s genes up-regulated, %s genes down-regulated", length(up), length(down))

12. To enable downstream analysis of the differentially expressed genes, save the gene IDs in two files, called "p-vs-b_up-regulated.txt" and "p-vs-b_down-regulated.txt" with the command below. In the GFF file the gene IDs contain the ending "_mRNA", which is removed using the R function sub() with the pattern "_mRNA" and an empty replacement "".

    write.table(sub(pattern = "_mRNA", replacement = "", x = up),

    file="p-vs-b_up-regulated.txt", col.names=FALSE,

    row.names=FALSE, quote = FALSE)

```
write.table(sub(pattern = "_mRNA", replacement = "", x =
down),
file="p-vs-b_down-regulated.txt", col.names=FALSE,
row.names=FALSE, quote = FALSE)
```

13. The DESeq2 package contains several methods to create overview plots of the differentially expressed genes. Use MA plot (plotMA(dds)) to create a plot of the log2 fold changes against mean normalized counts for each gene. Use the plotPCA function to create a Principal Component Analysis (PCA) plot. The number of genes that are taken into account for the distance calculation of the PCA can be specified. With "ntop=500" the 500 genes with the highest row variance, i.e., the highest variance of read counts per gene across all samples, will be used for the PCA.

```
plotPCA(dds, ntop=500)
```

14. To represent the log2 fold change distribution of all genes as a histogram highlighting differentially expressed genes for example in grey (as shown in Fig. 6), use the commands:

```
hist(res[!is.na(res$padj) & res$padj <= 0.01, ][,"log2Fold
Change"],
breaks=seq(-15,15,0.25),
col=c(rep("tomato", 56), rep("white", 8), rep("tomato", 56)),
xlab="Log2 Fold Change", main="Overall Log2 fold change,
p.adj<0.01") (see Note 17).
```

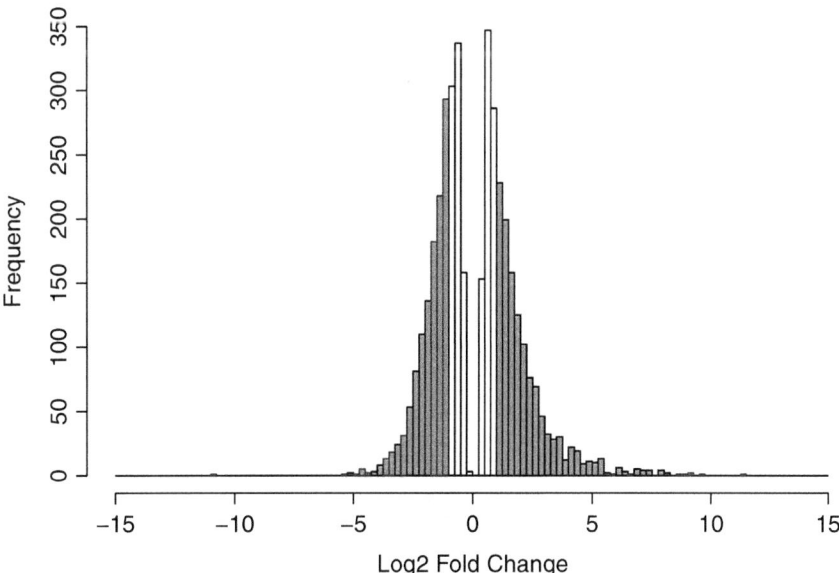

Fig. 6 Histogram of log$_2$ fold changes obtained by comparing the transcriptional profiles of *C. parapsilosis* cells grown in planktonic vs. biofilm conditions [22] using the Bioconductor package DESeq2. Significant log$_2$ fold changes are colored in *dark grey* (greater than 1, or less than −1), not significant log$_2$ fold changes are in *white*

3.4.6 Generating HTML Reports

As a last step, searchable HTML reports of the differentially expressed genes can easily be generated using the Bioconductor package ReportingTools [39].

1. Load ReportingTools into the R environment.

   library("ReportingTools")

2. Use the functions HTMLReport() and publish() to generate a website from the results data.frame created by DESeq2.

   htmlRep <- HTMLReport(shortName="P-vs-B_results", reportDirectory = "./reports")

   publish(cbind(GeneID=rownames(res),as.data.frame(res)), htmlRep)

3. Finally, create the report.

   finish(htmlRep)

3.4.7 Downstream Analysis of Differentially Expressed Genes

1. After the list of significantly differentially expressed genes is generated, several tools and websites can help to identify the biological mechanisms that drive the change between the two conditions, planktonic and biofilm transcriptomes in this example.

   Meta-information for *Candida* species is available from the *Candida* Genome Database (CGD, [40]). Equivalent sites for other genomes include *Saccharomyces* species from the *Saccharomyces* Genome Database (SGD, [41]) and *Aspergillus* species from the *Aspergillus* Genome Database (AspGD, [42]). For less characterized species, working with homologs from a closely related species can provide more biological insight. Homology information for *Candida* and *Saccharomyces* species is available from the *Candida* Gene Order Browser (CGOB, [35, 43]) and the *Yeast* Gene Order Browser (YGOB, [44]), as well as from AspGD.

2. Gene Ontology (GO) Analysis: Using the identified differentially expressed genes from the *C. parapsilosis* planktonic and biofilm samples, Gene Ontology terms that are enriched in upregulated or downregulated genes are identified using the GO Term Finder at CGD ([40], Table 4). In **step 1** on the GO Term Finder website, choose *Candida parapsilosis* as the target species. For **step 2** upload the file "p-vs-b_up-regulated.txt" or the equivalent file containing downregulated genes, which were saved at the end of the DESeq2 workflow. Choose one of the three Gene Ontologies in **step 3**, i.e., Biological Process, Molecular Function or Cellular Component. We want to know if our upregulated genes have any enriched GO terms in the Molecular Function ontology. Fig. 7 shows a screenshot of the GO Term Finder with sample settings. To start the analysis, click on the Search-button in the bottom left corner.

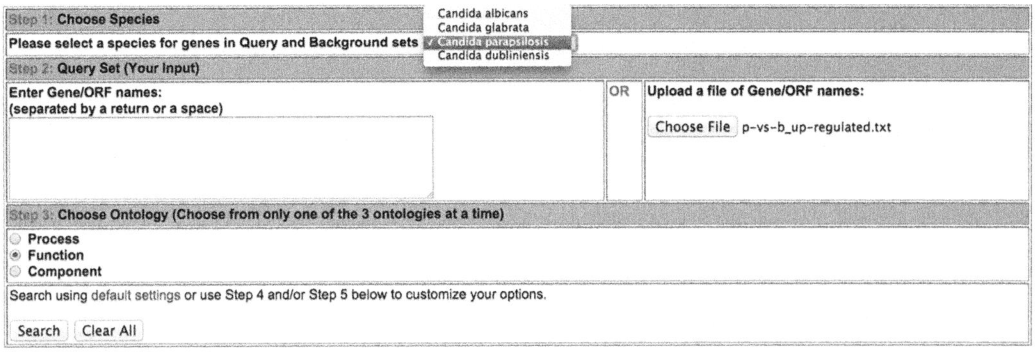

Fig. 7 Screenshot of the GO Term Finder from the *Candida* Genome Database. The species *C. parapsilosis* is selected in *Step 1* and a file containing gene names is specified in *Step 2*. The Molecular Function Gene Ontology is selected in *Step 3*

Results from the GO Term Finder can be downloaded in form of an Excel file at the bottom of the page. Alternatively the results can be saved by right-clicking on the page in the browser and selecting "Save As". This will also save the GO tree picture. In the GO tree for upregulated genes ("p-vs-b_up-regulated.txt") Molecular Function GO terms that are significantly shared between the list of upregulated genes include Oxidoreductase Activity, Transmembrane Transporter Activity, and Transition Metal Iron Binding.

3.5 Final Remarks

For relatively little bench time, RNA-seq can yield a large amount of data. With an established protocol and workflow, RNA-seq experiments can have a quick turnaround, with the longest waiting time being the actual sequencing itself. Even commercial companies are reducing this time, with some offering a turnaround of 4–6 weeks.

RNA-seq holds the potential to answer key research questions. With new strategies emerging for increased multiplexing and the availability of more whole genome sequencing data and reference genomes, the uses and benefits of RNA-seq and other NGS techniques are becoming widespread throughout the research community. They will soon be a staple technique in the lab environment.

4 Notes

1. We recommend using a commercial kit to isolate high quality RNA. The Ribopure Yeast RNA extraction kit from Ambion is particularly useful, but other kits or methods can be used providing the quality of the RNA produced is high.

2. It is now possible to design and synthesize long adapter sequences, using updated recommendations from Illumina

[21] that remove the necessity to add the P5 region during library amplification (Fig 2b [21]). The P5 sequence is included in the first oligonucleotide (the universal adapter). The indexing sequence is contained in the second oligonucleotide (indexed adapter), outside the short region that anneals with the universal adapter (Fig 2b). The sequence of twenty-seven 6 nucleotide indexes are provided by Illumina ([21], Oligonucleotide sequences, 2007–2013 Illumina, Inc. All rights reserved) and other home-made designs are described by Ford et al. [45]. Multiplexing of samples can be increased by also including an index sequence in the universal adapter (dual indexing). Recent kits from Illumina use six or eight nucleotide indexes, with up to eight different versions of the universal adapter, and 48 of the indexed adapter [21]. The libraries are amplified using regions derived from the P5 and P7 regions, and can be sequenced from either end using two sequencing primers. These adapters can also be combined with dUTP methodology for strand-specific sequencing [18].

3. AMPure XP beads (Beckman Coulter) are now recommended for clean-up procedures instead of gel purification.

4. We use 2-log DNA ladder from NEB, which allows visualization of DNA bands for 0.1–10 kb. Make 2-log DNA ladder mix by mixing 1 μl 2-log DNA ladder (NEB), 1 μl Blue loading dye (Promega), and 4 μl distilled water.

5. Consideration should be given to the length of the reads generated, and whether single-end or paired-end sequences are used. The adapters described here are for single read only, different sequences are required for paired end reads. It is possible to obtain increasingly long reads, but at a price. Long reads (>100 bases) are not necessary for RNA-seq. Paired-end reads can help in mapping, but are not strictly necessary. Strand-specific information however is strongly recommended as it enables identification of UTRs (untranslated regions) and antisense expression.

6. For users who are not yet proficient with the Terminal command line, a detailed Unix tutorial is available [46]. Commands and techniques described in the tutorial are applicable for both Linux and Mac OS X users. A large community of researchers that work with next-generation sequencing data can be reached on the SEQanswers forum [47]. BioStar [47, 48] is a more general and also very useful Bioinformatics forum.

7. If any error messages arise during the installation of Python, or if there are general questions about Unix environment variables or Unix commands, Stack Overflow [49] is a useful starting point to search for solutions.

8. More information on SRA formats are available in the SRA Knowledge Base [50] and the SRA Handbook [51].

9. More information about the GFF format is available at a dedicated website from the Sanger Institute [52].

10. There are several equivalently suitable tools available for trimming sequencing reads, some with extensive options to specify quality thresholds and filtering parameters. It is however for the user to decide which tool fits best for a specific task. The tool used in this RNA-seq workflow is Skewer [30]. Other popular tools include fastx_trimmer [27, 28], cutadapt [53], and TrimGalore! [54]. These tools are all capable of trimming sequencing reads based on quality thresholds.

11. There are different methods for generating strand-specific RNA-seq libraries [16]. To verify that the correct option for the RNA-seq data was used in TopHat, compare the aligned reads in a genome browser with the HTSeq count data for a specific gene. For unstranded RNA-seq data specify "fr-unstranded".

12. Other splice-aware aligners include GSNAP [55] or STAR [56]. STAR is a recently developed alignment tool that has significant speed advantages over other aligners. This is helpful for identifying the optimal parameters for aligning sequencing reads to a reference genome, e.g., number of allowed mismatches per read, number of times a read is allowed to map to the genome, or the length limitations for introns. A comparison of the different aligners was carried out by Engstrom et al. [57].

13. Other popular genome browsers are Artemis [58], the Integrative Genomics Browser [25], the web-browser based genome browsers JBrowse [59] and the UCSC Genome Browser [60].

14. Selecting a statistical package to identify differentially expressed genes from gene count data is an important step in an RNA-seq workflow. There is however not one method that is superior to all others. Commonly used packages include DESeq2, edgeR, baySeq, DEGSeq, NOISeq, tweeDEseq, and many more. The Bioconductor version 2.14 lists 138 packages used for Differential Expression analysis. DESeq2, edgeR, baySeq, and EBSeq assume that the count data follows a negative binomial distribution, which is the most commonly assumed distribution for RNA-seq data. Choosing any of the available methods will yield results, more important than the method itself however is that experiments are planned with enough replicates and that options recommended by the authors of each method are used. It is not ideal to switch between statistical methods when comparing different datasets, which could introduce additional biases. Detailed comparisons and analyses of several statistical methods for identifying differentially expressed genes from count data can be found at [61, 62].

15. A more sophisticated open source integrated development environment (IDE) for R is RStudio [63], which works on Windows, Mac, and Linux. Additionally, RStudio Server can be executed on a server and accessed from a web browser, which makes it easy to utilize the computing power of a server while using RStudio from a personal computer or laptop.

16. Specifying the levels for the group factor is important for running DESeq2, because it determines how the differentially expressed genes are reported. In this example "P" is the control condition and genes reported with a positive $\log_2$ fold change have increased expression in the test condition "B". If levels are not specified they will be assigned automatically in alphabetical order.

17. Several other useful plots for generating an overview of the results are mentioned in the DESeq2 package Vignette on Bioconductor, as well as extensive documentation of the DESeq2 methodology: http://www.bioconductor.org/packages/release/bioc/vignettes/DESeq2/inst/doc/DESeq2.pdf.

Acknowledgements

We are grateful to Dr Amanada Lohan, UCD, for helpful advice and for designing the original sequencing strategy. Work in the Butler lab is supported by Science Foundation Ireland and the Wellcome Trust.

References

1. Nagalakshmi U, Wang Z, Waern K et al (2008) The transcriptional landscape of the yeast genome defined by RNA sequencing. Science 320:1344–1349

2. Wang Z, Gerstein M, Snyder M (2009) RNA-Seq: a revolutionary tool for transcriptomics. Nat Rev Genet 10:57–63

3. Wilhelm BT, Landry JR (2009) RNA-Seq-quantitative measurement of expression through massively parallel RNA-sequencing. Methods 48:249–257

4. Bruno VM, Wang Z, Marjani SL et al (2010) Comprehensive annotation of the transcriptome of the human fungal pathogen *Candida albicans* using RNA-seq. Genome Res 20:1451–1458

5. Zupancic ML, Frieman M, Smith D et al (2008) Glycan microarray analysis of *Candida glabrata* adhesin ligand specificity. Mol Microbiol 68:547–559

6. Nobile CJ, Fox EP, Nett JE et al (2012) A recently evolved transcriptional network controls biofilm development in *Candida albicans*. Cell 148:126–138

7. Kumamoto CA, Vinces MD (2005) Contributions of hyphae and hypha-co-regulated genes to *Candida albicans* virulence. Cell Microbiol 7:1546–1554

8. Marioni JC, Mason CE, Mane SM et al (2008) RNA-seq: an assessment of technical reproducibility and comparison with gene expression arrays. Genome Res 18:1509–1517

9. R Code Team (2014) R: a language and environment for statistical computing. http://www.r-project.org

10. Gentleman RC, Carey VJ, Bates DM et al (2004) Bioconductor: open software development for computational biology and bioinformatics. Genome Biol 5:R80

11. CRAN (2014) The Comprehensive R Archive Network (CRAN). http://cran.r-project.org

12. Love MI, Huber W, Anders S (2014) Moderated estimation of fold change and dispersion for RNA-Seq data with DESeq2. Genome Biol 550

13. Soto T, Núñez A, Madrid M et al (2007) Transduction of centrifugation-induced gravity forces through mitogen-activated protein kinase pathways in the fission yeast Schizosaccharomyces pombe. Microbiology 153:1519–1529

14. Desjardins PR, Conklin DS (2011) Microvolume quantitation of nucleic acids. Curr Protoc Mol Biol Appendix:A.3J.1–A.3J.16

15. Zhang Z, Theurkauf WE, Weng Z et al (2012) Strand-specific libraries for high throughput RNA sequencing (RNA-Seq) prepared without poly(A) selection. Silence 3:9

16. Levin JZ, Yassour M, Adiconis X et al (2010) Comprehensive comparative analysis of strand-specific RNA sequencing methods. Nat Methods 7:709–715

17. Parkhomchuk D, Borodina T, Amstislavskiy V et al (2009) Transcriptome analysis by strand-specific sequencing of complementary DNA. Nucleic Acids Res 37, e123

18. Sultan M, Dokel S, Amstislavskiy V et al (2012) A simple strand-specific RNA-Seq library preparation protocol combining the Illumina TruSeq RNA and the dUTP methods. Biochem Biophys Res Commun 422:643–646

19. Craig DW, Pearson JV, Szelinger S et al (2008) Identification of genetic variants using barcoded multiplexed sequencing. Nat Methods 5:887–893

20. Weissenmayer BA, Prendergast JG, Lohan AJ et al (2011) Sequencing illustrates the transcriptional response of Legionella pneumophila during infection and identifies seventy novel small non-coding RNAs. PLoS One 6, e17570

21. Illumina (2014) Illumina customer sequence letter. http://support.illumina.com/downloads/illumina-customer-sequence-letter.html

22. Holland LM, Schröder MS, Turner SA et al (2014) Comparative phenotypic analysis of the major fungal pathogens Candida parapsilosis and Candida albicans. PLoS Pathog 10, e1004365

23. Barrett T, Wilhite SE, Ledoux P et al (2013) NCBI GEO: archive for functional genomics data sets--update. Nucleic Acids Res 41: D991–D995

24. Kodama Y, Shumway M, Leinonen R et al (2012) The sequence read archive: explosive growth of sequencing data. Nucleic Acids Res 40:D54–D56

25. Thorvaldsdottir H, Robinson JT, Mesirov JP (2013) Integrative genomics viewer (IGV): high-performance genomics data visualization and exploration. Brief Bioinform 14:178–192

26. The SAM/BAM Format Specification Working Group (2014) Sequence alignment/map format specification. https://github.com/samtools/hts-specs

27. Blankenberg D, Gordon A, Von Kuster G et al (2010) Manipulation of FASTQ data with Galaxy. Bioinformatics 26:1783–1785

28. Hannon G (2014) FASTX-Toolkit: FASTQ/A short-reads pre-processing tools. http://hannonlab.cshl.edu/fastx_toolkit

29. Andrews S (2010) FastQC: a quality control tool for high throughput sequence data. http://www.bioinformatics.babraham.ac.uk/projects/fastqc

30. Jiang H, Lei R, Ding SW et al (2014) Skewer: a fast and accurate adapter trimmer for next-generation sequencing paired-end reads. BMC Bioinformatics 15:182

31. Trapnell C, Pachter L, Salzberg SL (2009) TopHat: discovering splice junctions with RNA-Seq. Bioinformatics 25:1105–1111

32. Butler G, Rasmussen MD, Lin MF et al (2009) Evolution of pathogenicity and sexual reproduction in eight Candida genomes. Nature 459:657–662

33. Fitzpatrick DA, Butler G (2010) Comparative genomic analysis of pathogenic yeasts and the evolution of virulence. In: Ashbee HR, Bignell E (eds) Pathogenic yeasts. Springer, Heidelberg, pp 1–18

34. Guida A, Lindstadt C, Maguire SL et al (2011) Using RNA-seq to determine the transcriptional landscape and the hypoxic response of the pathogenic yeast Candida parapsilosis. BMC Genomics 12:628

35. Maguire SL, Oheigeartaigh SS, Byrne KP et al (2013) Comparative genome analysis and gene finding in Candida species using CGOB. Mol Biol Evol 30:1281–1291

36. Langmead B, Salzberg SL (2012) Fast gapped-read alignment with Bowtie 2. Nat Methods 9:357–359

37. Li H, Handsaker B, Wysoker A et al (2009) The sequence alignment/map format and SAMtools. Bioinformatics 25:2078–2079

38. Anders S, Huber W (2010) Differential expression analysis for sequence count data. Genome Biol 11:R106

39. Huntley MA, Larson JL, Chaivorapol C et al (2013) ReportingTools: an automated result processing and presentation toolkit for high-throughput genomic analyses. Bioinformatics 29:3220–3221

40. Inglis DO, Arnaud MB, Binkley J et al (2012) The Candida Genome Database incorporates multiple Candida species: multispecies search

and analysis tools with curated gene and protein information for *Candida albicans* and *Candida glabrata*. Nucleic Acids Res 40:D667–D674

41. Costanzo MC, Engel SR, Wong ED et al (2014) *Saccharomyces* genome database provides new regulation data. Nucleic Acids Res 42:D717–D725

42. Cerqueira GC, Arnaud MB, Inglis DO et al (2014) The *Aspergillus* genome database: multispecies curation and incorporation of RNA-Seq data to improve structural gene annotations. Nucleic Acids Res 42:D705–D710

43. Fitzpatrick DA, O'Gaora P, Byrne KP et al (2010) Analysis of gene evolution and metabolic pathways using the Candida Gene Order Browser. BMC Genomics 11:290

44. Byrne KP, Wolfe KH (2005) The Yeast Gene Order Browser: combining curated homology and syntenic context reveals gene fate in polyploid species. Genome Res 15:1456–1461

45. Ford E, Nikopoulou C, Kokkalis A et al (2014) A method for generating highly multiplexed ChIP-seq libraries. BMC Res Notes 7:312

46. Stonebank M (2001) UNIX tutorial for beginners. http://www.ee.surrey.ac.uk/Teaching/Unix

47. Li JW, Schmieder R, Ward RM et al (2012) SEQanswers: an open access community for collaboratively decoding genomes. Bioinformatics 28:1272–1273

48. Parnell LD, Lindenbaum P, Shameer K et al (2011) BioStar: an online question & answer resource for the bioinformatics community. PLoS Comput Biol 7, e1002216

49. Overflow S (2014) Question and answer site for professional and enthusiast programmers. http://www.stackoverflow.com

50. SRA SS (2011) Using the SRA Toolkit to convert .sra files into other formats. http://www.ncbi.nlm.nih.gov/books/NBK158900/. 2014

51. Stack Overflow (2009, updated in 2014) Download guide. National Center for Biotechnology Information (US). http://www.ncbi.nlm.nih.gov/books/NBK47540/

52. GFF: an exchange format for feature description (1999) Sanger Institute. https://www.sanger.ac.uk/resources/software/gff/. 2014

53. Martin M (2011) Cutadapt removes adapter sequences from high-throughput sequencing reads. EMBnet J 17:10–12

54. Krueger F (2014) Trim Galore! a wrapper tool around Cutadapt and FastQC http://www.bioinformatics.babraham.ac.uk/projects/trim_galore

55. Wu TD, Nacu S (2010) Fast and SNP-tolerant detection of complex variants and splicing in short reads. Bioinformatics 26:873–881

56. Dobin A, Davis CA, Schlesinger F et al (2013) STAR: ultrafast universal RNA-seq aligner. Bioinformatics 29:15–21

57. Engstrom PG, Steijger T, Sipos B et al (2013) Systematic evaluation of spliced alignment programs for RNA-seq data. Nat Methods 10:1185–1191

58. Rutherford K, Parkhill J, Crook J et al (2000) Artemis: sequence visualization and annotation. Bioinformatics 16:944–945

59. Skinner ME, Uzilov AV, Stein LD et al (2009) JBrowse: a next-generation genome browser. Genome Res 19:1630–1638

60. Kent WJ, Sugnet CW, Furey TS et al (2002) The human genome browser at UCSC. Genome Res 12:996–1006

61. Kvam VM, Liu P, Si Y (2012) A comparison of statistical methods for detecting differentially expressed genes from RNA-seq data. Am J Bot 99:248–256

62. Soneson C, Delorenzi M (2013) A comparison of methods for differential expression analysis of RNA-seq data. BMC Bioinformatics 14:91

63. RStudio (2014) RStudio: Integrated development environment for R (Version 0.98.1062). http://www.rstudio.org/

64. Saito R, Smoot ME, Ono K et al (2012) A travel guide to cytoscape plugins. Nat Methods 9:1069–1076

65. Foundation PS (2014) Python language reference, version 2.7. http://www.python.org

Chapter 2

Enhancing Structural Annotation of Yeast Genomes with RNA-Seq Data

Hugo Devillers, Nicolas Morin, and Cécile Neuvéglise

Abstract

The number of fully sequenced genomes of yeasts is dramatically increasing but both structural and functional annotation quality are usually neglected, as most frequently based on automatic annotation transfer tools from reference genomes. RNA sequencing technologies offer the possibility to better characterize yeast transcriptomes and to correct or improve the prediction of mRNA, ncRNA, or miscellaneous RNA. We describe a computational approach to enhance structural annotation of yeast genomes based on RNA-Seq data exploitation. The proposed pipeline is primarily based on read mapping with TopHat2. Mapping outputs are then used for various applications such as: (1) validation of exon–exon junctions of predicted transcripts, (2) definition of new transcribed features, (3) prediction of 3' UTR, and (4) identification of extra features absent from the genome assembly. We strongly encourage curators to proceed to a manual validation and editing of the reference genome. Releasing genomes with high-quality annotation is an important issue, as they will be considered as references for further predictions.

Key words Transcriptome, Genome annotation, Intron, Genome curation, RNA-Seq, Yeast

1 Introduction

Most of the genomic studies directly rely on the annotation associated with assembled sequences. As a consequence, the reliability and the accuracy of these analyses critically depend on the quality of the provided annotations. With the rise of next-generation sequencing (NGS), the number of complete genomes increases exponentially while each of them contains thousands of genetic feature descriptions in their relative annotations. However, the quality and the correctness of these data are particularly heterogeneous and most of the time, rather low [1].

This quality issue is explained by the limits of the different methods that are employed to generate genome annotation. Indeed, with the flood of newly sequenced genomes, it is necessary to use automated methods to infer gene structures. However, although these methods can predict a majority of gene structures,

Frédéric Devaux (ed.), *Yeast Functional Genomics: Methods and Protocols*, Methods in Molecular Biology, vol. 1361, DOI 10.1007/978-1-4939-3079-1_2, © Springer Science+Business Media New York 2016

they generally failed to identify "complex" gene structures (e.g., multi-intronic genes) or "new" gene structures. Briefly, automated methods can be classified into two categories. The first one consists in inferring genes according to a theoretical model that detects the possible evidence of protein-coding genes along the genome sequences. These methods are referred to ab initio. They are particularly efficient for prokaryote genomes where genes do not have spliceosomal introns, but they rapidly failed to infer exon–exon junctions in eukaryote genomes. The second kind of method is the annotation transfer from reference genome(s). It consists in mapping known features from a closely related species on a newly sequenced genome. This method can be particularly efficient and fast if the reference genome is close enough and if it has a high quality annotation. Unfortunately, these requirements are generally overlooked and therefore, the reference annotation often contains mistakes themselves inferred or transferred from another genome annotation.

As a consequence, to obtain a good quality annotation, a particular effort has to be made to expertise and validate predicted annotations. Today, too few genome annotations have benefit from such an effort [2]. Among the available yeast genomes, only a few are considered as reference genomes (e.g., *Saccharomyces cerevisiae*, *Candida albicans*, *Schizosaccharomyces pombe*, *Neurospora crassa*) and are consequently used even though the target species is not closely related. This highlights the critical needs to fill this gap by providing more genomes with validated annotations.

With the development of second and third generation sequencing techniques, it is now possible to easily sequence the complete transcriptome of a given organism. RNA sequencing (RNA-Seq) offers an invaluable opportunity to experimentally identify all the transcripts and hence highly facilitates the detection of genetic features. However, RNA-Seq data are complex, voluminous, and often biased. Therefore, they require specific methods and procedures to extract usable and fruitful information.

In this chapter, we present different protocols that aim at retrieving gene structure information from RNA-Seq data to complete and correct annotations obtained by automated methods.

2 Materials

2.1 Genomic Data

For illustration purpose, we considered the genome of *Yarrowia lipolytica* strain E150/CLIB122 [3] with two different RNA-Seq studies (ENA project accession numbers PRJEB7323 and PRJEB7354), and the genome of *Rhodosporidium toruloides* strain CECT1137 [4] with one RNA-Seq study (unpublished data). EMBL format is required for the reference genome files. More details about these datasets are provided in **Note 1**. RNA sequencing design strategy is discussed in **Note 2**, while reference genome selection is addressed in **Note 3**.

2.2 Bioinformatic Tools

1. FastQC [5]: read quality evaluation.
2. Trimmomatic [6]: read trimmer.
3. Bowtie2 [7]: read mapping on reference genome.
4. TopHat2 [8]: read mapping on reference genome.
5. Cufflinks [9]: mapped read assembling.
6. Trinity [10]: de novo assembler tool for transcripts.
7. RSEM [11]: estimation of gene isoform expression level.
8. Samtools [12]: manipulation of BAM/SAM files.
9. BLAST+ suite [13]: local alignment tool suite.

2.3 Data Visualization and Edition

Data visualization is a key aspect for refining structural annotation. Thus, it is crucial to use a tool able to display and to handle the different kind of data (e.g., EMBL, GFF, BAM) and to facilitate the edition of the structure of the different genetic features. In this context, we recommend the genome browser Artemis [14] developed in java by the Sanger Institute.

2.4 Scripting Language

Some pretreatments and/or posttreatments on the inputs and outputs of the tools presented in Subheading 2.2 require the use of scripts to facilitate and automate analyses. In the different protocols presented in the next section, three scripting languages are used for that purpose:

1. Bash: for basic file handling.
2. Perl: for more advanced script, often rooted on the BioPerl library [15].
3. R: for matrix treatments and statistics [16].

2.5 System Requirements

Most of the methods and tools used in this paper are dedicated to UNIX systems although some of them can run under Windows. The different procedures presented in Subheading 3 were developed on a 12-core server under RedHat 6.5 with 96 GB of RAM (*see* **Note 4**).

3 Methods

All the protocols described in this section are summarized in Fig. 1.

3.1 Read Quality Evaluation

Global evaluation of the quality of reads provides fruitful information for further analyses and treatments, especially for the trimming step (*see* Subheading 3.2). Quality assessment is performed

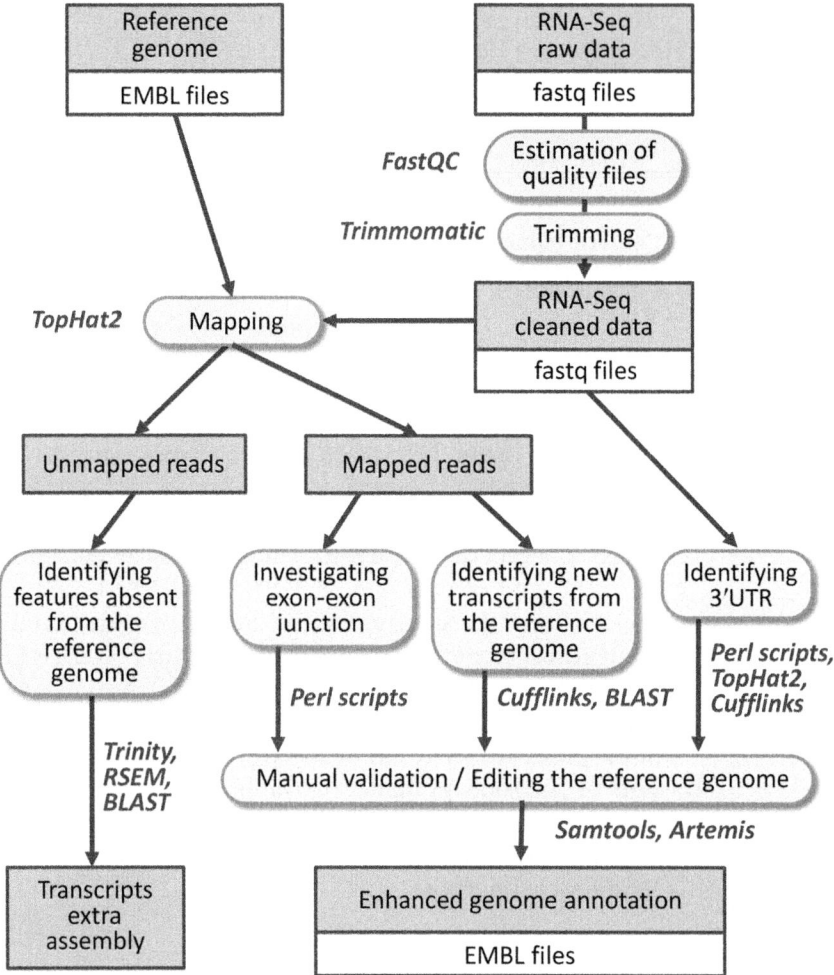

Fig. 1 Global representation of the workflow (tools, data, and file formats) presented in this chapter

with the FastQC tool. This software takes raw fastq files as input, and generates an html report with a variety of diagnostic plots (*see* **Note 5**).

3.2 Trimming RNA-Seq Data

Quality trimming is performed with the Trimmomatic tool. It allows the definition of a complete trimming pipeline for fastq files. For Illumina read data, we propose the following pipeline:

–   ILLUMINACLIP is used as a first step to clip adaptor sequences from reads. It takes 4 arguments as input: (1) a fasta file describing adaptor sequences, (2) the maximum number of mismatches (set to 2 in our cases) and (3–4) two score thresholds, namely *palindromeClipThreshold* and *simpleClipThreshold* (reciprocally set to 15 and 5). For more details, *see* [6] and **Note 6**.

- LEADING and TRAILING subsequently allow to crop reads in 5′ and 3′ if the quality score provided by the fastq file is below a given threshold (set to 5 for both sides).

- SLIDINGWINDOW follows as a next step. It is a more advanced quality trimming algorithm, based on sliding windows. It considers two arguments, (1) the window size (set to 5) and (2) the minimum average quality score for the window (set to 20).

- MINLEN is then called to finalize the trimming. It is a length filter tool, which deletes cropped reads with a length above a given threshold (we generally use 36 for 50 bp reads and 50 bp for 100 bp reads).

The definition of this pipeline is discussed in **Note 7**. The usefulness of read trimming is discussed in **Note 8**.

3.3 RNA-Seq Mapping

After the trimming step, RNA-Seq data are mapped on a reference genome. As we consider eukaryote genomes whose genes may contain introns, we use TopHat2 for mapping reads on our yeast genomes (*see* [17] for a comprehensive comparison of mapping tools). The procedure is the following:

1. Preparing input data: TopHat2 requires reads data in fastq, a fasta file of the chromosomes/scaffolds of the reference genome and eventually a GFF file describing the known transcripts and the coordinates of their exon–exon junctions (*see* **Note 9**). The GFF file of transcripts can be easily obtained either with a BioPerl script, or by using the data selector of a genome viewer such as Artemis.

2. Preprocessing the reference genome: To run TopHat2 a pretreatment is required on the reference chromosomes. To do so, we use the "bundle" function from Bowtie2, denoted bowtie2-build, with default options. Running Bowtie2-build produces an index of the sequences of the genome.

3. Running TopHat2: Here are the options and parameter settings used for mapping reads on yeast chromosomes with TopHat2 (*see* **Note 10**):

   --microexon-search (boolean)

   --min-intron-length 30 (integer, bp)

   --min-coverage-intron 30 (integer, bp)

   --min-segment-intron 30 (integer, bp)

   --max-intron-length 4000 (integer, bp)

   --max-multihits 1 (integer)

   --library-type fr-firststrand (for oriented reads)

Input data are provided as follows:

--GTF<transcript.gff>--transcriptome-index=<path_to_transcriptome_index>

<path_to_bowtie_index><list_of_read_files>.

Log files and results are written into an output directory. In the following protocols, three output files will be considered (*see* **Note 11**):

- *accepted_hits.bam*, which contains the alignment of reads on the reference genome.

- *junctions.bed*, which lists the exon–exon junctions found by the mapping.

- *unmapped.bam*, which enumerates the reads that do not map on the reference genome.

3.4 Investigating Exon–Exon Junctions

As described in Subheading 3.3, TopHat2 provides among other outputs a specific file describing all the exon–exon junctions identified through the mapping of RNA-Seq reads on a reference genome. This file, named "*junctions.bed*" uses the UCSC BED format (*see* **Note 12**). Each splice junction identified by mapping is represented by two connected blocks, where each block is as long as the maximal overhang of any read spanning the junction. Using a scripting language such as BioPerl, one can parse the "*junctions.bed*" file to list coordinates and supporting information for each junction identified. Here is the procedure to extract information from the *junctions.bed* file and compare it to the structural annotations described in the reference embl file(s), in order to validate or correct or identify exon–exon junctions (including alternative splicing events).

1. Parsing the "*junctions.bed*" file: for each junction, store the following information into a hashtable:

   - The name of the chromosome (i.e., first field of the BED table) on which the junction has been identified.

   - The start and stop coordinates of the alignment spanning the junction (i.e., respectively second and third fields of the BED table).

   - The score of the junction (i.e., fourth field of the BED table): TopHat2 gives a score for each junction, representing the number of reads validating the junction. It can be further used to discriminate splicing events strongly supported by RNA-Seq data from alignment artifacts.

   - The block sizes (i.e., 11th field of the BED table): this particular field contains two values that can be described as the maximal portion of reads aligned on both sides of the junction.

- The coordinates of the splice junction, which can be calculated using the alignment coordinates and the block sizes (*see* **Note 13**):

junction start = alignment start + left block size + 1

junction end = alignment end – right block size

All that information should be stored using a unique hashkey that could be used to quickly compare junctions. As such, we use a hashkey combining the name of the chromosome and the start and stop coordinates of the junctions, as calculated above.

2. Parsing the embl file(s) of the reference genome: in a similar fashion, store the junction information that can be extrapolated from an annotated embl file. For each mRNA or misc_RNA features, store the following information into a hashtable:

- The name of the chromosome (i.e., first field of the BED table) on which the feature can be found.
- The locus_tag of the feature.
- The location operator, which contains the coordinates of the various blocks (i.e., exons or UTR regions) that compose the feature. Junction coordinates can be extrapolated by parsing the coordinates of two consecutive blocks:

junction start = end of block 1 + 1

junction end = start of block 2 – 1

As for the "*junctions.bed*" file, information extracted from the embl file(s) should be stored using a unique hashkey, using the same structure (e.g., combining the name of the chromosome and the start and stop coordinates of the junctions, as calculated above).

3. Compare the two hashtables: by using similar hashkey structures, one can quickly compare the tables, solely based on the hashkeys.

- If a key is present in the two tables, then there are evidences in the RNA-Seq data to validate the presence of an already known intron. We name the corresponding junction as "KNOWN".—If a key is only present in the "*junctions.bed*" hashtable, then there are evidences in the RNA-Seq data to highlight the presence of a novel intron. We name the corresponding junction as "NEW". Expert validation is further necessary to decide if the intron is indeed new, or if it corresponds to a correction of a wrongly defined intron, or to an alternative splicing event.
- If a key is not present in the "*junctions.bed*" hashtable, then there are no evidences in the RNA-Seq data to support the presence of an intron described in the embl file(s). We name the corresponding junction as "MISSED".

Expert validation is further necessary to decide whether or not the intron was indeed wrongly defined in the reference genome, or if the junction could not be identified due to the lack of coverage (e.g., non-expressed gene).

While comparing the two hashtables we produce an output containing the information acquired for each junction (i.e., chromosome location, coordinates, and KNOWN/NEW/MISSED status).

3.5 Identifying New Transcripts of the Reference Genome

To identify transcripts or fragment of transcripts that are not included in a known annotation of the reference genome, a possible solution is to assemble mapped reads and to compare the resulting transcripts with those from the annotation. To do so, we propose the following protocol using the Cufflinks tools:

1. Preparing input data: Cufflinks can work directly with the bam files produced by TopHat2. A GFF file with the complete list of known transcripts is also required in this protocol.

2. Cufflinks assembly: For each condition/medium tested, transcript assembly of mapped reads is obtained by running Cufflinks on BAM files retrieved from TopHat2 output directory. RNA-Seq replicates are pooled together (Cufflinks accepts one or more input bam file(s)). We recommend using the option *--mask-file<transcript.gff>* that masks known transcripts from the assembly process (*see* **Note 14**). Cufflinks produces a GTF file (named *transcripts.gtf*) containing transcripts with associated potential isoforms.

3. Filtering transcript isoforms: It is possible to filter transcript isoforms with low covering support. Different statistics of covering and relative abundance for each proposed isoform are available in the *transcripts.gtf* file (*see* **Note 15**). For example, it is possible to select a transcript validated by a minimal number of reads (*cov* parameter). The filtering procedure can be easily implemented with a perl script, based on the use of regular expressions (*see* **Note 16**). We named the filtered GTF file *transcripts_filtered.gtf*.

4. Merge transcripts from the different conditions/media: If RNA-Seq experiments include different growth conditions/media, it can be interesting to merge the different transcripts together in order to obtain consensus transcripts reflecting the whole experiments. To do so, we run the *cuffmerge* tool on a list of predicted transcripts GTF files (in a simple text file, one file path per line). A relative abundance filter threshold can be setup, optionally, for example, *--min-isoform-fraction 0.25* (meaning that only isoforms whose relative proportion is over 25 % are kept). Comments about this step are available in **Note 17**.

5. BLAST search on potential new transcripts: This is an optional step. In order to facilitate the analysis of the obtained transcripts, BLAST search can be performed, for example with BLASTX on RefSeq database (protein sequences from NCBI Reference Sequence project; www.ncbi.nlm.nih.gov/refseq/) or on another selection of protein sequences.

The resulting list of transcripts allows the identification of different kind of elements and structures:

– Transcripts distant from known features: new transcribed features (coding DNA sequence (CDS), non-coding RNA, miscellaneous RNA).

– Transcripts joined to known features: 5′ and 3′ untranslated regions (UTR). *See* **Note 18** for comments.

3.6 Identifying 3′ UTR

The aim of the following protocol is to identify the polyadenylation sites of transcripts, and hence to find their 3′UTR. This procedure relies on TopHat2, cufflinks, and a couple of perl scripts:

1. Filtering reads ending with a poly-A monomer (*see* **Note 19**): This can be done with a perl script and one regular expression. Poly-A tails are removed from the reads, which are then retained and saved in a separated fastq file.

2. Mapping the selected reads: Using the same procedure given in Subheading 3.3, the selected reads are mapped to the reference genome with TopHat2.

3. Mapped read assembly: the mapped reads from **step 2** (*accepted_hits.bam*) are then assembled with cufflinks (*see* **step 2** from the Subheading 3.5, without the masking option).

4. Identifying poly-A sites: Fragments of transcripts generated in the previous **step 3** are analyzed to determine if a poly-A exists in the genome just upstream or downstream from the mapped transcript (depending on the strand):

– If a poly-A is found, the fragment is not a poly-A site.

– If not, the fragment is a poly-A site. Then, the coordinates of the mRNA/ncRNA/misc_RNA can be extended.

3.7 Identifying Features Absent from the Reference Genome

It is noteworthy that the reference genome used to map RNA-Seq reads can be incomplete (e.g., gapped scaffolds, missing subtelomeric regions). RNA-Seq data offer an invaluable opportunity to fill this gap as it can help to identify transcripts that are missing from the reference sequences. We define the following protocol using the de novo assembler Trinity:

1. Preparing input data: Trinity assembler needs a fastq file of the unmapped read. These reads are retrieved from the output directory of TopHat2 (*see* Subheading 3.3), from the file *unmapped.bam*. The latter is converted into fastq format with the bundle function from the TopHat2 tool suite, *bam2fastx*.

2. De novo assembly of transcripts: The obtained reads are assembled with Trinity, using the Perl wrapper Trinity.pl provided by the Trinity tool suite (*see* **Note 20**). Assembled transcripts and the associated isoforms are written in the fasta file *trinity.fasta*.

3. Preparing trinity outputs for filtering: Transcripts from *trinity.fasta* are indexed in order to recomputed covering and abundance statistics. We recommend to use the following command line: *bowtie2-build trinity.fasta trinity.fasta.TRANS* (*see* **Note 21**).

4. Computing statistics: For each transcript, covering and abundance statistics are computed with the RSEM tool. We use the provided Perl wrapper run_RSEM_align_n_estimate.pl. It requires the fasta file of transcripts and the fastq file(s) of the RNA-Seq reads. This tool must be run in the same directory as for the previous step. Statistics are written in the output directory, in the text file *RSEM.isoforms.results*.

5. Filtering isoforms: We use the Perl wrapper script filter_fasta_by_rsem_value.pl from the Trinity tool suite to filter transcript isoforms according to the statistics computed in the previous step. Three cutoffs can be set:

   --tmp-cutoff: Transcripts per million.

   --fpkm-cutoff: Fragment per kilobase per million.

   --isopct-cutoff: relative abundance of isoforms (in %).

   From a trial and error procedure, we selected the two following cutoffs: *--tmp-cutoff 10* and *--isopct-cutoff 30*.

6. BLAST search against reference genome: The aim of this step is to identify transcripts that do not match on the reference genome (*see* **Note 22**). Thus, a BLASTN against the reference genomic sequences or a BLASTX against the reference proteome is performed with the selected transcripts from the previous steps as queries. Only "no hit" queries are kept as potential new transcripts.

7. BLAST search against external references: This is an optional step. To go further, the identified transcripts can be "blasted" against different resources, such as other yeast proteomes or a more general protein database (e.g., RefSeq or nr non-redundant database ftp://ftp.ncbi.nih.gov/blast/db/nr), in order to determine if these transcripts correspond to protein-coding sequences (*see* **Note 23**).

The above procedure allows the identification of unknown CDS and/or untranslated transcripts but also to reveal potential biological contaminations (*see* **Note 24**).

3.8 Manual Validation

All the protocols described above lead to identify features or regions that can be amended in the reference genome. However, this last step cannot be done without human expertise. Ultimately, all the potential amendments identified by the previous protocols must be

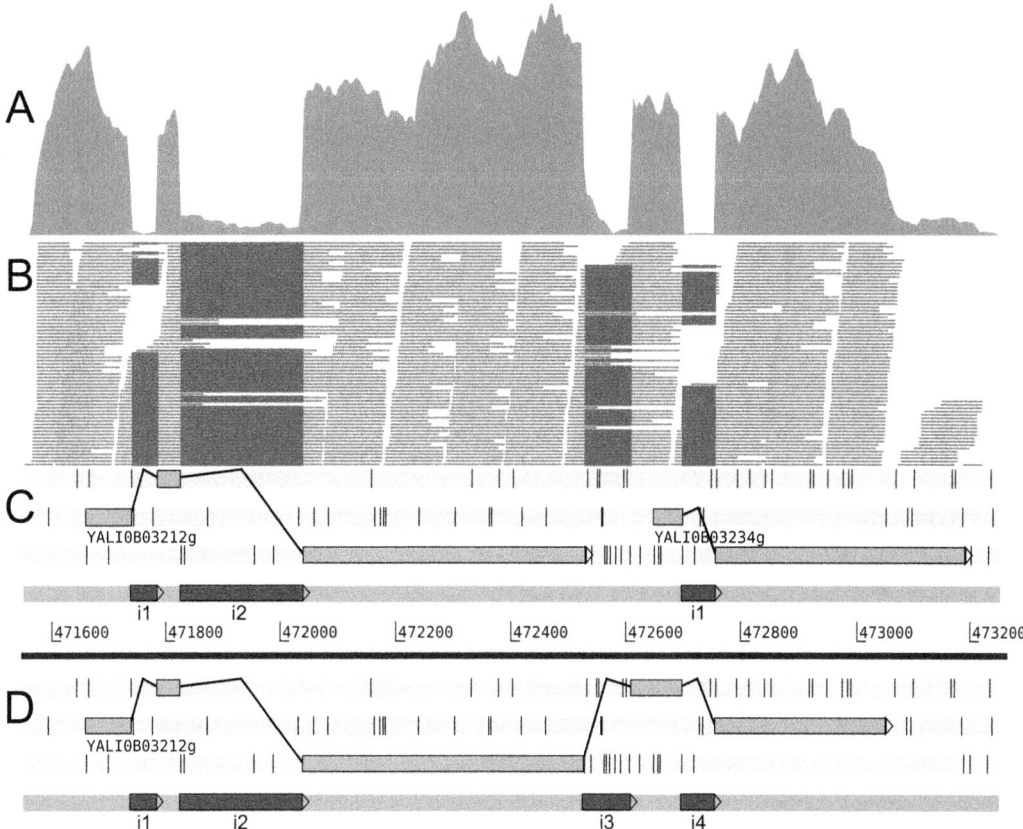

Fig. 2 Illustration of annotation amendments retrieved from the Artemis genome browser on a genomic region of the chromosome B of *Yarrowia lipolytica*: (**a**) RNA-Seq read global coverage; (**b**) "stack" representation of read mapping, light fragments are matching reads, *dark rectangles* are introns deduced from exon–exon junction prediction; (**c**) Initial (predicted) annotation. *Light rectangles* depict CDS and *dark rectangles* predicted introns; (**d**) corrected annotation. In this example, the investigation of RNA-Seq reveals that the two genes YALI0B03212g and YALI0B03234g are in fact parts of a unique gene. A new intron is identified and the 3′ end of the gene is on another frame. Inspection of read coverage shows potential intron retention for the second intron (i2)

manually validated. To do so, we use a genome browser, Artemis, able to handle all the different outputs obtained from the different procedures (i.e., bam, gtf, gff, embl, fasta), as well as to easily edit the reference annotation. *See* Fig. 2 for illustration.

3.9 Editing Reference Sequences

RNA-Seq can reveal sequencing errors in the reference genome. Indeed, during the mapping procedure (Subheading 3.3), a certain number of mismatches can be highlighted between reads and the reference genome. Substitutions and/or insertions/deletions (indels) supported by a high number of reads for a given position strongly suggest that this position is wrong in the reference sequence (*see* **Note 25**). To identify these positions, it is possible to use the Samtools suite. The following command lines allow the identification of the most supported sequence variations:

samtools faidx<ref.fasta>

samtools mpileup -uf<ref.fasta><list-of-bam-files>| bcftools view -bvcg ->var.raw.bcf

bcftools view var.raw.bcf>var.raw.vcf

vcfutils.pl varFilter -d 50>var.flt.vcf

Where *<ref.fasta>* is the fasta file of the reference chromosomes and *<list-of-bam-files>* a space separated list of bam files containing the mapped reads from TopHat2. The most supported variations are then given in the text file *var.flt.vcf* (*see* **Note 26**).

4 Notes

1. Three RNA-Seq datasets were used to illustrate the different processes exposed in this paper. The first one is from *Yarrowia lipolytica* CLIB122 on six different conditions/media: oleic acid, alkane, YPD (yeast extract, peptone, and glucose, 10 g/L each) at 28 °C, YPD at 18 °C, YPD at pH 4, YPD and H_2O_2. Sequencing was performed with Solexa Illumina HiSeq 2000, in paired-end 100 bp (clusters of 300 bp) with two replicates per condition/medium. The second dataset is also from *Yarrowia lipolytica* CLIB122, on YPD at 28 °C without replicate. Sequencing technology was Solexa Illumina with HiSeq 2000 sequencing system. Strand-specific single reads of 50 bp were obtained. The last dataset is also from Solexa Illumina HiSeq 2000 sequencing, 50 bp single reads, non-stranded, one experiment on YNB medium (yeast nitrogen base, 17 g/L, NH_4Cl 5.3 g/L, NaK buffer 50 mM, no yeast extract) supplemented with 1 % glucose, without replicate.

2. In addition to the sequencing technology employed, different "parameters" can vary in the design of RNA-Seq experiments, such as the read length, the library depth (coverage), the number of replicates, the use of single or paired reads, stranded or non-stranded, the number of media/condition tested, and so on. Making appropriate choices is crucial and directly depends on the aim of the study. Thus, it is well admitted that for differential gene expression analysis, the number of replicates is a critical parameter and must be as high as possible to ensure the statistical evaluation of gene expression. Alternatively, if RNA-Seq data are used to amend gene structural annotations, the two most important parameters are the depth of libraries, to catch genes with low expression level and alternative transcripts,

and the number of media/conditions, as some genes can be expressed only under specific conditions.

3. To amend a reference genome with RNA-Seq, it is highly recommended to use the same strain for RNA-Seq experiments. Using a different strain may produce many single nucleotide polymorphisms (SNP) when mapping on genome and therefore many read losses.

4. As mentioned in Subheading 2.5, UNIX systems are highly recommended for the different procedures described in this paper. However, all the mentioned tools can be more or less easily installed on Windows systems, generally requiring compilation from the source code. A traditional desktop configuration can be enough to do the job. However, most of the tools discussed in this chapter propose optional multi-threading optimization reducing significantly computational time. As a consequence, multi-core servers or cluster of computers present a great advantage.

5. Quality evaluation of reads is often provided by the sequencing platform, which highly facilitates the parameter setting of trimming algorithms (e.g., quality thresholds, region to crop, minimal length to keep reads). In addition, fastQC shows eventual read contaminations (e.g., overrepresented sequences) such as sequencing adaptors or rRNA fragments.

6. The match score used in the module ILLUMINACLIP is based on the number of matches between the read and the adaptor sequences coupled with the quality of read sequencing provided in the fastq files. The *palindromeClipThreshold* parameter is used when treating paired-end reads (as an adaptor can be found in the both strands), the *simpleClipThreshold* is used for single reads.

7. The parameter values, presented in the Trimmomatic pipeline, were calibrated on the first dataset (*see* **Note 1**). Obviously, depending on the dataset to trim, some adjustments can be required.

8. It is well admitted that mapping tools are generally not influenced by quality issues and adaptor contamination as they compare reads to a reference genome. However, if the quality of sequencing is particularly low or if the cover (number of reads) is low too, the use a specific trimming tool such as Trimmomatic provides better results.

9. One of the great advantages of TopHat2 is to consider known transcripts to facilitate intron detection and avoid spurious transcript isoforms supported by too few reads. This is optional but highly recommended.

10. The proposed options and parameter settings were calibrated with different RNA-Seq datasets including those presented in Subheading 2.1.

11. A fourth output file from TopHat2 should be checked before going further, it is the *align_summary.txt* file. It contains basic statistics about the number of mapped reads and unmapped reads. This allows to identify potential issues during the mapping step. Thus, for example, it is admitted that if less than 80 % of reads map the genome (for Illumina sequencing) there is probably a problem on the reads or on the parameter setting.

12. A complete description of the BED format can be found on the University of California Santa Cruz Genome Browser website (http://genome.ucsc.edu/FAQ/FAQformat.html#format1).

13. When working with oriented reads, one can use the strand information (i.e., fifth field of the BED table) to determine the orientation of the junction alignment. Knowing this information, one can then identify the junction coordinates calculated in Subheading 3.3 as 5′ or 3′ splice sites.

14. Masking known transcripts during the cufflinks assembly is not necessary but this allows focusing only on the detection of unknown transcripts and hence, is highly recommended.

15. Three statistics can be used for filtering transcripts: the read covering, the reads per kilobase per million (RPKM) and the fragments per kilobase per million (FPKM). It is not necessary to set threshold on the three parameters at the same time, but filtering allows reducing significantly the number of spurious transcript isoforms. Last, it is noteworthy that there are no optimal values for filtering these transcripts. Too stringent filtering will produce smaller and fewer transcripts while too permissive filtering will lead to many inconsistent fragments. The best is to try different settings and mix the results.

16. Here are some examples of regular expressions (in Perl) that can be used to retrieve the different statistical values from the *transcripts.gtf* file produced by cufflinks:
    - To get the covering value: */cov\"([\d\.]+)\"/*
    - To get the FPKM value: */FPKM\"([\d\.]+)\"/*

17. Merging is recommended but not mandatory. It allows reducing the number of transcripts by joining the different fragments that overlap between the different experiments. Most of the time, larger and more consistent transcripts are obtained after merging. Note that the *cuffquant* tool allows the recomputation of the relative read coverage of each merged transcript according to the different conditions/media.

18. Part of the obtained transcripts can be unusable. Indeed, especially for non-stranded reads, overlapping features (5′ and 3′ UTR) or small intergenic regions will lead to large predicted transcripts without clear feature junctions.

19. The protocol described in Subheading 3.6 is for stranded reads. For non-stranded reads, it is necessary to consider poly-T monomers at the same time.

20. Trinity de novo assembly can be rather time-consuming and hence should be run using multi-threading options, (*--CPU X*) as well as the memory allocation limit (*--JM XG*), whenever possible.

21. We recommend using the suffix pattern ".TRAN" as it is used by default in the next step.

22. Without BLAST filtering, most of the identified transcripts are in fact chimerical transcripts due to repeat regions or palindromic sequences, and hence, must be deleted. Note that the *e*-value threshold of the BLAST search can change the number of "no hit" transcripts. We recommend the use of a stringent threshold (low value) to eliminate spurious transcripts.

23. This procedure provides only transcripts and not gene structures. For instance, exon–exon junctions of intron-containing genes will not be predicted.

24. The transcripts that do not match to the reference genome may have two possible origins: either they come from a non-contigated part of the genome or from contaminations during RNA-Seq production. To differentiate between these two origins, further investigations have to be done taking into account non-assembled reads from the reference genome assembly.

25. Reference genome sequences may contain wrong bases or indels. This can come from the sequencing or the assembly. For example, it is well admitted that 454 sequencing often fails for sequencing homopolymers, inducing indels and hence frameshifts in coding sequences.

26. Sequence variations or SNP (for single nucleotide polymorphism) between reads and the reference genome have different origins:
    - RNA-Seq sequencing error: In that case, the SNP frequency is expected to be low.
    - Error on reference sequence: In that case, the SNP frequency should be high.
    - Repeated regions present in several conserved genes: In that case, the SNP frequency should have an average range of values and several other SNP are expected on the same feature.

References

1. Eilbeck K, Moore B, Holt C et al (2009) Quantitative measures for the management and comparison of annotated genomes. BMC Bioinformatics 10:67

2. Guida A, Lindstädt C, Maguire SL et al (2011) Using RNA-seq to determine the transcriptional landscape and the hypoxic response of the pathogenic yeast Candida parapsilosis. BMC Genomics 12:628

3. Dujon B, Sherman D, Fischer G et al (2004) Genome evolution in yeasts. Nature 430:35–44

4. Morin N, Calcas X, Devillers H et al (2014) Draft genome sequence of Rhodosporidium toruloides CECT1137, an oleaginous yeast of biotechnological interest. Genome Announc 2:e00641–14

5. Andrews S. FastQC: a quality control tool for high throughput sequence data http://www.bioinformatics.babraham.ac.uk/projects/fastqc/. Accessed 6 Feb 2015

6. Bolger AM, Lohse M, Usadel B (2014) Trimmomatic: a flexible trimmer for Illumina sequence data. Bioinformatics 30:2114–2120

7. Langmead B, Salzberg SL (2012) Fast gapped-read alignment with Bowtie 2. Nat Methods 9:357–359

8. Kim D, Pertea G, Trapnell C et al (2013) TopHat2: accurate alignment of transcriptomes in the presence of insertions, deletions and gene fusions. Genome Biol 14:R36

9. Trapnell C, Roberts A, Goff L et al (2012) Differential gene and transcript expression analysis of RNA-seq experiments with TopHat and Cufflinks. Nat Protoc 7:562–578

10. Grabherr MG, Haas BJ, Yassour M et al (2011) Full-length transcriptome assembly from RNA-Seq data without a reference genome. Nat Biotechnol 29:644–652

11. Li B, Dewey CN (2011) RSEM: accurate transcript quantification from RNA-Seq data with or without a reference genome. BMC Bioinformatics 12:323

12. Li H, Handsaker B, Wysoker A et al (2009) The sequence alignment/map format and SAMtools. Bioinformatics 25:2078–2079

13. Camacho C, Coulouris G, Avagyan V et al (2009) BLAST+: architecture and applications. BMC Bioinformatics 10:421

14. Rutherford K, Parkhill J, Crook J et al (2000) Artemis: sequence visualization and annotation. Bioinformatics 16:944–945

15. Stajich JE, Block D, Boulez K et al (2002) The Bioperl Toolkit: Perl modules for the life sciences. Genome Res 12:1611–1618

16. Core Team R (2013) R: a language and environment for statistical computing. R Foundation for Statistical Computing, Vienna

17. Engström PG, Steijger T, Sipos B et al (2013) Systematic evaluation of spliced alignment programs for RNA-seq data. Nat Methods 10:1185–1191

Chapter 3

Pathogen Gene Expression Profiling During Infection Using a Nanostring nCounter Platform

Wenjie Xu, Norma V. Solis, Scott G. Filler, and Aaron P. Mitchell

Abstract

NanoString nCounter is a recently developed platform that can make direct multiplexed measurement of gene expression using color-coded probe pairs (Geiss et al., Nat Biotechnol 26(3):317–325, 2008; Malkov et al., BMC Res Notes 2:80, 2009). We have found that this platform is uniquely suitable for quantification of pathogen gene expression during infection, where pathogen RNA comprises a tiny portion of total RNA isolated from the infected tissue. Here, we describe a protocol that we have successfully applied to a number of pathogens across multiple infection models, including both invasive and mucosal infection by *Candida albicans*, and lung infection by *Aspergillus fumigatus* and *Cryptococcus neoformans*.

Key words Pathogen, Infection, RNA isolation, Gene expression, In vivo profiling, NanoString

1 Introduction

What genes does a pathogen express during infection to cope with the host environment? Which regulatory pathways are responsible for the onset and retreat of such responses? These are among the most important questions in microbial pathogenesis. However, for most mammalian pathogens, gene expression profiling studies have been limited by the technical difficulty to accurately quantify pathogen gene transcripts from infected tissues, even in light of new genome-wide technologies [1–3]. Host RNA constitutes an overwhelming portion (usually >99 %) of the total RNA isolated from infected tissue samples. This poses a challenge for most expression profiling technologies: it contributes to high background on microarrays, and it dominates sequence reads from RNA-Seq. Pathogen cell isolation from infected tissue, in theory, can help to enrich for pathogen RNA, but it poses additional problems: it requires large quantities of infected tissue; the procedure can be tedious; and most importantly, it is difficult to conserve the native state or integrity of RNA during the lengthy process. Here we describe an in vivo gene expression profiling protocol that is

Frédéric Devaux (ed.), *Yeast Functional Genomics: Methods and Protocols*, Methods in Molecular Biology, vol. 1361, DOI 10.1007/978-1-4939-3079-1_3, © Springer Science+Business Media New York 2016

fast, extremely sensitive and highly reproducible (*see* **Note 1**). We developed this protocol during our investigation of the fungal pathogen *Candidaalbicans* in a murine model of hematogenously disseminated candidiasis. Using this protocol, we have documented time courses of dynamically regulated *C. albicans* gene expression during kidney infection, and discovered unexpected features of the gene expression response to antifungal drug treatment in vivo [4]. We have successfully applied this protocol to a number of other tissue types, pathogens, and infection models [5–7].

2   Materials

2.1   RNA Isolation from Infected Tissue

1. GentelMACS dissociator and M-type homogenization tube (Miltenyi Biotec).
2. Minibeadbeater.
3. Tabletop centrifuge.
4. Zirconia beads.
5. Phenol–chloroform–isoamyl alcohol 25:24:1.
6. RNeasy kit, including buffers RLT, RW, and RPE, and RNeasy spin columns.
7. BioPhotometer.

2.2   NanoString Profiling

1. NanoString nCounter system (nanoString).
2. NanoString nCounter master kit (nanoString).
3. NanoString custom-built codeset: The investigator selects the high priority genes and nanoString will design the probes and synthesize the codeset reaction mix. For in vivo profiling studies, it is important to remind the nanoString codeset design team to avoid probe sequences that could cross-hybridize to host genes.
4. Thermal cycler.
5. MultiExperiment Viewer 4 software [8].

3   Methods

3.1   RNA Isolation from Infected Tissue (Using Mouse Kidney as an Example)

1. Male Balb/c mice weighting 20–22 g were used for all studies. Three mice per experimental group were inoculated intravenously with 1×10^6 yeast-phase *C albicans* cells. The animals were sacrificed and kidneys were harvested at specific time points post-infection. The right kidney was snap frozen in liquid nitrogen and stored at −80 °C in a screw cap tube for later RNA extraction (*see* **Note 2**).

2. Prepare the following reagents before removing kidneys from the −80 °C freezer: add 2-mercaptoethanol (1% V/V) to buffer RLT; label M-tubes and chill on ice; label 2 ml screw cap tubes, and add approximately 300 μl Zirconia beads.

3. Remove kidneys from −80 °C freezer and put on ice. Add 1.2 ml of buffer RLT with 2-mercaptoethanol to each kidney (*see* **Note 3**). Decant kidney with buffer into an M-tube.

4. Homogenize the kidney in M-tube using gentelMACS dissociator on pre-loaded setting RNA_02.01.

5. Centrifuge the M-tube at $1000 \times g$ for 1 min in a tabletop centrifuge at room temp.

6. Transfer 600 μl of homogenate from M-tube to the screw cap tube containing Zirconia beads. Save the remaining homogenate on ice.

7. Add 600 μl phenol–chloroform–isoamyl alcohol 25:24:1 to the tubes from the previous step.

8. Close the lids tightly and vortex on mini-beadbeater for 3 min in a 4 °C cold room.

9. Centrifuge the tubes at $15,000 \times g$ for 5 min in a 4 °C cold room.

10. Carefully transfer the aqueous phase to a new 1.5 ml microfuge tube, mix well with equal volume of 70 % ethanol, then load onto the RNeasy spin column.

11. Wash the spin column once with 700 μl buffer RW, followed by twice with 500 μl buffer RPE. Centrifuge 1 extra minute in a dry collection tube to remove remaining liquid in the spin column.

12. Elute RNA with 50 μl of H_2O, and measure the RNA concentration using a BioPhotometer (*see* **Note 4**).

3.2 NanoString Profiling (12 Reactions)

1. Thaw the nanoString reporter codeset (green cap tube) and capture codeset (gray cap tube) on ice.

2. Add 130 μl hybridization buffer to the reporter codeset (green), invert to mix and spin down.

3. Add 20 μl of the mix to each of the 12 reaction tubes.

4. Add 10 μg of total tissue RNA (in a volume of 5 μl) to each tube (*see* **Note 5**). Mix by pipetting.

5. Add 5 μl capture codeset to each tube. Mix by pipetting.

6. Incubate the reaction at 65 °C in a thermal cycler overnight (12–18 h).

7. Take out one sealed sample cartridge (−20 °C) and two prep plates (4 °C). Let them warm to room temperature.

8. Centrifuge the prep plates at $670 \times g$ for 2 min in a tabletop centrifuge.

9. Set up the nanoString prep station following on-screen instructions.

10. Remove the reactions from the thermal cycler and immediately load on the prep station. Select to run the high sensitivity program (3 h).

11. When the prep station program is complete, remove the cartridge, seal the lanes with a clear tape (provided by nanoString), apply mineral oil to the bottom of the cartridge (for generation one nCounter only, later generations do not require this step), then load the cartridge onto the nanoString digital analyzer.

12. Set up the nanoString digital analyzer following on-screen instructions.

13. Select the high resolution (600 fields) option, run the scanning program (~4.5 h).

14. Upon receiving the results (by email), import raw data into nSolver software (provided by nanoString). The software will automatically check data quality and raise flags if the quality of the data falls out of the normal range. Perform technical adjustment using the built-in function (optional, follow instructions in the software), then export the data as an excel file (*see* **Note 6**).

15. Normalize the data using one of the following methods (*see* **Note 7**): total counts from all genes in the codeset; one or a few internal control genes; geometric mean of highly expressed genes.

16. Calculate the mean expression values for each gene (if the experiment was done in replicates or triplicates), then calculate the ratio of expression levels among different experimental groups (*see* **Note 8**).

17. Visualize the datasets in heap maps by a clustering program such as MultiExperiment Viewer [8].

4   Notes

1. This protocol requires total processing time of less than 48 h, from tissue to expression profiling data. The hands-on time is around 4 h for 12 samples. While it is feasible to perform RNA extraction for 12 samples at the same time, we recommend doing six samples at a time to ensure quality. On the other side, we recommend to run a full nanoString cartridge (12 samples) at a time when possible, both to save material costs and to improve consistency of data.

2. We first tried using RNAlater (Life Technologies) to preserve tissue samples before RNA isolation. We found RNAlater

solution adds viscosity to tissue homogenate and adversely affects quality and quantity of RNA recovery. We then tried snap-freezing tissues in liquid nitrogen followed by storage at −80°C before RNA isolation. We were able to consistently recover RNA with high quality and high yield from snap-frozen tissues.

3. One key variable in this protocol is the amount of buffer RLT added to the tissue at the very first step. Too much buffer will dilute the homogenate and lead to lower RNA concentration, while too little buffer may lead to formation of viscous gels after the phenol–chloroform extraction step and leave no aqueous phase for RNA recovery. We have empirically determined the optimal volume of buffer to use for the following tissues (assuming typical sizes): one mouse kidney (1.2 ml), one mouse tongue (1.0 ml), one mouse lung (2.0 ml).

4. To improve the recovery of RNA, add the first round eluate back to the column, and elute again. We can routinely recover approximately 100 μg of total tissue RNA from one RNeasy spin column (~2 μg/μl × 50 μl). If larger amount of RNA is needed, a second prep can be made from the remaining tissue homogenate. Do not dispose of the remaining homogenate until RNA concentration has been measured, just in case.

5. Depending on the infection model, the inoculum size and the pathogen strain, the percentage of pathogen RNA in total tissue RNA varies from 0 to 2 %, and typically falls within the 0.05–0.5 % range. For example, 10 μg of total tissue RNA from *C. albicans* infected kidney could generate nanoString raw counts equal to that of 10 ng of pure *C. albicans* RNA from an in vitro culture (10 ng/10 μg = 0.1 %).

6. One main challenge for pathogen gene expression profiling in vivo is to get enough reads from pathogen RNA. Given the low percentage of pathogen transcripts in total RNA, we have to use large amount of total RNA. The nanoString platform has a unique advantage in this perspective: it is so specific in recognizing the target RNA that the overwhelming amount of host RNA does not cause a significant level of noise. As shown in Fig. 1 (adapted from ref. [7]), raw counts from uninfected tissue sample were all below 10, while raw counts from infected tissue samples ranged between <10 and $>10^5$. Only four out of 135 genes fell below the noise levels.

7. Because the percentage of pathogen RNA in total tissue RNA varies in a wide range, we do not really know the quantity of pathogen RNA in a given amount of total RNA that we use for hybridization with nanoString probes. Therefore, normalization using pathogen genes is a critical step before the expression profiles can be compared among different

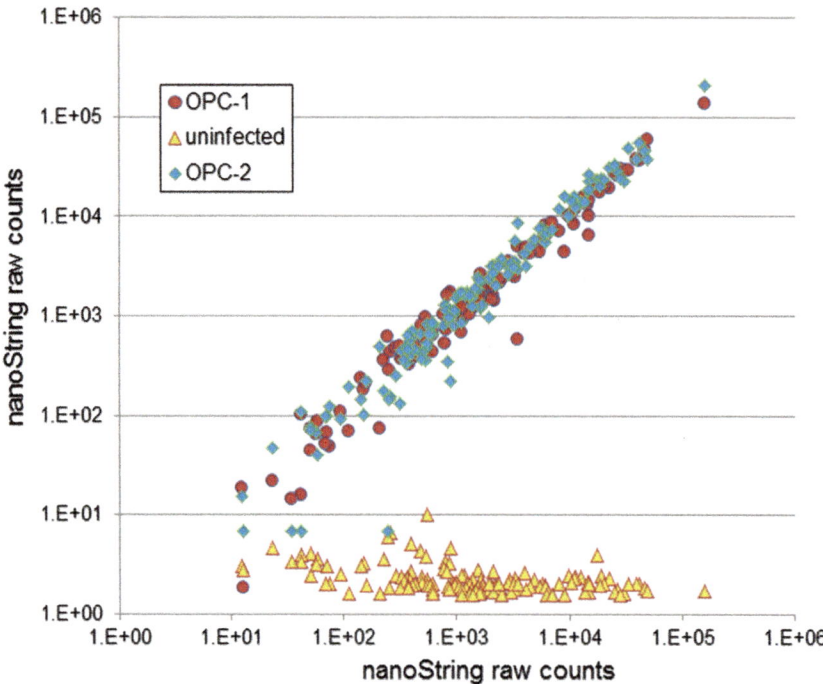

Fig. 1 Raw nanoString counts for 135 *C. albicans* environmental response genes. NanoString probe counts for infected (*red* and *blue* data points) and uninfected (*yellow*) mouse tongue samples are presented as scatter plots for the 2 days postinfection time point. Adapted from ref. [7]

samples. There are three commonly used methods for normalization. (1) Use total counts from all genes in the codeset. (2) Use one or a few "housekeeping" genes. (3) Use the geometric mean (Nth root of the product of N numbers) of highly expressed genes. Each method has its pros and cons. For a large codeset (>100 genes) containing "randomly" selected probes (such as all genes in the genome that encode a protein with a DNA-binding domain), using total counts for normalization can be a good choice, because the total counts of a large number of unrelated genes may faithfully reflect the amount of RNA input. For a small codeset (<100 genes) containing probes focused on a specific process (such as hyphal growth, given that hyphal growth genes tend to be co-regulated), choosing one or a few housekeeping genes as control for normalization is essential. *TDH3*, a robustly expressed metabolic gene, has served well as control for many of our experiments. The third method is a hybrid of method 1 and 2, with an emphasis for equal contribution from highly

Table 1

Three methods to calculate normalization factors

Methods	OPC sample 1	OPC sample 2	OPC sample 3	Sample 1–3 average
Total counts for all genes	196,429	469,794	152,306	272,843
Normalization factor based on total counts (method 1)	1.39	0.58	1.79	
Counts for TDH3	10,813	21,362	6677	12,951
Normalization factor based on TDH3 (method 2)	1.20	0.61	1.94	
Geometric mean for 20 most highly expressed genes	5602	14292	4172	8022
Normalization factor based on geometric mean (method 3)	1.43	0.56	1.92	

expressed genes. In Table 1, we use oropharyngeal candidiasis (OPC) experiment data from Fanning et al. [7] to demonstrate how the normalization factors are calculated and to what extent these different normalization methods could affect the outcome of expression profiling. In Table 1, row 2, total counts for all genes for sample 1–3 are 196,429, 469,794 and 152,306, and the average for the three samples is 272,843. Divide the average total counts by sample 1 total counts, we get the normalization factor for sample 1 as 273,943/196,429 = 1.39. Following the same calculations we can generate normalization factors for each sample by all three methods. In the case shown in Table 1, normalization factors based on the three methods are within 15 % difference from each other, hence unlikely to have a significant influence on the interpretation of the profiling data.

8. Expression data generated using this protocol are highly reproducible (Fig. 1, comparing red and blue dots; and Fig. 2, comparing among biological triplicates in a heat map). The nanoString platform is extremely sensitive and has a dynamic range encompassing the biological expression levels. We were able to quantify transcripts level for >95 % of genes in our codeset in the OPC model (Figs. 1 and 2), and were able to discern dynamic gene expression changes in a time course study of kidney infection [4].

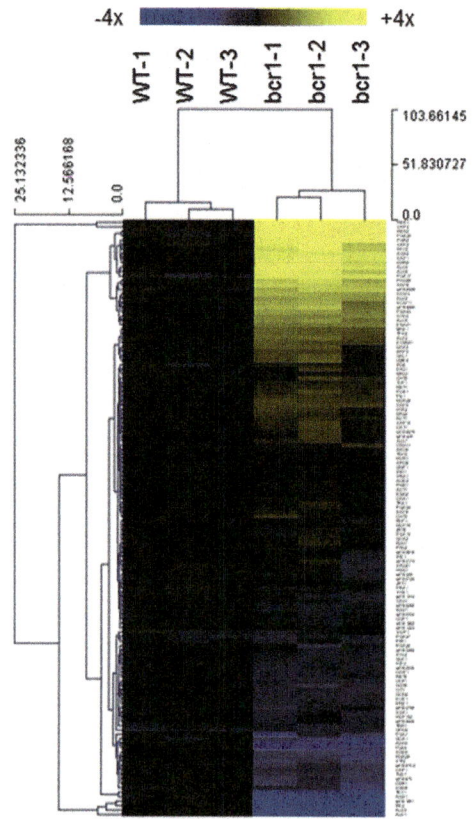

Fig. 2 Expression of *C. albicans* environmentally responsive genes during OPC. Expression levels for 135 *C. albicans* genes at 2 days postinfection in the mouse OPC model are presented in a heat map format. Expression levels were normalized to the mean value of the wild type. Color saturation represents the extent of the expression change, with full saturation at fourfold upregulation or downregulation. Adapted from ref. [7]

Acknowledgement

This work was supported in part by NIH grants R21 DE023311 (A.P.M.), R56 AI111836 (A.P.M. and S.G.F.), R01 AI054928 (S.G.F.), and R01 DE017088 (S.G.F.).

References

1. Geiss GK, Bumgarner RE, Birditt B, Dahl T, Dowidar N, Dunaway DL, Fell HP, Ferree S, George RD, Grogan T, James JJ, Maysuria M, Mitton JD, Oliveri P, Osborn JL, Peng T, Ratcliffe AL, Webster PJ, Davidson EH, Hood L, Dimitrov K (2008) Direct multiplexed measurement of gene expression with color-coded probe pairs. Nat Biotechnol 26(3):317–325

2. Malkov VA, Serikawa KA, Balantac N, Watters J, Geiss G, Mashadi-Hossein A, Fare T (2009) Multiplexed measurements of gene signatures in different analytes using the Nanostring nCounter Assay System. BMC Res Notes 2:80

3. Westermann AJ, Gorski SA, Vogel J (2012) Dual RNA-seq of pathogen and host. Nat Rev Microbiol 10(9):618–630

4. Xu W, Solis NV, Ehrlich RL, Woolford CA, Filler SG, Mitchell AP (2015) Activation and alliance of regulatory pathways in *C. albicans* during mammalian infection. PLoS Biol 13(2), e1002076

5. O'Meara TR, Xu W, Selvig KM, O'Meara MJ, Mitchell AP, Alspaugh JA (2013) The Cryptococcus neoformans Rim101 transcription factor directly regulates genes required for adaptation to the host. Mol Cell Biol 34(4):673–684

6. Cheng S, Clancy CJ, Xu W, Schneider F, Hao B, Mitchell AP, Nguyen MH (2013) Profiling of Candida albicans gene expression during intra-abdominal candidiasis identifies biologic processes involved in pathogenesis. J Infect Dis 208(9):1529–1537

7. Fanning S, Xu W, Solis N, Woolford CA, Filler SG, Mitchell AP (2012) Divergent targets of Candida albicans biofilm regulator Bcr1 in vitro and in vivo. Eukaryot Cell 11(7):896–904

8. Saeed AI, Sharov V, White J, Li J, Liang W, Bhagabati N, Braisted J, Klapa M, Currier T, Thiagarajan M, Sturn A, Snuffin M, Rezantsev A, Popov D, Ryltsov A, Kostukovich E, Borisovsky I, Liu Z, Vinsavich A, Trush V, Quackenbush J (2003) TM4: a free, open-source system for microarray data management and analysis. Biotechniques 34(2):374–378

Chapter 4

Comparative Transcriptomics in Yeasts

Dawn A. Thompson

Abstract

Comparative functional genomics approaches have already shed an important light on the evolution of gene expression that underlies phenotypic diversity. However, comparison across many species in a phylogeny presents several major challenges. Here, we describe our experimental framework for comparative transcriptomics in a complex phylogeny.

Key words Evolution, Gene regulation, Comparative expression profiling

1 Introduction

Divergence in gene regulation can play a major role in evolution. Among eukaryotes, the *Ascomycota* fungi provide an excellent model to study the evolution of gene regulation [1–11]. They include the model organisms *Saccharomyces cerevisiae*, *Schizosaccharomyces pombe*, and *Candidaalbicans*, as well as many non-model, genetically tractable species with sequenced genomes. Their physiology is also well understood, but it is also surprisingly diverse: for example, different species of yeast colonize different ecological niches, utilize a range of different carbon sources, and differ in their preference for oxidative phosphorylation vs. a more fermentative lifestyle. Species in the phylogeny diverged before and after a whole genome duplication event [12, 13] (WGD), allowing us to study the consequences of this evolutionary mechanism [12–14].

Despite these advantages, collecting experimental data across species and implementation of appropriate analytical approaches is a complex problem. An important challenge is to collect experimental data in such a way that would minimize irrelevant differences, for example, due to growth conditions, and allow focusing on true evolutionary distinctions (Fig. 1). To address these challenges we developed an experimental and computational framework to understand the evolution of modular gene

Frédéric Devaux (ed.), *Yeast Functional Genomics: Methods and Protocols*, Methods in Molecular Biology, vol. 1361, DOI 10.1007/978-1-4939-3079-1_4, © Springer Science+Business Media New York 2016

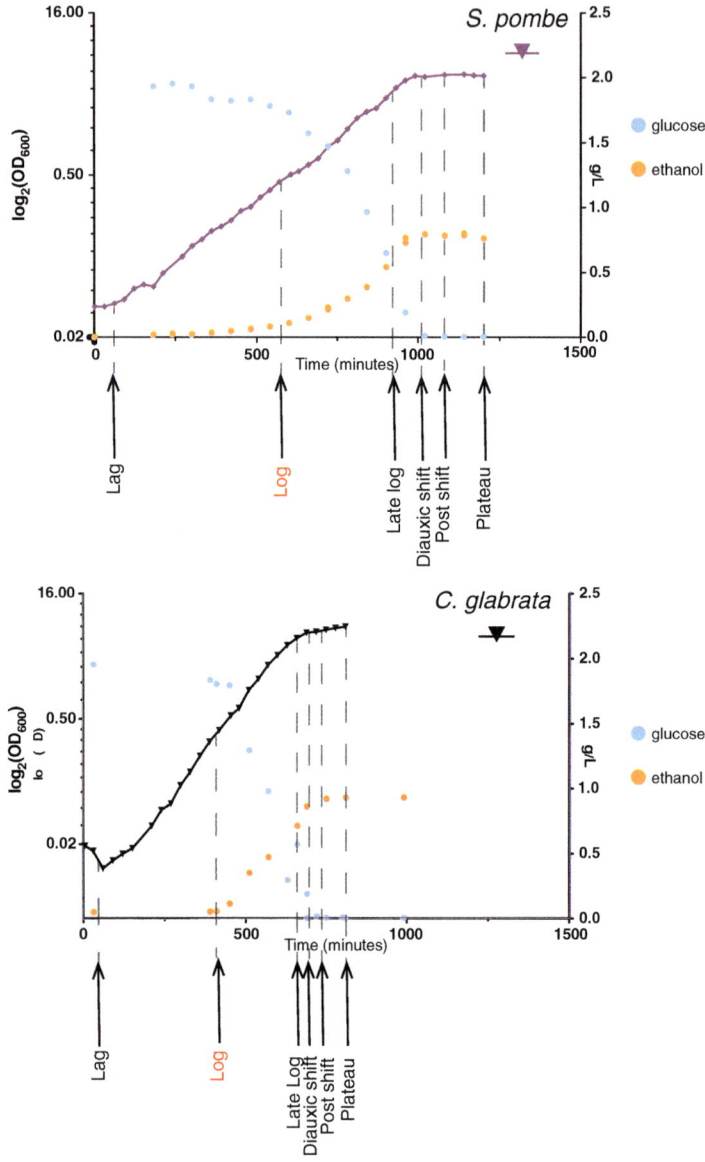

Fig. 1 Choosing "physiologically comparable" time points. Our experiments compare "physiologically analogous" time points across all species. For example, shown is the growth curve (x axis: time, minutes; y axis: growth rate, in $\log_2(OD_{600})$ and glucose levels (g/L, *blue*) and ethanol levels (g/L, *orange*) for the relative slow growing species *S. pombe* (*top*) vs. the growth curve for the faster growing *C. glabrata* (*bottom*). Biological samples from each species were taken at the time points indicated by *arrows*

regulation [15]. The experimental framework presented here includes: growth conditions, methods for phenotypic profiling needed for physiologically comparable sampling, collection and fixation of cells and RNA isolation. Expression profiling is conducted using standard protocols for either custom Agilent microarrays [15] or strand-specific RNA seq [16]. To analyze the evolution of regulatory modules (groups of co-expressed genes), we use a new algorithm, Arboretum [17], to identify expression modules within and across species and to reconstruct their evolutionary history. Our framework for comparative functional genomics is applicable to any complex phylogeny, and can reveal principles of regulatory evolution in many responses and species.

2 Materials

2.1 Growth Media

1. BMW (Rich medium): 1.5 % yeast extract, 1 % peptone, 2 % dextrose, 2 g/L SC amino acid mix, 100 mg/L adenine, 100 mg/L tryptophan, 100 mg/L uracil. When fully dissolved, filter-sterilize using a 1liter filtration unit (0.22 μm) designed for cell culture medium (*see* **Notes 1–4**).

2. Minimal medium: 3000 mg/L potassium phosphate, monobasic, 6.75 mg/L calcium chloride, dihydrate, 244 mg/L magnesium sulfate, anhydrous, 22.5 mg/L EDTA, disodium, 0.08 mg/L D-biotin, 1.5 mg/L calcium pantothenate, 37.5 mg/L inositol, 1.5 mg/L nicotinic acid, 0.3 mg/L 4-aminobenzoic acid (PABA), 1.5 mg/L pyridoxine hydrochloride, 1.5 mg/L thiamine hydrochloride, 1.5 mg/L boric acid, 0.45 mg/L copper (II) sulfate pentahydrate, 0.15 mg/L potassium iodide, 0.95 mg/L manganese chloride, anhydrous, 0.51 mg/L disodium molybdate, anhydrous, 4.21 mg/L Zinc sulfate, monohydrate, 4.5 mg/L iron sulfate (II) heptahydrate, 0.25 mg/L cobalt (II) chloride, anhydrous, 6.7 g/L yeast nitrogen base w/o amino acids, 6.16 g/L EMM—dextrose, 20 g/L dextrose. When fully dissolved, filter-sterilize using a 1-L filtration unit (0.22 μm) designed for cell culture medium (*see* **Notes 1–4**).

2.2 Phenotypic Analysis

1. Spectrophotometer.

2. Multi-mode plate reader with temperature control and orbital shaking.

3. New Brunswick Scientific Edison model TC-7 roller drum.

4. New Brunswick Scientific Edison water bath model C76 shaker.

5. Nexcelom Cellometer Auto M10.

6. YSI Biochemistry Analyzer Model 2700.

2.3 Sample Collection

1. Either methanol (100 %) or ethanol (95 %).

2. Liquid nitrogen.

3. Dry ice.

4. 2 mL tubes with O-ring cap.

5. Wire rack to hold 50 mL conical tubes.

6. rectangular ice tray large enough to hold wire racks.

7. 50 mL conical tubes (*see* **Note 5**).

8. RNAlater (Life Technologies).

9. DNase/RNase-free distilled water.

2.4 RNA Isolation

1. Zirconia beads.

2. 2ml tubes with O-ring cap.

3. Bead beater (Biospec).

4. RNeasy plus mini kits (Qiagen).

5. DNase/RNase-free distilled water.

6. RNA 6000 Nano ll kit (Agilent).

7. Bioanalyzer 2100 instrument (Agilent).

---

3 Methods

3.1 Phenotypic Profiling

3.1.1 Media and Experimental Parameter Screening

1. For each strain, cells are plated onto BMW plates from frozen glycerol stocks.

2. After 2 days, cells are taken from plates, resuspended into liquid BMW, and counted using the Cellometer Auto M10.

3. Next, an appropriate volume of this culture is used to inoculate a 3 mL BMW (or other appropriate medium) culture at 1×10^6 cells/mL.

4. The resulting 3 mL culture is placed in a New Brunswick Scientific Edison model TC-7 roller drum on the fastest rotation until saturated (1–2 days).

5. The cells were then counted and diluted back to 1×10^6 cells/mL. One hundred and fifty microliters of culture was placed in a 96-well plate and growth curves were generated using the Synergy H1 plate reader (Biotek) in the desired conditions to be tested. The instrument was set to continuous low shaking and OD_{600} was measured every 15 min. Growth curves were measured in two biological replicates (*see* **Note 6**).

3.1.2 High-Resolution Growth Curve Data Are Collected for Each Species

1. For each strain, cells are plated onto BMW plates from frozen glycerol stocks. After 2 days, cells are taken from plates and resuspended into liquid BMW (or other medium of choice), and counted using a Cellometer Auto M10 (Nexcelom).

2. A 3 mL BMW culture was inoculated at 1×10^6 cells/mL and placed in a New Brunswick Scientific Edison model TC-7 roller drum on the highest speed until saturated (1–2 days) at the optimal growth temperature for particular species.

3. The saturated cultures are counted as described above and then used to inoculate 300 mL BMW (or other medium of choice) at 1×10^6 cells/mL.

4. Flasks are then transferred to New Brunswick Scientific Edison water bath model C76 shakers set to 200 rpm. The OD_{600} was measured every 15–60 min using a Thermo Spectronic Genesys 20 spectrophotometer, and 1 mL media samples were taken to measure extracellular glucose and ethanol levels on a YSI Biochemistry Analyzer Model 2700 according to manufacturer's instructions (*see* **Note 7**).

3.2   Physiologically Comparable Sampling

1. The data from the high resolution phenotypic profiling is plotted for each species (Fig. 1) and physiologically comparable time points (e.g., log phase) for sampling are determined by visual inspection of the data for each species (*see* **Note 8**).

3.2.1   Finalizing Sample Collection

1. Once the initial selection of appropriate experimental parameters and determination of sampling time points is conducted based on the methodology described above, a pilot expression study should be performed (using the methods described below) for validation. This involves higher resolution sampling in one biological replicate for each species to either confirm transcriptional response of growth stage (e.g., log phase), timing of an environmental response, and/or appropriate experimental parameter for a robust response (*see* **Note 9**).

3.3   Growth Curve Alignments

Since sampling time points are selected in real time during the experiment (based on the previous growth curve data and data collected concurrently), after the data was collected it is important to confirm that the sample time points indeed matched their expected categorization. To this end, two methods are used to align the measured growth curves.

1. In the first method, data collected from replicate experiments in each species are manually aligned by overlaying growth curves for each experiment.

2. Samples are then categorized into time point classes (e.g., growth phase) by their position on the growth curve and their correlation in expression profiles.

3. In the second method, two transformations are performed to align growth curves. First, in each species sampling times for growth curves of biological replicates are shifted in order to align the exponential growth phase. The doubling time for each replicate should be consistent.

4. Next, a line is fitted to the exponential growth phase using all replicate data in order to get an average growth curve.

5. This average growth curve for each species is then aligned to a reference species (e.g., *S. cerevisiae*) growth curve, adjusting for the doubling time (slope) and speed (shift along *x*-axis) during exponential growth.

6. Finally, the plotted glucose consumption is overlaid on the growth curve and used to manually align a particular growth phase (e.g., log phase) such that it matches that in the reference species. Sampling times are then extracted from the aligned growth curve. Sampling is deemed correct if the two approaches match and are consistent with the original sampling choice based on the phenotypic analysis and pilot expression studies described above.

3.4 Sample Collection and Storage

3.4.1 Methanol/Ethanol

Sample collection volumes should be calculated such that each will have an appropriate number of cells ($2–5 \times 10^7$) to yield ample amount of total RNA with some cells remaining for permanent storage (see description below).

1. Samples are collected in 50 mL conicals filled with the appropriate amount of 100 % methanol (or 95 % ethanol) to produce a 60/40 methanol–culture mixture once the sample is added. The methanol-filled tubes were stored at –80 °C until ready for use (*see* **Note 10**).

2. During sample collection tubes are placed in a rack in a dry ice–ethanol bath kept at approximately –40 °C. Once the sample is added to the methanol, the methanol and media are separated from the cells by centrifugation ($3700 \times g$ for 5 min at 4 °C and poured off (*see* **Note 10**).

3. The conicals containing the cell pellet are flash frozen in liquid nitrogen and then stored at –80 °C until processed for permanent storage.

4. For permanent storage, the cell pellets are thawed on ice and then washed in 5 mL of ice-cold nuclease-free water and spun for 5 min at $3700 \times g$ at 4 °C.

5. The supernatant is discarded and the pellet resuspended in 2 mL of RNAlater (Ambion) and transferred to a 2 mL tube with O-ring cap for storage.

6. The samples should be put at 4 °C for 24 h before being moved to a –80 °C freezer as per manufacturers instructions for RNAlater application to yeast.

3.5 RNA Isolation

1. Samples are removed from –80 °C and thawed on ice.

2. Remove an appropriate volume containing the number of cells needed for total RNA isolated using the RNeasy plus Mini Kit (Qiagen) and transferred to a 2 ml tube with O-ring cap and spun down in a RNase free microfuge at top speed for 5 min at 4 °C.

3. Pour off RNAlater solution and remove remaining liquid with a pipet.

4. Resuspend cells in the appropriate volume RLT lysis buffer, provided in the RNeasy kit and prepared according to the Qiagen instructions for mechanical lysis of yeast cells.

5. Add an equal volume of zirconia glass beads.

6. Samples are lysed in a bead beater (Biospec) for 3 min. at top speed.

7. Process samples according to the provided instructions in the RNeasy kit. All steps should be done at room temperature. Including the on column DNase treatment to remove genomic DNA contamination.

8. Samples are then quality control tested for yield and integrity with the RNA 6000 Nano ll kit (Agilent) and analyzed on the Bioanalyzer 2100 instrument (Agilent).

Expression analysis is performed using either species-specific microarrays (Agilent) according to manufacturer's instructions [15] or strand-specific RNA seq with standard protocols [16]

4   Notes

1. Due to lifestyle differences some of the species do not grow well in typical media formulations (e.g., YPD). We therefore first optimized our growth medium to minimize growth differences between species. Our formulations, one rich (BMW) and one minimal, boosts the growth of otherwise slow growers, without substantially impacting the growth of fast growers. It is also necessary to test optimal growth temperature of each species. For example, *S. cerevisiae* grows optimally at 30 °C whereas optimal growth temperature for *N. castellii* is 25 °C.

2. It is important for the reproducibility of gene expression studies that care is taken in medium preparation such that it is made following the exact same procedure each time (e.g., same volume of water added). It is also critical that the medium be filter-sterilized and not autoclaved. Therefore it must be fully dissolved prior to filtration. Heat can be used during preparation to aid dissolution but no more than 30 °C.

3. The preparation of the complex minimal medium can be simplified by ordering a custom formulation from Sunrise Science.

4. It is important for experimental design to understand as much as possible about the lifestyle of each species. For example, *Debaromyces Hansenii* is a halophilic yeast that requires the growth medium both rich and minimal be made up in filtered sea water purchased from a pet store for salt water aquariums.

The book "The Yeast a Taxonomic Guide" [18] is an excellent resource.

5. It is important to cover the sample labels on the 50 mL conicals used for sample collection with Scotch brand magic tape to prevent the labels from being dissolved by the methanol or the ethanol in the ethanol–dry ice bath.

6. In addition to the basic growth mediums described above it is essential to conduct high-resolution phenotypic characterization for each experimental condition (e.g., low glucose) or environmental response (e.g., heat shock) in each species to be compared. Experimental parameters can be easily screened using a plate reader (Synergy H1 from Biotek) with temperature control and orbital shaking capabilities. Alternatively, screening can be done in 3 mL tubes grown for in a TC-7 roller drum to quantify the "saturation coefficient." The saturation coefficient was determined by inoculating a 3 mL culture with 1×10^6 cells/mL of each species and measuring the OD_{600} using a Thermo Spectronic Genesys 20 spectrophotometer after 24 hours of growth. Once the experimental parameters have been determined by screening, high resolution growth curves (OD_{600}) are performed in the exact same conditions (e.g., flask and media volume) to be used in subsequent profiling experiments where samples will be collected. Glucose and ethanol measurements are also taken at this time (Fig. 1).

7. For species difficult to measure by OD_{600} such as clumpy or filamentous fungi, glucose concentration is an excellent proxy to determine growth phase.

8. It is important to identify physiologically comparable time points across species. This is critical for distinguishing true inter-specific variation from temporal shifts due to physiological parameters, e.g., growth rate differences. For example, in a comparative expression study across 15 ascomycota species to compare the response to batch growth on glucose and its depletion [15] the following time points were chosen from visual inspection of the high resolution phenotypic profiling (Fig. 1). The Lag phase time point was taken 30 min after inoculation for all species. Log phase was defined as the midpoint of exponential growth. Diauxic Shift is the point at which glucose levels reached 0. Two time points before the diauxic shift, Early Late Log, Late Log, and two time points after, Post Shift, Late Post Shift, were chosen for each species at times proportional to the maximum growth rate in exponential phase in each species. Finally, the Plateau time point was defined as approximately 2 hours after the growth had plateaued.

9. In the Thompson et al. [15] paper, a pilot study was conducted where expression profiles were measured for each of the eight

time points for one biological replicate for each species. These data were used to finalize the six time points used in the larger study. Likewise, in a study to compare the transcriptional response to various stress conditions [17, 19] pilot expression studies were conducted to assess the response to various experimental parameters initially chosen based on the growth curve response. These included confirmation of the timing of each stress response across species and the observation that for the human commensal species, *C. glabrata* and *C. albicans*, the appropriate heat shock regimen was a shift from 22 to 42 °C as opposed to 37 °C in the other species.

10. Either cold methanol or ethanol can be used during sample collection. Methanol should be used if protein and/or metabolites in addition to RNA will be isolated. The cold alcohol quenching immediately kills the cells and eliminates the contribution to the expression signature of downstream manipulation (e.g., centrifugation). This greatly improves reproducibility of the expression profiles.

Acknowledgement

This work was supported by NIH grant 2R01CA119176-01 and a SPARC grant from the Broad Institute.

References

1. Tsong AE, Miller MG, Raisner RM, Johnson AD (2003) Evolution of a combinatorial transcriptional circuit: a case study in yeasts. Cell 115:389

2. Tsong AE, Tuch BB, Li H, Johnson AD (2006) Evolution of alternative transcriptional circuits with identical logic. Nature 443:415

3. Tanay A, Regev A, Shamir R (2005) Conservation and evolvability in regulatory networks: the evolution of ribosomal regulation in yeast. Proc Natl Acad Sci U S A 102:7203

4. Field Y et al (2008) Distinct modes of regulation by chromatin encoded through nucleosome positioning signals. PLoS Comput Biol 4, e1000216

5. Ihmels J et al (2005) Rewiring of the yeast transcriptional network through the evolution of motif usage. Science 309:938

6. Hogues H et al (2008) Transcription factor substitution during the evolution of fungal ribosome regulation. Mol Cell 29:552

7. Tirosh I, Barkai N (2008) Evolution of gene sequence and gene expression are not correlated in yeast. Trends Genet 24:109

8. Tsankov AM, Thompson DA, Socha A, Regev A, Rando OJ (2010) The role of nucleosome positioning in the evolution of gene regulation. PLoS Biol 8, e1000414

9. Tsankov A, Yanagisawa Y, Rhind N, Regev A, Rando OJ (2011) Evolutionary divergence of intrinsic and trans-regulated nucleosome positioning sequences reveals plastic rules for chromatin organization. Genome Res 21(11):1851–1862

10. Baker CR, Tuch BB, Johnson AD (2011) Extensive DNA-binding specificity divergence of a conserved transcription regulator. Proc Natl Acad Sci U S A 108:7493

11. Habib N, Wapinski I, Margalit H, Regev A, Friedman N (2012) A functional selection model explains evolutionary robustness despite plasticity in regulatory networks. Mol Syst Biol 8:619

12. Kellis M, Birren BW, Lander ES (2004) Proof and evolutionary analysis of ancient genome duplication in the yeast *Saccharomyces cerevisiae*. Nature 428:617

13. Wolfe KH, Shields DC (1997) Molecular evidence for an ancient duplication of the entire yeast genome. Nature 387:708

14. Wapinski I, Pfeffer A, Friedman N, Regev A (2007) Natural history and evolutionary principles of gene duplication in fungi. Nature 449:54

15. Thompson DA et al (2013) Evolutionary principles of modular gene regulation in yeasts. ELife 2, e00603

16. Levin JZ et al (2010) Comprehensive comparative analysis of strand-specific RNA sequencing methods. Nat Methods 7:709

17. Roy S et al (2013) Arboretum: reconstruction and analysis of the evolutionary history of condition-specific transcriptional modules. Genome Res 23:1039

18. Kurtzman CP (ed) (2000) The yeasts a taxonomic study, 4th edn. Elsevier, New York, NY, p 1055

19. Wapinski I et al (2010) Gene duplication and the evolution of ribosomal protein gene regulation in yeast. Proc Natl Acad Sci U S A 107:5505

Chapter 5

Mapping the Transcriptome-Wide Landscape of RBP Binding Sites Using gPAR-CLIP-seq: Experimental Procedures

Ting Han and John K. Kim

Abstract

An estimated 5–10 % of protein-coding genes in eukaryotic genomes encode RNA-binding proteins (RBPs). Through dynamic changes in RNA recognition, RBPs posttranscriptionally regulate the biogenesis, transport, inheritance, storage, and degradation of RNAs. Understanding such widespread RBP-mediated posttranscriptional regulatory mechanisms requires comprehensive discovery of the in vivo binding sites of RBPs. Here, we describe the experimental procedures of the gPAR-CLIP-seq (global photoactivatable-ribonucleoside-enhanced cross-linking and precipitation followed by deep sequencing) approach we recently developed for capturing and sequencing regions of the transcriptome bound by RBPs in budding yeast. Unlike the standard PAR-CLIP method, which identifies the bound RNA substrates for a single RBP, the gPAR-CLIP-seq method was developed to isolate and sequence all mRNA sites bound by the cellular "RBPome." The gPAR-CLIP-seq approach is readily applicable to a variety of organisms and cell lines to profile global RNA–protein interactions underlying posttranscriptional gene regulation. The complete landscape of RBP binding sites provides insights to the function of all RNA *cis-regulatory* elements in an organism and reveals fundamental mechanisms of posttranscriptional gene regulation.

Key words RNA binding proteins, 4-Thiouracil, UV cross-linking, Posttranscriptional gene regulation

1 Introduction

The transmission of genetic information, from DNA to RNA to protein, is regulated at multiple levels to ensure accuracy, robustness, and adaptability of gene expression programs [1]. According to this central dogma, RNA serves as a critical intermediate. However, RNAs are not naked in a cell but rely on protein partners to dictate their stability, storage, translational efficiency, and subcellular residence [2, 3]. Compared to DNA-binding proteins that regulate transcription, little is known about the roles of RNA-binding proteins (RBPs). Out of hundreds of RBPs encoded in a eukaryotic genome, less than 10 % have had their RNA targets identified, leaving a major gap in our understanding of the role of

Frédéric Devaux (ed.), *Yeast Functional Genomics: Methods and Protocols*, Methods in Molecular Biology, vol. 1361,
DOI 10.1007/978-1-4939-3079-1_5, © Springer Science+Business Media New York 2016

RNA–protein interactions governing gene expression programs [4–10]. Furthermore, recent proteomic surveys of proteins crosslinked to mRNAs have expanded the definition and repertoire of RBPs to include RNA-binding metabolic enzymes, kinases, cytoskeletal proteins, and many other factors with no known RNA binding domains [6–8, 11]. How these newly identified RBPs regulate gene expression posttranscriptionally is not known.

In order to study the functions of RBPs, several labs have developed methods to map transcriptome-wide binding sites of RBPs in vivo. Robert Darnell's lab pioneered the cross-linking immunoprecipitation (CLIP) protocol that couples UV crosslinking with immunopurification of RBPs [12]. In the CLIP procedure, live cells are irradiated with 254 nm UV to cross-link RBPs and their interacting RNAs. The formation of covalent cross-links allows purification of RBP-RNA complexes using specific antibodies against RBPs. Cross-linked RNA fragments are then isolated and converted to cDNA for sequencing. cDNA mapping to the transcriptome enables the identification of binding sites of RBPs in vivo. Recently, Thomas Tuschl's lab developed a modified CLIP technique, PAR-CLIP (photoactivatable-ribonucleoside-enhanced CLIP) [13]. The PAR-CLIP procedure starts with the incorporation of photoreactive ribonucleoside analogs into the transcriptome of live cells, followed by 365 nm UV irradiation to cross-link photoreactive nucleoside-labeled RNAs to interacting RBPs. PAR-CLIP generates frequent and non-random nucleotide substitutions at cross-linking sites to reveal specific RBP-RNA contact sites with single nucleotide resolution. Since these original studies using mammalian cells, other labs have applied the CLIP and PAR-CLIP procedures to study diverse RBPs in a variety of systems, greatly advancing our understanding of RBP functions [14–16]. However, all of these studies are limited to investigation of individual RBPs.

In order to identify all RBP-bound RNA sequences in a cell and monitor their dynamics under different perturbations, we recently developed the gPAR-CLIP-seq methodology using the budding yeast *S. cerevisiae* as a model [17]. gPAR-CLIP-seq stems from the PAR-CLIP technique. The major difference between PAR-CLIP and gPAR-CLIP-seq is that PAR-CLIP identifies RNAs bound to a specific protein while gPAR-CLIP-seq captures RNAs bound to the entire "RBPome." In gPAR-CLIP-seq, we first utilize 4-thiouridine-enhanced UV cross-linking technology to promote covalent bond formation between closely interacting nucleotides and amino acid side chains, essentially "freezing" all RNA–protein interactions in vivo [13, 17]. We then implement three biochemical strategies to capture RNA regions bound by the "RBPome" (Fig. 1): (1) sucrose gradient centrifugation to reduce ribosome abundance; (2) oligo(dT) selection to deplete abundant structural non-coding RNAs (e.g., rRNAs); and (3) chemical biotinylation of proteins via primary amines to enable purification of

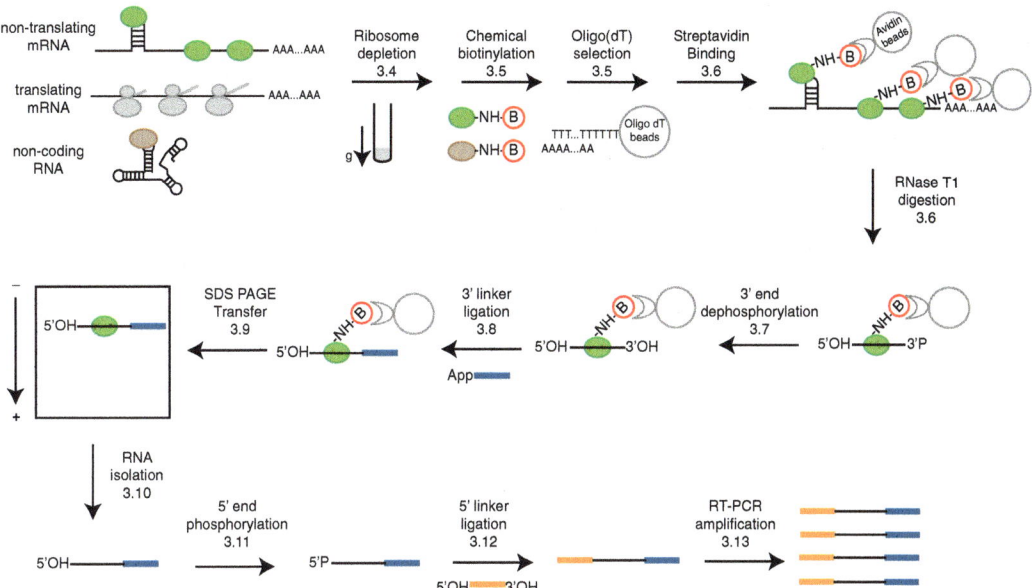

Fig. 1 Overview of the major steps in the gPAR-CLIP-seq protocol

all RNA–protein complexes. All RNAs purified from RNA–protein complexes are converted into cDNA libraries for next generation sequencing. Parallel to our effort, Markus Landthaler's group used oligo(dT) selection followed by ammonium sulfate precipitation to capture protein-bound mRNAs from cultured mammalian cells and generated a transcriptome-wide map of potential *cis-regulatory* elements in mammalian cells [11].

gPAR-CLIP-seq is a useful method for examining the dynamics of RNA–protein interactions under different physiological and pathological conditions. For example, using gPAR-CLIP-seq, we elucidated over 13,000 RBP binding sites in untranslated regions (UTR) covering 73 % of protein-coding transcripts encoded in the genome. In addition, we found 25 % of RBP binding sites respond to glucose or nitrogen deprivation, with major impacts on metabolic pathways as well as mitochondrial and ribosomal gene expression [17]. In addition to mapping protein-binding sites on messenger RNAs, gPAR-CLIP-seq can also reveal binding sites on non-coding RNAs [18]. For example, by omitting the step of oligo(dT) selection, gPAR-CLIP-seq unveils binding sites on many non-coding RNAs. As most noncoding RNAs require protein co-factors for their function, the visualization of protein-binding sites will enable experimentation to study non-coding RNA structure and function. In this protocol, we describe the detailed experimental procedures of gPAR-CLIP-seq, from 4sU incorporation to preparation of next generation sequencing libraries (Fig. 1). A companion chapter (Chapter 6) describes the bioinformatics pipeline for analyzing gPAR-CLIP-seq data to generate biological insights.

2   Materials

Perform RNA molecular biology in an RNase-free environment. Always wear gloves and use RNase-free plastic wares and reagents. RNase-free filter tips are recommended.

2.1 Yeast Strain and Culture

1. WT yeast strain BY4742 (*MATα his3Δ1 leu2Δ0 lys2Δ0 ura3Δ0*).

2. Yeast Synthetic Defined (SD) media: dissolve a pouch of SD/-Ura Broth powder in 0.5 L of deionized H_2O, supplemented with 10 mg of uracil. Autoclave before use.

2.2 Chemicals and Reagents

1. 4-thiouracil (4sU, Sigma-Aldrich).

2. Sucrose (Sigma-Aldrich).

3. 10 mM EZ-Link NHS-SS-Biotin (Pierce) dissolved in dimethylformamide (*see* **Note 1**).

4. Oligo(dT)$_{25}$ magnetic beads (NEB).

5. Streptavidin M280 Dynabeads (Life Technologies).

6. 1 M DTT (*see* **Note 2**).

7. NuPAGE 4–12 % Bis-Tris gel (Life Technologies).

8. Full-range rainbow molecular weight markers (GE Healthcare).

9. Protran BA 85 nitrocellulose membrane (pore size 0.45 µm, Whatman).

10. Phenol–chloroform–isoamyl alcohol 25:24:1 (Sigma-Aldrich).

11. 3 M NaOAc, pH 5.5.

12. 100 % ethanol.

13. 15 mg/mL GlycoBlue (Life Technologies).

14. 6 % TBE UREA gel (Life Technologies).

15. Low range ssRNA ladder (NEB).

16. 10 bp DNA ladder (Life Technologies).

17. SYBR Gold Stain (Life Technologies).

18. Costar Spin-X centrifuge tube filters (cellulose acetate membrane, pore size 0.22 µm, non-sterile) (Corning).

19. 10 mM dNTP.

20. 10 % TBE gel (Life Technologies).

21. DNA Clean & Concentrator-5 (Zymo).

2.3 Buffers and Enzymes

1. HBSS (Life Technologies).

2. Polysome lysis buffer: 20 mM HEPES pH 7.5, 140 mM KCl, 1.5 mM $MgCl_2$, 1 % Triton X-100, 1× Complete Mini Protease Inhibitor EDTA-free (Roche), 0.2 U/µL SUPERase·In.

3. Polysome gradient buffer: 20 mM HEPES pH 7.5, 140 mM KCl, 5 mM $MgCl_2$.

4. Hybridization buffer: 10 mM HEPES pH 7.5, 0.5 M NaCl, 1 mM EDTA.

5. Elution buffer: 10 mM HEPES pH 7.5, 1 mM EDTA.

6. 10× PBS (Life Technologies).

7. RNase T1 (Fermentas).

8. Wash buffer: 1× PBS, 0.1 % SDS, 0.5 % deoxycholate, 0.5 % NP-40.

9. High-salt wash buffer: 5× PBS, 0.1 % SDS, 0.5 % deoxycholate, 0.5 % NP-40.

10. 1× PNK buffer: 50 mM Tris-HCl pH 7.4, 10 mM $MgCl_2$, 0.5 % NP-40.

11. CIP mix: 50 mM Tris-HCl pH 7.9, 100 mM NaCl, 10 mM $MgCl_2$, 0.5 U/μL calf intestinal alkaline phosphatase (CIP) (NEB).

12. 1× PNK+EGTA buffer: 50 mM Tris-HCl pH 7.4, 20 mM EGTA, 0.5 % NP-40.

13. 3′ Ligation mix: 50 mM Tris-HCl pH 7.4, 10 mM $MgCl_2$, 0.5 mM DTT, 2 μM Pre-adenylated 3′ DNA linker, 25 % PEG-8000, 10 U/μL T4 RNA ligase 2, truncated K227Q (NEB) (*see* **Note 3**).

14. 4× NuPAGE LDS sample buffer (Life Technologies).

15. NuPAGE MOPS SDS Running Buffer (20×) (Life Technologies).

16. 4 mg/mL Proteinase K prepared in 1× PK buffer: 100 mM Tris-HCl pH 7.5, 50 mM NaCl, 10 mM EDTA.

17. 7 M urea prepared in 1× PK buffer (*see* **Note 4**).

18. PNK mix: 70 mM Tris-HCl pH 7.6, 10 mM $MgCl_2$, 5 mM DTT, 1 mM ATP, 1 U/μL T4 polynucleotide kinase (NEB), 1 U/μL SUPERase·In.

19. 5′ Ligation mix: 50 mM Tris-HCl pH 7.5, 10 mM $MgCl_2$, 10 mM DTT, 1 mM ATP, 0.1 mg/mL BSA, 2 μM 5′ RNA linker, 1 U/μL T4 RNA ligase (Fermentas), 1 U/μL SUPERase·In, 10 % DMSO.

20. 2× formamide gel loading buffer (Life Technologies).

21. 10× TBE (Life Technologies).

22. SuperScript III Reverse transcriptase (Life Technologies).

23. AccuPrimeTaq High Fidelity (Life Technologies).

24. 6× DNA Loading Dye.

2.4   Equipment

1. 30 °C shaker incubators.

2. UVP CL-1000L UV cross-linker (*see* **Note 5**).

3. TLS-55 rotor and Optima MAX-E ultracentrifuge (Beckman Coulter); polycarbonate centrifugation tubes (11 × 34 mm) (Beckman Coulter).

4. Thermomixer (Eppendorf Thermomixer Comfort).

5. Magnetic Particle Concentrator (Life Technologies).

2.5   Oligonucleotides for Constructing gPAR-CLIP-seq Libraries

All the 3′ DNA linker oligonucleotides were ordered from Integrated DNA technologies with two modifications: 5′ phosphorylation and 3′ block with inverted deoxythymidine (*see* **Note 6**). The first six nucleotides (underlined sequences below) of the oligonucleotides represent the barcode sequences.

2.5.1   Barcoded 3′ DNA Linker Oligonucleotides

Perform pre-adenylation of 3′ DNA linker oligonucleotides with 5′ DNA adenylation kit (NEB) by mixing 100 pmol of 5′ phosphorylated DNA oligonucleotide with 100 µM ATP and 100 pmol of *M*th RNA ligase in 1× 5′ DNA Adenylation reaction buffer (total volume of 20 µL). Incubate at 65 °C for 1 h followed by heat inactivation at 85 °C for 5 min. To purify adenylated oligonucleotides, mix the adenylation reaction with 80 µL of H_2O and 100 µL of phenol–chloroform–isoamyl alcohol 25:24:1. Vortex and spin for 5 min at 20,000 × *g*. Transfer the liquid phase (90 µL) into a new tube, mix with 10 µL of 3 M NaOAc, 250 µL of 100 % ethanol, and 1 µL of 15 mg/mL GlycoBlue, and precipitate for 2 h at −80 °C. Collect oligonucleotides by centrifugation for 20 min at 20,000 × *g* at room temperature followed by two washes with cold 75 % ethanol. After brief air-drying, resuspend pellet in 10 µL of H_2O. The final concentration of adenylated oligonucleotide is around 10 µM.

Index 1: 5′ pATCACGTCGTATGCCGTCTTCTGCTTGidT 3′.

Index 2: 5′ pCGATGTTCGTATGCCGTCTTCTGCTTGidT 3′.

Index 3: 5′ pTTAGGCTCGTATGCCGTCTTCTGCTTGidT 3′.

Index 4: 5′ pTGACCATCGTATGCCGTCTTCTGCTTGidT 3′.

Index 5: 5′ pACAGTGTCGTATGCCGTCTTCTGCTTGidT 3′.

Index 6: 5′ pGCCAATTCGTATGCCGTCTTCTGCTTGidT 3′.

Index 7: 5′ pCAGATCTCGTATGCCGTCTTCTGCTTGidT 3′.

Index 8: 5′ pACTTGATCGTATGCCGTCTTCTGCTTGidT 3′.

2.5.2   5′ RNA Linker (No Modification Required, PAGE Purified)

5′ GUUCAGAGUUCUACAGUCCGACGAUC 3′.

2.5.3   Barcoded RT Primers (No Modification Required, PAGE Purified)

Index 1: 5′ CAAGCAGAAGACGGCATACGACGTGAT 3′.

Index 2: 5′ CAAGCAGAAGACGGCATACGAACATCG 3′.

Index 3: 5′ CAAGCAGAAGACGGCATACGAGCCTAA 3′.

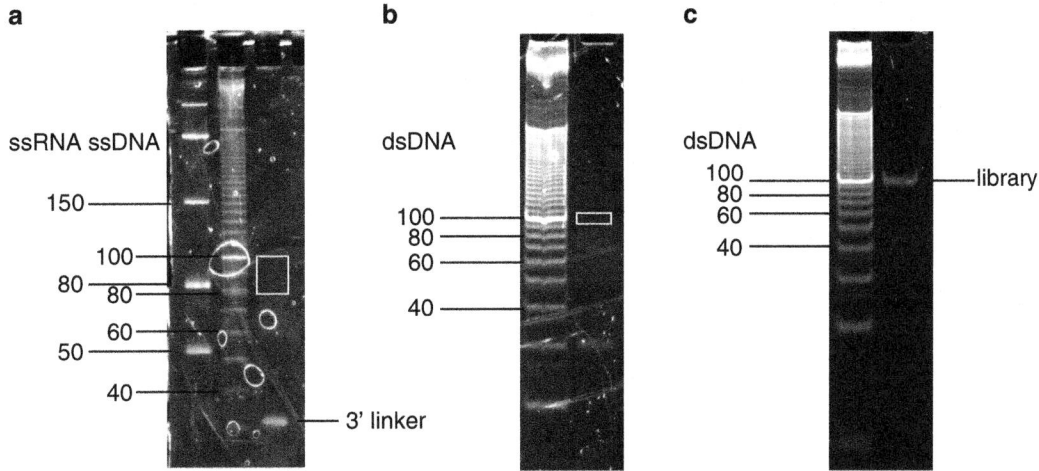

Fig. 2 Representative gel images for size-selection and quality assessment of libraries. (**a**) Size selection of 70–90 nt single-stranded RNA from a 10 % TBE UREA gel (related to Subheading 5′ RNA Linker Ligation and RNA Size Selection). (**b**) Size selection of 96–112 bp double-stranded DNA after the first round of PCR from a 10 % TBE gel (related to Subheading Preparation of Sequencing Libraries). (**c**) Final library displayed on a 10 % TBE gel (related to Subheading Preparation of Sequencing Libraries)

Index 4: 5′ CAAGCAGAAGACGGCATACGA<u>TGGTCA</u> 3′.

Index 5: 5′ CAAGCAGAAGACGGCATACGA<u>CACTGT</u> 3′.

Index 6: 5′ CAAGCAGAAGACGGCATACGA<u>ATTGGC</u> 3′.

Index 7: 5′ CAAGCAGAAGACGGCATACGA<u>GATCTG</u> 3′.

Index 8: 5′ CAAGCAGAAGACGGCATACGA<u>TCAAGT</u> 3′.

2.5.4 PCR Primers (No Modification Required, PAGE Purified)

P7 primer: 5′ CAAGCAGAAGACGGCATACGA 3′.

P5 long primer:

5′ AATGATACGGCGACCACCGACAGGTTCAGAGTTCTAC AGTCCGA 3′.

Illumina primer A: 5′ AATGATACGGCGACCACCGA 3′.

Illumina primer B: 5′ CAAGCAGAAGACGGCATACGA 3′.

3   Methods

3.1   Yeast Growth and 4sU Incorporation

1. Inoculate a 3 mL starter culture with a fresh colony of WT strain BY4742 (*see* **Note 7**). Grow at 30 °C with vigorous shaking (250 rpm) in synthetic defined (SD) media overnight.

2. Supplement 50 mL of SD media with 200 μM 4sU (*see* **Note 8**). Inoculate with 0.1 mL of starter culture. Grow at 30 °C with vigorous shaking (250 rpm) to $OD_{600} = 0.7$–0.8 (*see* **Note 9**).

3.2 UV Cross-Linking

1. Pellet 50 mL of mid-log phase cultures for 5 min at $3000 \times g$ at room temperature, resuspend in 2 mL of HBSS, and transfer to a 60 mm cell culture dish.

2. Place the culture dish on ice, and irradiate with 365 nm UV at 150 mJ/cm^2 four times using a UVP CL-1000 L UV cross-linker.

3. Pellet the cells for 2 min at $5000 \times g$ at 4 °C. Remove HBSS and quickly freeze the cells in liquid nitrogen (*see* **Note 10**).

3.3 Extract Preparation

1. Resuspend cross-linked cells in polysome lysis buffer (1 mL/g of pellets), mix with ½ volume of acid-washed glass beads, and lyse cells by vortexing four times at 4 °C, 1 min each with 1 min incubation on ice in between.

2. Remove cell debris by centrifugation for 5 min at $1300 \times g$ at 4 °C. Transfer supernatant to a new tube chilled on ice.

3. Spin at $20,000 \times g$ for 10 min at 4 °C. Transfer supernatant to a new tube chilled on ice.

3.4 Ribosome Depletion Using Sucrose Density Gradients

1. Prepare 50, 41.25, 32.5, 23.75, and 15 % sucrose (w/v) dissolved in polysome gradient buffer. Prepare 15–50 % (w/v) sucrose density gradients in Beckman polycarbonate centrifugation tubes (11×34 mm) by sequentially layering and freezing 0.24 mL of 50, 41.25, 32.5, 23.75, and 15 % sucrose solutions. Before use, thaw gradients overnight at 4 °C.

2. Carefully load 100 µL of clarified yeast extract on top of a sucrose gradient, centrifuge for 1 h at 200,000 g at 4 °C using a TLS-55 rotor in an Optima MAX-E ultracentrifuge (Beckman Coulter).

3. Recover the top 600 µL of the gradient and supplement with 2 µL of SUPERase·In (20 U/µL) (*see* **Note 11**).

3.5 Chemical Biotinylation and polyA Selection

1. Add 60 µL of freshly prepared 10 mM EZ-Link NHS-SS-Biotin (dissolved in dimethylformamide) to the recovered 600 µL of ribosome-depleted lysate and incubate on a rotating wheel for 2 h at 4 °C.

2. Add 50 µL of 5 M NaCl to the lysate to increase the total salt concentration to 0.5 M. Mix the lysate with 1 mg of oligo(dT)$_{25}$ magnetic beads, then incubated on a rotating wheel for 30 min at 4 °C.

3. Pellet beads with Magnetic Particle Concentrator (MPC). Wash the beads four times with ice-cold hybridization buffer.

4. Elute the RNAs by incubating beads with 500 µL of elution buffer and heating at 65 °C for 3 min. Transfer the eluted sample to a new tube and mix with 55 µL of 10× PBS.

3.6 Streptavidin Binding and RNase T1 Digestion

1. Mix polyA-selected samples with 1 mg of streptavidin M280 Dynabeads and incubate on a rotating wheel for 30 min at 4 °C.

2. Pellet beads with MPC. Wash the beads three times with 1× PBS, then incubate with 20 µL of 50 U/µL RNase T1 at 22 °C for 15 min on an Eppendorf Thermomixer (15 s shaking at 1000 rpm followed by a 2 min rest interval), followed by 5 min incubation on ice.

3. Pellet beads with MPC. Wash beads twice with wash buffer, twice with high-salt wash buffer, and twice with 1× PNK buffer.

3.7 On-Bead CIP Treatment

1. Incubate beads with 20 µL of CIP mix at 37 °C for 15 min, with 15 s shaking at 1000 rpm followed by a 2 min rest interval on a Thermomixer.

2. Pellet beads with MPC. Wash beads twice with 1× PNK + EGTA buffer and twice with 1× PNK buffer.

3.8 On-Bead 3′ DNA Linker Ligation

1. Incubate beads with 20 µL of 3′ ligation mix at 16 °C overnight (≥16 h), with 15 s shaking at 1000 rpm followed by a 2 min interval on a Thermomixer (see **Note 12**).

2. Pellet beads with MPC. Wash beads three times with 1× PNK + EGTA buffer.

3.9 SDS-PAGE and Transfer to Nitrocellulose Membrane

1. Mix beads with 12 µL of 1× PNK + EGTA buffer, 3 µL of freshly made 1 M DTT and 15 µL of 4× NuPAGE LDS sample buffer, and incubate at 70 °C for 10 min in a Thermomixer (Eppendorf).

2. Pellet beads with MPC. Load the supernatant onto NuPAGE 4–12 % Bis-Tris gel and run at 150 V for 35 min using 1× MOPS SDS running buffer (see **Note 13**). Run 5 µL of full-range rainbow markers as size standards.

3. Transfer proteins from the gel to Protran BA 85 nitrocellulose membrane using Novex wet transfer at 30 V for 1 h.

4. Use a clean razor blade to excise a broad band from 31 kDa up to the top of the gel, cut into small pieces, and transfer into a microfuge tube.

3.10 RNA Isolation and Purification

1. Incubate excised membranes with 500 µL of 4 mg/mL Proteinase K prepared in 1× PK buffer for 20 min at 37 °C on a Thermomixer.

2. Add 500 µL of 7 M urea prepared in 1× PK buffer to the tube followed by another 20 min incubation at 37 °C in a Thermomixer.

3. Mix the Proteinase K digestion reaction with 1 mL of phenol–chloroform–isoamyl alcohol 25:24:1 by vortexing and spin for 5 min at 20,000 × g.

4. Transfer the liquid phase into a new tube, mix with 125 μL of 3 M NaOAc, 2.5 mL of 100 % ethanol and 1 μL of 15 mg/mL GlycoBlue, and precipitate for 2 h at –80 °C. Collect RNAs by centrifugation for 20 min at $20,000 \times g$ at room temperature followed by two washes with cold 75 % ethanol.

3.11 RNA 5′ End Phosphorylation

1. Air-dry RNA pellets briefly, resuspend in 10 μL of PNK mix and incubate at 37 °C for 30 min in a Thermomixer.

2. Add 90 μL of H_2O and 100 μL of phenol–chloroform–isoamyl alcohol 25:24:1 to the reaction, mix well and spin for 5 min at $20,000 \times g$.

3. Mix the liquid phase with 12.5 μL of 3 M NaOAc, 250 μL of 100 % ethanol, 1 μL of 15 mg/mL GlycoBlue and precipitate for 2 h at –80 °C. Collect RNAs by centrifugation for 20 min at $20,000 \times g$ at room temperature, followed by two washes with cold 75 % ethanol.

3.12 5′ RNA Linker Ligation and RNA Size Selection

1. Resuspend RNA pellets in 10 μL of ligation mix and incubate at 15 °C for 2 h in a Thermomixer.

2. Terminate ligation reaction by adding 10 μL of 2× formamide gel loading buffer, heat for 2 min at 70 °C and then quickly chill on ice.

3. Load samples onto a 6 % TBE UREA gel together with 500 ng of low range ssRNA ladder and 250 ng of 10 bp DNA ladder (prepared in 1× formamide gel loading buffer, heated for 2 min at 70 °C and then quickly chilled on ice). Run the gel at 150 V for 45 min.

4. Stain the gel with 1× SYBR Gold Stain (diluted in 1× TBE). Visualize stain under a UV lamp. Excise a gel piece corresponding to 70–90 nt RNA (80–100 nt ssDNA) (Fig. 2a).

5. Crush and soak gels in 400 μL of 0.3 M NaOAc overnight at room temperature (*see* **Note 14**).

6. Remove gel pieces by passing through Costar Spin-X centrifuge tube filters. Mix the solution with 1 mL of 100 % EtOH and 1 μL of 15 mg/mL GlycoBlue and precipitate for 2 h at –80 °C.

7. Collect RNA by centrifugation for 20 min at $20,000 \times g$ at room temperature, followed by two washes with cold 75 % ethanol. After brief drying, dissolve RNA in 15 μL of H_2O.

3.13 Reverse Transcription and Test PCR Amplification

1. In a PCR tube, mix 10 μL of the ligated RNA with 2 μL of 5 μM RT primer, heat at 65 °C for 5 min, and then quickly chill on ice.

2. Per reaction, add 8 μL of reverse transcriptase mix (1 μL of 10 mM dNTP, 1 μL of 0.1 M DTT, 4 μL of 5× first strand buffer,

1 μL of SUPERase·In (20 U/μL), and 1 μL of SuperScript III reverse transcriptase). Incubate the reactions in a thermocycler at 50 °C for 45 min, 55 °C for 15 min and 90 °C for 5 min.

3. Perform a test PCR with 2.5 μL of reverse transcription product in 50 μL PCR mix (1× AccuPrime PCR buffer I, 0.5 μM P5 long primer, 0.5 μM P7 primer, 0.2 μL AccuPrimeTaq High Fidelity). Use the cycling program with an initial 3 min denaturation at 98 °C, followed by 14–22 cycles of 80 s denaturation at 98 °C, 90 s annealing and extension at 65 °C, and termination with a final 5 min extension at 65 °C. Collect 15 μL PCR product after 14, 18, and 22 cycles, add 3 μL of 6× DNA loading dye, and analyze on a 10 % TBE gel at 150 V for 1 h to determine the optimal amplification cycles (the lowest cycle number required to generate 96–116 bp amplicons detected by SYBR Gold staining).

3.14 Preparation of Sequencing Libraries

1. Perform a 50 μL PCR reaction with the determined cycle number using the condition listed in Subheading Reverse Transcription and Test PCR amplification.

2. Purify amplicons using Zymo DNA Clean & Concentrator-5. Elute amplicons in 6 μL of H$_2$O, add 1 μL of 6× DNA loading dye, run on 10 % TBE gels at 150 V for 1 h, and stain with SYBR Gold. Load 250 ng of 10 bp DNA ladder (prepared in 1× DNA loading dye) as size markers. Excise a gel piece corresponding to 96–116 bp DNA (Fig. 2b).

3. Crush and soak gel pieces overnight in 400 μL 0.3 M NaOAc at room temperature.

4. Remove gel pieces by passing through Spin-X filters. Mix the solution with 1 mL of 100 % EtOH and 1 μL of 15 mg/mL GlycoBlue and precipitate for 2 h at –80 °C.

5. Collect DNAs by centrifugation for 20 min at 20,000 × g at room temperature, followed by two washes with cold 75 % ethanol. After brief drying, resuspend amplicons in 20 μL of H$_2$O.

6. Use 5 μL of purified amplicons to seed a second round of PCR in 50 μL: 1× AccuPrime PCR buffer I, 0.5 μM Illumina Primer A, 0.5 μM Illumina Primer B, 0.2 μL AccuPrimeTaq High Fidelity for 6–12 cycles (*see* **Note 15**) using the same cycling conditions as in Subheading Reverse Transcription and Test PCR Amplification.

7. Purify second PCR amplicons with Zymo DNA Clean & Concentrator-5. Elute in 25 μL of H$_2$O. Mix 5 μL of PCR amplicons with 1 μL of 6× DNA loading dye, and run a 10 % TBE gel to check library quality. A single band centered on 100 bp should be seen (Fig. 2c). Sequence the libraries on an Illumina HiSeq 2000 sequencer (*see* **Notes 16** and **17**).

4   Notes

1. Store EZ-Link NHS-SS-Biotin at −20 °C with desiccant, and equilibrate to room temperature before opening. Make fresh solution with dimethylformamide (DMF) before use. Discard unused solution.

2. Reconstitute DTT in H_2O before use. Discard unused solution.

3. T4 RNA ligase 2, truncated K227Q ligates preadenylated DNA oligonucleotides to the 3′ ends of cross-linked RNA fragments in an ATP-independent buffer [19]. 25 % PEG-8000 is included to enhance ligation efficiency.

4. Make fresh 7 M urea solution before use. Discard unused solution.

5. For efficient cross-linking of 4sU labeled RNAs to proteins, a 365 nm UV light source is needed.

6. Oligonucleotides can be ordered from Integrated DNA Technologies with the corresponding modifications. PAGE purification is recommended. Preadenylation and 3′ blocking of 3′ DNA linkers ensures the correct directionality of ligation reactions and prevents self-ligation of the 3′ DNA linkers.

7. Other strains defective in uracil synthesis (*ura3Δ*) can be used. *ura3Δ* strains readily take up 4sU from the media. Inside the cell, 4sU is converted by Fur1p (uracil phosphoribosyltransferase) to 4-thiouridine monophosphate that can be incorporated during RNA synthesis [20].

8. 4sU incorporation rates can be estimated using spectrophotometry. Dissolve RNA samples isolated from cells grown in the presence or absence of 4sU in 100 μL of 12 mM Tris-HCl buffer, pH 7. Adjust the A260 absorption to the same value. Measure A330 for both samples using a Q6 quartz cuvette with 1 mm light path in a Thermo Scientific BioMate 3 UV-Vis spectrophotometer. 4sU incorporation rates per kilobase of RNA can be calculated as $500 \times [(A330(+4sU)) - (A330(-4sU))]/A260$. Using 200 μM 4sU, the incorporation rate was roughly four 4sU per kilobase of transcript.

9. To perform gPAR-CLIP-seq under starvation conditions, after OD_{600} reaches 0.7–0.8, pellet cells for 5 min at $3000 \times g$ at room temperature, discard all media, rinse once with H_2O, and resuspend cells in an equal volume of SD without glucose or nitrogen (supplemented with 200 μM 4sU). Return cells to 30 °C with shaking for 2 h.

10. Finish the whole cross-linking procedure within 5 min to minimize exposure to non-physiological conditions (e.g., nutrient-free media, ice incubation, and UV irradiation).

11. This protocol only recovers the top part of the sucrose gradient, which is composed of non-translated mRNAs. This step is performed to reduce the representation of ribosome binding sites on both mRNAs and rRNAs. However, we also had success analyzing the lower part of the gradient, which contains translating mRNAs. Because ribosomes occupy 5′ UTR and coding sequences (CDS) of mRNAs during translation, binding sites on 3′ UTRs are likely derived from RNA-binding proteins.

12. 3′ linker ligation is performed under an optimized condition [19, 21].

13. The NUPAGE Bis-Tris gel system operates under a neutral pH, to ensure RNA stability during electrophoresis.

14. Crush the gel slice by forcing it through a small opening. Use a hot syringe needle to make a hole in the bottom of a 0.5 mL tube. Place the gel slice in this tube, then place this tube into a 1.5 mL tube. Spin the assembly at 15,000 g for 2 min to collect crush gel pieces in the lower tube.

15. A test PCR is recommended to determine the optimal amplification cycles. Set up a 50 μL PCR reaction, and collect 12 μL of PCR product after 6, 8, 10, and 12 cycles. Run PCR product on a 10 % TBE gel, and pick the cycle number that is in the linear range of amplification.

16. The sequencing constructs are:

    5′-AATGATACGGCGACCACCGACAGGTTCAGAGTTCT ACAGTCCGACGATC-(N)$_{20-40}$-(XXXXXX)- TCGTATGCCGTCTTCTGCTTG-3′

    (N)$_{20-40}$ is the 20–40 nt insert. XXXXXX is the 6 nt barcode.

    Use 50 nt single-end (50SE) sequencing on the Illumina platform with the sequencing primer: 5′-CGACAGGTTCAGAG TTCTACAGTCCGACGATC-3′.

17. Because of the high-throughput nature of the Illumina platform, typically eight yeast gPAR-CLIP-seq libraries each with a unique barcode sequence can be sequenced in one lane. This typically results in 10–12 million reads obtained per library. Higher sequencing depth is recommended for organisms with larger genomes.

Acknowledgements

This research was supported by National Institute of General Medical Sciences (NIGMS) R01GM088565 and the Pew Charitable Trusts. The authors thank Mallory Freeberg and Danny Yang for helpful comments on the manuscript.

References

1. Crick F (1970) Central dogma of molecular biology. Nature 227(5258):561–563

2. Moore MJ (2005) From birth to death: the complex lives of eukaryotic mRNAs. Science 309(5740):1514–1518. doi:10.1126/science.1111443

3. Mitchell SF, Parker R (2014) Principles and properties of eukaryotic mRNPs. Mol Cell 54(4):547–558. doi:10.1016/j.molcel.2014.04.033

4. Tsvetanova NG, Klass DM, Salzman J, Brown PO (2010) Proteome-wide search reveals unexpected RNA-binding proteins in Saccharomyces cerevisiae. PLoS One 5(9):pii:e12671, doi:10.1371/journal.pone.0012671

5. Scherrer T, Mittal N, Janga SC, Gerber AP (2010) A screen for RNA-binding proteins in yeast indicates dual functions for many enzymes. PLoS One 5(11):e15499. doi:10.1371/journal.pone.0015499

6. Castello A, Fischer B, Eichelbaum K, Horos R, Beckmann BM, Strein C, Davey NE, Humphreys DT, Preiss T, Steinmetz LM, Krijgsveld J, Hentze MW (2012) Insights into RNA biology from an atlas of mammalian mRNA-binding proteins. Cell 149(6):1393–1406. doi:10.1016/j.cell.2012.04.031

7. Kwon SC, Yi H, Eichelbaum K, Fohr S, Fischer B, You KT, Castello A, Krijgsveld J, Hentze MW, Kim VN (2013) The RNA-binding protein repertoire of embryonic stem cells. Nat Struct Mol Biol 20(9):1122–1130. doi:10.1038/nsmb.2638

8. Mitchell SF, Jain S, She M, Parker R (2013) Global analysis of yeast mRNPs. Nat Struct Mol Biol 20(1):127–133. doi:10.1038/nsmb.2468

9. Hogan DJ, Riordan DP, Gerber AP, Herschlag D, Brown PO (2008) Diverse RNA-binding proteins interact with functionally related sets of RNAs, suggesting an extensive regulatory system. PLoS Biol 6(10), e255. doi:10.1371/journal.pbio.0060255

10. Riordan DP, Herschlag D, Brown PO (2010) Identification of RNA recognition elements in the Saccharomyces cerevisiae transcriptome. Nucleic Acids Res 39(4):1501–1509. doi:10.1093/nar/gkq920

11. Baltz AG, Munschauer M, Schwanhausser B, Vasile A, Murakawa Y, Schueler M, Youngs N, Penfold-Brown D, Drew K, Milek M, Wyler E, Bonneau R, Selbach M, Dieterich C, Landthaler M (2012) The mRNA-bound proteome and its global occupancy profile on protein-coding transcripts. Mol Cell 46(5):674–690. doi:10.1016/j.molcel.2012.05.021

12. Ule J, Jensen KB, Ruggiu M, Mele A, Ule A, Darnell RB (2003) CLIP identifies Nova-regulated RNA networks in the brain. Science 302(5648):1212–1215. doi:10.1126/science.1090095

13. Hafner M, Landthaler M, Burger L, Khorshid M, Hausser J, Berninger P, Rothballer A, Ascano M Jr, Jungkamp AC, Munschauer M, Ulrich A, Wardle GS, Dewell S, Zavolan M, Tuschl T (2010) Transcriptome-wide identification of RNA-binding protein and microRNA target sites by PAR-CLIP. Cell 141(1):129–141. doi:10.1016/j.cell.2010.03.009

14. Zisoulis DG, Lovci MT, Wilbert ML, Hutt KR, Liang TY, Pasquinelli AE, Yeo GW (2010) Comprehensive discovery of endogenous Argonaute binding sites in Caenorhabditis elegans. Nat Struct Mol Biol 17(2):173–179. doi:10.1038/nsmb.1745

15. Creamer TJ, Darby MM, Jamonnak N, Schaughency P, Hao H, Wheelan SJ, Corden JL (2011) Transcriptome-wide binding sites for components of the Saccharomyces cerevisiae non-poly(A) termination pathway: Nrd1, Nab3, and Sen1. PLoS Genet 7(10), e1002329. doi:10.1371/journal.pgen.1002329

16. Lebedeva S, Jens M, Theil K, Schwanhausser B, Selbach M, Landthaler M, Rajewsky N (2011) Transcriptome-wide analysis of regulatory interactions of the RNA-binding protein HuR. Mol Cell 43(3):340–352. doi:10.1016/j.molcel.2011.06.008

17. Freeberg MA, Han T, Moresco JJ, Kong A, Yang YC, Lu ZJ, Yates JR, Kim JK (2013) Pervasive and dynamic protein binding sites of the mRNA transcriptome in Saccharomyces cerevisiae. Genome Biol 14(2):R13. doi:10.1186/gb-2013-14-2-r13

18. Yang Y, Umetsu J, Lu ZJ (2014) Global signatures of protein binding on structured RNAs in Saccharomyces cerevisiae. Sci China Life Sci 57(1):22–35. doi:10.1007/s11427-013-4583-0

19. Viollet S, Fuchs RT, Munafo DB, Zhuang F, Robb GB (2011) T4 RNA ligase 2 truncated active site mutants: improved tools for RNA analysis. BMC Biotechnol 11:72. doi:10.1186/1472-6750-11-72

20. Kern L, de Montigny J, Lacroute F, Jund R (1991) Regulation of the pyrimidine salvage pathway by the FUR1 gene product of Saccharomyces cerevisiae. Curr Genet 19(5):333–337

21. Munafo DB, Robb GB (2010) Optimization of enzymatic reaction conditions for generating representative pools of cDNA from small RNA. RNA 16(12):2537–2552. doi:10.1261/rna.2242610

Chapter 6

Mapping the Transcriptome-Wide Landscape of RBP Binding Sites Using gPAR-CLIP-seq: Bioinformatic Analysis

Mallory A. Freeberg and John K. Kim

Abstract

Protein–RNA interactions are integral components of posttranscriptional gene regulatory processes including mRNA processing and assembly of cellular architectures. Dysregulation of RNA-binding protein (RBP) expression or disruptions in RBP–RNA interactions underlie a variety of human pathologies and genetic diseases including cancer and neurodegenerative diseases (reviewed in (Cooper et al., Cell 136(4):777–793, 2009; Darnell, Cancer Res Treat 42(3):125–129, 2010; Lukong et al., Trends Genet 24 (8):416–425, 2008)). Recent studies have uncovered only a small proportion of the extensive RBP–RNA interactome in any organism (Baltz et al., Mol Cell 46(5):674–690, 2012; Castello et al., Cell 149(6):1393–1406, 2012; Freeberg et al., Genome Biol 14(2):R13, 2013; Hogan et al., PLoS Biol 6(10):e255, 2008; Mitchell et al., Nat Struct Mol Biol 20(1):127–133, 2013; Tsvetanova et al. PLoS One 5(9): pii: e12671, 2010; Schueler et al., Genome Biol 15(1):R15, 2014; Silverman et al., Genome Biol 15(1):R3, 2014). To expand our understanding of how RBP–RNA interactions govern RNA-related processes, we developed gPAR-CLIP-seq (global photoactivatable-ribonucleoside-enhanced cross-linking and precipitation followed by deep sequencing) for capturing and sequencing all regions of the *Saccharomyces cerevisiae* transcriptome bound by RBPs (Freeberg et al., Genome Biol 14(2):R13, 2013). This chapter describes a pipeline for bioinformatic analysis of gPAR-CLIP-seq data. The first half of this pipeline can be implemented by running locally installed programs or by running the programs using the Galaxy platform (Blankenberg et al., Curr Protoc Mol Biol. Chapter 19:Unit 19 10 11–21, 2010; Giardine et al., Genome Res 15 (10):1451–1455, 2005; Goecks et al., Genome Biol 11(8):R86, 2010). The second half of this pipeline can be implemented by user-generated code in any language using the pseudocode provided as a template.

Key words Bioinformatics, High-throughput sequencing, RNA-binding proteins, Global PAR-CLIP-seq, Posttranscriptional gene regulation

1 Introduction

RNA-binding proteins (RBPs) are responsible for regulating a variety of processes including storage, transport, inheritance, and degradation of RNAs. The identification of both RBP-specific and general RBP interactions with RNA is necessary for understanding the mechanisms underlying these key biological processes. Recently, techniques utilizing UV light to induce covalent bond

Frédéric Devaux (ed.), *Yeast Functional Genomics: Methods and Protocols*, Methods in Molecular Biology, vol. 1361, DOI 10.1007/978-1-4939-3079-1_6, © Springer Science+Business Media New York 2016

formation between directly interacting nucleotides and amino acid side chains followed by purification of a protein of interest and deep sequencing of the bound RNAs (CLIP-seq) have been successfully implemented to identify the precise sites on target mRNAs bound by RBPs [1]. In one of the first studies to pioneer this approach, distinct mRNA binding sites were identified for neuron-specific RNA-binding Nova proteins [2], which are associated with paraneoplastic neurologic degenerations [3] and involved in regulating alternative splicing through direct binding of targets [4]. A modified CLIP-seq technique, PAR-CLIP-seq, was subsequently developed that incorporated photoactivatable ribonucleoside analogs into nascent transcripts to improve cross-linking efficiency and create a mismatch signature in resulting deep sequencing reads to more accurately pinpoint RBP–RNA contact sites [5].

Expanding upon these recent studies, we developed an approach to identify a comprehensive set of sites on the *Saccharomyces cerevisiae* transcriptome that interact with any RBP under normal or environmentally stressed conditions. Similar to traditional PAR-CIP-seq, our global PAR-CLIP-seq (gPAR-CLIP-seq) approach, described in the previous chapter (Chapter 5), utilizes 4-thiouridine (4sU)-enhanced UV cross-linking to promote covalent bond formation in vivo between closely interacting nucleotides and amino acid side chains. Instead of immunopurifying a protein of interest, we biochemically biotinylated all proteins, purified RBP–RNA complexes, and sequenced the RBP-bound RNA fragments. Our protocol captures binding patterns of all RBPs, so a novel bioinformatic analysis approach is required that was different from published methods for analyzing single-RBP PAR-CLIP-seq data. Similar to these published methods, we take advantage of the nucleotide mismatch signature resulting from cross-linking [6–8]; however, unlike these published methods, we incorporate sequencing error from mRNA-seq libraries to assign a false-discovery rate to our identified RBP binding sites, thus enabling an accurate measure of confidence that we are identifying biologically relevant RBP binding sites from background noise.

We describe below our novel pipeline developed for bioinformatic analysis of high-throughput sequencing data derived from the gPAR-CLIP-seq protocol. The basic steps of the protocol, outlined in Fig. 1, include: processing sequencing reads (Subheading 3.1), mapping reads to a reference genome (Subheading 3.2), generating binding sites and per-nucleotide cross-linking scores from mapped reads (Subheadings 3.3 and 3.5), assessing binding site quality (Subheading 3.4), and functionally characterizing binding sites (Subheading 3.6).

In parallel to performing the gPAR-CLIP-seq protocol to identify RBP-bound sites on mRNAs, we recommend performing traditional mRNA-seq to quantify transcript abundance. This

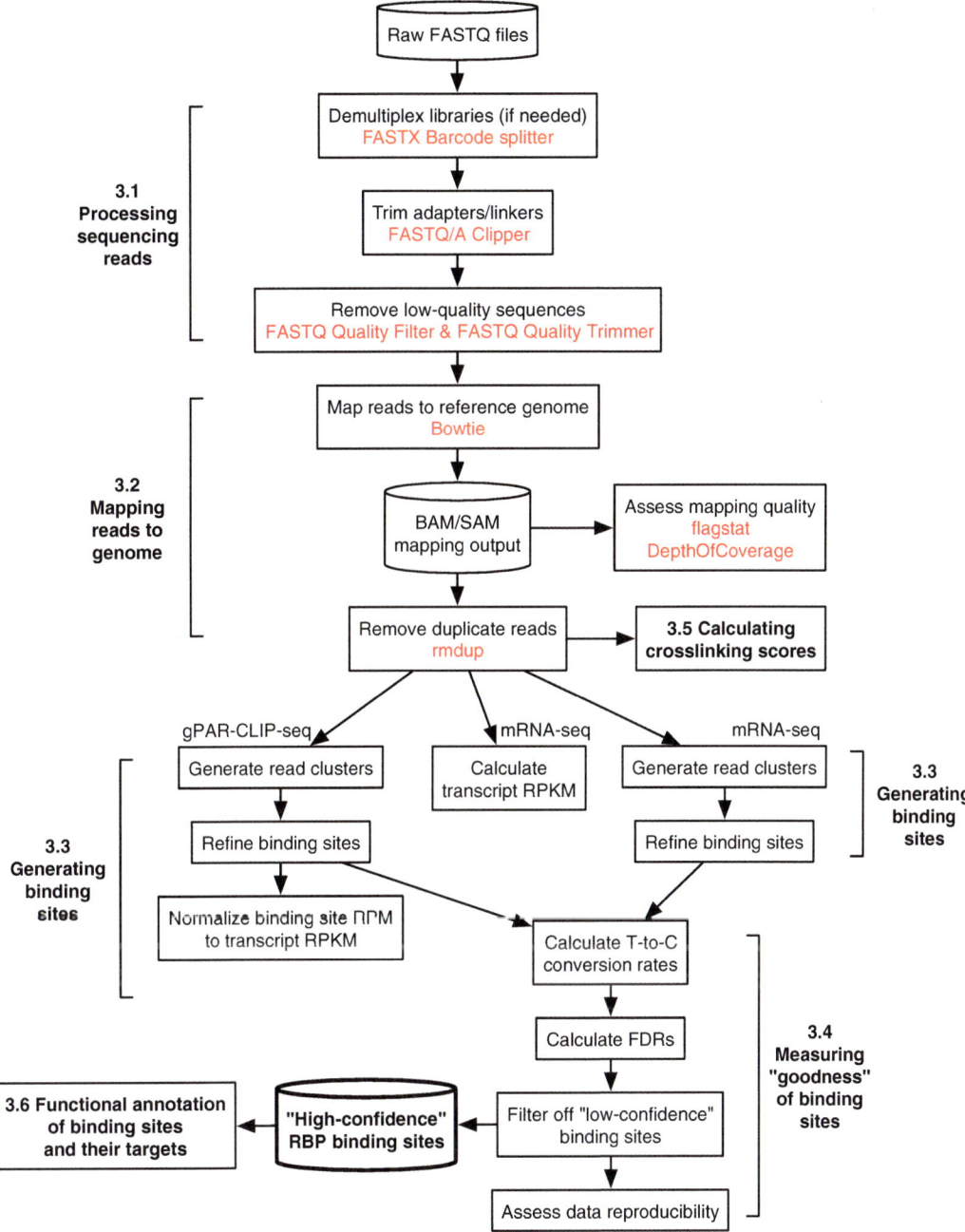

Fig. 1 Pipeline for analysis of gPAR-CLIP-seq data. Steps corresponding to the bioinformatic analysis pipeline are displayed as a flowchart. Programs available for download or through Galaxy are highlighted in *red*

allows comparisons of the relative strength of RBP binding across different transcripts using read coverage as a proxy for binding strength (Subheadings 3.3 and 3.5). mRNA-seq reads are also used in Subheading 3.4 to calculate a false-discovery rate for each gPAR-CLIP-seq-derived binding site. If transcript abundance data are not available, global binding sites can still be calculated, but users must be careful when comparing read coverage of binding sites located on different transcripts as more gPAR-CLIP-seq reads will be recovered and sequenced from the most highly abundant transcripts [9, 10].

2   Materials

The methods presented here can be run on any operating system (Max OS X, Windows, or Linux) depending on user preference and algorithm dependencies. We implemented our methods using custom Perl (v5.10.1) scripts and code developed in R (v2.15.2) [11] or using downloadable programs, which are indicated at each step in Subheading 3. Our work was executed on a single RedHat Enterprise Linux 6 machine with 256GB of RAM and two Intel Xeon E5-2680v2 10-core processors capable of hyperthreading. The computer was attached to the network and 3TB of fast NFS-based storage via 10Gbit Ethernet. Manual parallelization can speed up performance at individual steps of the computational pipeline depending on the exact programs used. For example, read mapping (Subheading 3.2) can be parallelized by breaking up a raw sequencing read file into multiple input files for the Bowtie algorithm.

3   Methods

3.1   Processing Sequencing Reads

Prior to mapping, sequencing reads must be sorted into their respective samples, if libraries were multiplexed, and processed to remove undesirable sequences. Sequencing centers may offer to perform these steps before returning files of sequencing reads. If not, users should perform the following steps.

1. *De-multiplex libraries.* In the case of multiplexed libraries, reads need to be sorted into their respective samples based on barcode sequences added during cDNA library preparation. This can be accomplished using the *FASTX Barcode splitter* algorithm from the FASTX-Toolkit available through the Galaxy platform or for download at http://hannonlab.cshl.edu/fastx_toolkit/download.html. Recommended parameters for *FASTX Barcode splitter*: [--mismatches 1] for 6-nt barcodes (*see* **Note 1**).

2. *Trim adapters/linkers.* Sequencing reads need to be trimmed of artificial adapter sequences added during cDNA library preparation. This can be accomplished using the *FASTQ/A Clipper*

algorithm from the FASTX-Toolkit. Recommended parameters for *FASTQ/A Clipper*: [-l 15] [-C] (*see* **Note 2**).

3. *Remove low-quality sequences.* To increase mapping efficiency, low-quality reads and low-quality nucleotides from 3′ ends of reads should be removed (*see* **Note 3**). This can be accomplished using the *FASTQ Quality Filter* (for removing low-quality reads) and *FASTQ Quality Trimmer* (for removing low-quality nucleotides from 3′ ends of reads) algorithms from the FASTX-Toolkit. Recommended parameters for *FASTQ Quality Filter*: [-q 30]. Recommended parameters for *FASTQ Quality Trimmer*: [-t 30] [-l 15].

3.2 Mapping Reads to the Genome

Many programs are available for mapping sequencing reads to reference genomes. Users are encouraged to use a mapping program with which he or she is most familiar.

1. *Map gPAR-CLIP-seq reads to a reference genome.* Use Bowtie [12], or an alternate mapping algorithm, to map gPAR-CLIP-seq reads to a reference genome. Bowtie is available through Galaxy or for download at https://github.com/BenLangmead/bowtie. Mapping output can be saved in BAM and SAM formats, which are commonly used as input to a variety of downstream analysis programs. Recommended Bowtie parameters: [-v 3] [--best] [--strata] (*see* **Note 4**).

2. *Map mRNA-seq reads to a reference genome.* Use Bowtie, or an alternate mapping algorithm, to map mRNA-seq reads to the same reference genome used above. Reads mapping with 0 mismatches will be used for transcript quantification (Subheading 3.3); reads mapping with 0-2 T-to-C mismatches will be used for FDR calculations (Subheading 3.4).

3. *Assess mapping quality.* Results of mapping should be assessed for quality and efficiency. Users can choose from a variety of programs including: *flagstat* (reports total number of reads, number of duplicate reads, percentage of reads mapped, etc.) from the SAMtools [13] package (available through Galaxy or for download at http://sourceforge.net/projects/samtools/files/samtools/) and *DepthOfCoverage* (reports read coverage per interval, gene, etc.) from the Genome Analysis Toolkit [14, 15] software package (available through Galaxy or for download at http://www.broadinstitute.org/gatk/download). If mapping quality or efficiency is unacceptable (this will depend on the organism and sample being analyzed), refer to the manual for the algorithm used to improve mapping results.

4. *Remove duplicate reads.* To eliminate amplification bias introduced during PCR in the library preparation step (*see* Chapter 5 Experimental Procedures chapter), duplicated read artifacts should be removed [14, 13]. This can be accomplished using

rmdup from the SAMtools package (available through Galaxy or for download). Alternatively, users can use the *MarkDuplicates* algorithm from Picard, which is available for download at http://sourceforge.net/projects/picard/files/picard-tools/. Recommended parameters for *MarkDuplicates*: [REMOVE_DUPLICATES=true] to prevent duplicate reads from being written to a new file.

3.3 Generating Binding Sites

For this step, code was generated in-house and is not available through Galaxy or for download as a stand-alone program. The original code was written in R, but pseudocode is provided (Figs. 2 and 4) so that users may implement the algorithm in any language.

1. *Generate read clusters.* Read clusters are defined as continuous stretches of nucleotides covered by at least one read harboring 0, 1, or 2 T-to-C conversion events only (*see* **Note 4**; Fig. 2). To differentiate between true RBP binding sites and noise in the data, read clusters that do not contain any T-to-C conversions (suggesting these RNA regions are not actually bound by an RBP) are treated as "low-confidence" and removed in a final filtering step (Subheading 3.4).

Algorithm 1   Generate read clusters

```
 1: for each chromosome chr do
 2:     for each read rd on chromosome chr do
 3:         for each genomic position pos in rd do
 4:             readcount(pos) ←readcount(pos) + 1
 5:         end for
 6:     end for
 7:     i ← 1                                    ▷ Initialize clusterID to 1
 8:     for each genomic position pos on chromosome chr do
 9:         if readcount(pos) > 0 then
10:             clusterID(pos) ← i
11:             while readcount(nextpos) > 0 do
12:                 clusterID(nextpos) ← i
13:             end while
14:             i ← i + 1
15:         end if
16:     end for
17: end for
end
```

Fig. 2 Pseudocode describing how to generate read clusters from mapped gPAR-CLIP-seq data. A read cluster is defined as a continuous stretch of nucleotides covered by at least 1 gPAR-CLIP-seq read with 0-2 T-to-C conversion events. Input: chromosome, start position, end position, and strand information for reads mapping to the genome with 0-2 T-to-C mismatches. Output: a list of every chromosomal position, how many unique reads map to each position, and a cluster ID denoting the distinct read cluster to which each position belongs. Users can obtain genomic start and end coordinates of each read cluster by calculating the minimum and maximum position for each cluster ID. In the case of stranded sequencing libraries, separately analyze reads mapping to the plus and minus strands of the reference.

2. *Refine binding sites.* Some read clusters span hundreds of nucleotides and contain one or more distinct peaks indicative of unique RBP–RNA binding events (Fig. 3). To isolate distinct peaks within long read clusters, we fit a Gaussian curve (normal kernel function) to each read cluster and used the inflection points of this curve to define the boundaries of individual binding sites (Fig. 4). The bandwidth parameter for the normal kernel function was chosen to reflect the expected size of an RBP binding site (e.g., 21 nt); however, this parameter can be empirically determined by the user and adjusted as

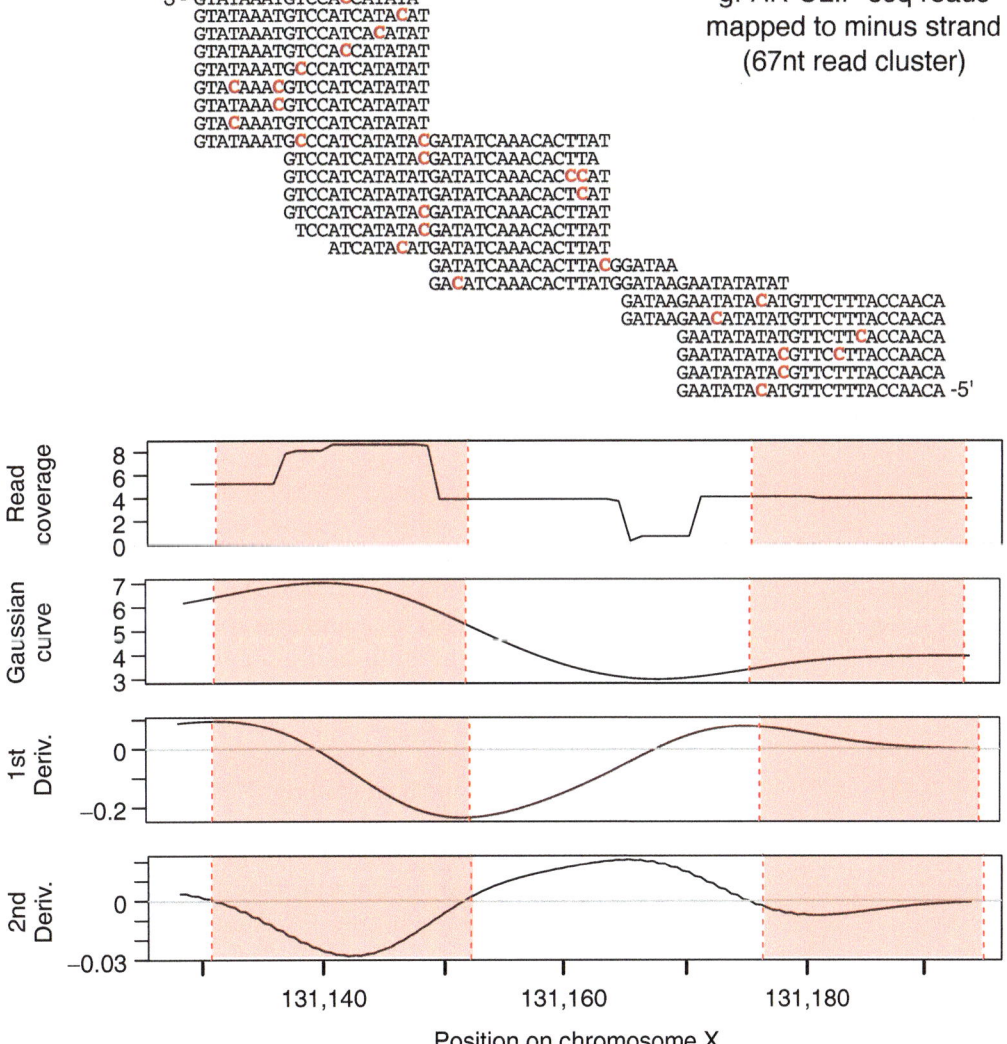

Fig. 3 Distinct binding events identified within long read clusters. Shown are sequencing reads mapping to the minus strand of chromosome X with T-to-C conversion events highlighted in *red*. These read form a read cluster 67 nt long. Two distinct binding events are determined by identifying the inflection points of the second derivative of the fitted Gaussian curve. *Red shaded blocks* indicate the new, refined binding sites.

Algorithm 2   Refine binding sites

```
 1: for each read cluster rc do
 2:     f(rc) ← Gaussian curve fit to read count data across rc
 3:     Calculate f'(rc) (first derivative)
 4:     Calculate f''(rc) (second derivative)
 5:     for each position pos in rc do
 6:         if f''(pos) = 0 and slope of f''(pos) < 0 then
 7:             append pos to startcoordinates
 8:         else if f''(pos) = 0 and slope of f''(pos) > 0 then
 9:             append pos to endcoordinates
10:         end if
11:     end for
12: end for
end
```

Fig. 4 Pseudocode describing how to identify distinct binding events in long read clusters. Read cluster start and end coordinates are refined to break long read clusters into smaller, distinct peaks. The general approach is to fit a Gaussian curve to the read counts across each binding site and define start and end coordinates as the inflections points of this curve. Input: output from Algorithm 1. Output: refined binding site boundary start and end coordinates. While read count data are discrete, the data are treated as continuous for this analysis. In lines 6 and 8, the points at which $f''(pos) = 0$ will likely not be integers, so users should round to the nearest integer to get chromosomal coordinates

needed (*see* **Note 5**). From this new set of refined binding sites, read coverage is determined by averaging the reads per million mapped read (RPM) values at each position across each refined binding site.

3. *Calculate transcript abundance.* Using perfectly mapped mRNA-seq reads obtained in Subheading 3.2, calculate transcript reads per million mapped reads per kilobase of transcript (RPKM; also called FPKM) using an established method such as Cufflinks [16] available through Galaxy or for download at http://cufflinks.cbcb.umd.edu/downloads/. Additional methods and documentation describing best practices for quantifying transcript levels are readily available [17], so details will not be described here. Alternatively, published RPKM values of transcripts can be obtained and used in Subheading 3.3; however, (1) the sample conditions must be similar so that the published RPKM values are an accurate proxy for transcript levels in the samples used to generate gPAR-CLIP-seq libraries, and (2) published mRNA-seq libraries cannot be used for FDR calculation as the rate of sequencing error varies from machine to machine.

4. *Normalize binding site RPM to transcript abundance.* To allow comparison of RBP binding sites on different transcripts, binding site read coverage must be normalized by transcript abundance by dividing binding site RPM by the RPKM of the

associated transcript and multiplying by 1000 to account for the kilobase normalization of RPKM values. Some transcripts with no mRNA-seq reads contain gPAR-CLIP-seq binding site (typically very few; often with no T-to-C conversion events); these site are treated as "low-confidence" and removed in a final filtering step (Subheading 3.4).

3.4 Defining High-Quality RBP Binding Sites from gPAR-CLIP-seq Data

A small fraction of T-to-C mismatches in gPAR-CLIP-seq reads likely represent sequencing error instead of true RBP–RNA interaction events, so binding sites derived from this error need to be identified and removed. The general approach is to calculate an FDR for each gPAR-CLIP-seq-derived binding site by performing binding site generation (Subheadings 3.1–3.3) using mRNA-seq reads with 0-2 T-to-C mismatches, which is a proxy for the rate of T-to-C sequencing error. This is followed by comparison of T-to-C conversion rates between gPAR-CLIP-seq-derived and mRNA-seq-derived binding sites, removal of "low-confidence" binding sites, and assessment of data reproducibility.

1. *Generate mRNA-seq binding sites.* Repeat steps in Subheadings 3.1–3.3 (through **step 2**) using mRNA-seq reads and the same reference genome and algorithm parameters that were used for generating binding sites from gPAR-CLIP-seq data.

2. *Calculate T-to-C conversion rates.* For each gPAR-CLIP-seq- and mRNA-seq-derived binding site, calculate the T-to-C conversion rate as the number of reads with at least 1T-to-C conversion event divided by the total number of reads covering at least 1 thymine.

3. *Assign FDRs.* Bin gPAR-CLIP-seq- and mRNA-seq-derived binding sites separately into groups based on total read coverage. Because total read coverage values are approximated by a negative binomial distribution, we recommend binning such that roughly equal numbers of binding sites are in each bin (Fig. 5). For each gPAR-CLIP-seq-derived binding site within each bin, calculate the proportion of mRNA-seq binding sites in that bin with a higher T-to-C conversion rate. This proportion represents the FDR for that gPAR-CLIP-seq binding site.

4. *Filter off "low-confidence" binding sites.* To determine a final set of "high-confidence" gPAR-CLIP-seq-derived binding sites, we removed binding sites that met any of the following criteria: (1) contain no T-to-C conversion events, (2) map to transcripts with no mRNA-seq reads, (3) have low RPM coverage, or (4) have an FDR above 1 % (strict) or 5 % (conservative).

5. *Assess data reproducibility.* If replicate gPAR-CLIP-seq libraries are generated, reproducibility can be assessed by calculating a correlation coefficient for binding site RPM values.

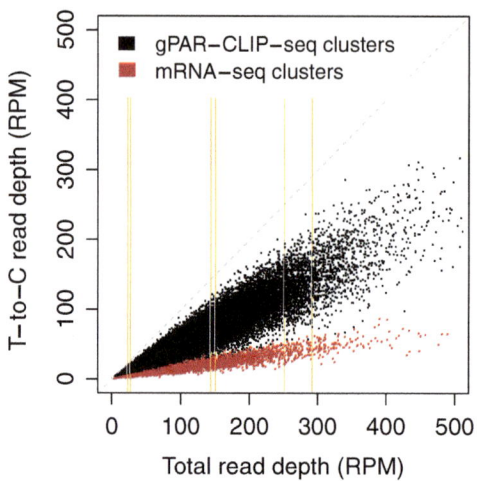

Fig. 5 Total versus T-to-C read coverage for determining gPAR-CLIP-seq binding site FDRs. Plotted are total read coverage versus T-to-C read coverage of 50,000 random gPAR-CLIP-seq (*black*) and 50,000 random mRNA-seq (*red*) read clusters. gPAR-CLIP-seq read clusters were grouped into 50 bins with ~1000 clusters in each bin. The 5th, 40th, and 49th bins are demarcated by *orange lines* and show that although the bins contain roughly the same number of gPAR-CLIP-seq binding sites, they cover a varied range of total read depth RPMs.

3.5 Calculating Per-Nucleotide Cross-Linking Scores (CLSs)

In addition to identifying discrete, transcriptome-wide RBP binding sites, users can calculate a measure of cross-linking, or binding, strength on a per-nucleotide level (*see* **Note 6**). The general approach is similar to how FDR values are assigned to binding sites in Subheading 3.4.

1. A CLS for each transcriptomic uracil (represented by thymines in our cDNA sequencing libraries) is calculated as the number of reads covering that position that contain a T-to-C conversion event divided by the transcript RPKM and multiplied by a factor of 1000 to account for the kilobase normalization in RPKM values.

3.6 Functional Annotation of Binding Sites and Their Targets

After obtaining a set of "high-confidence" RBP binding sites across the transcriptome, binding sites and their mRNA targets can be further analyzed to obtain biologically functional information. Below are some common analysis tools. Specific information about how to run these tools and interpret the results is beyond the scope of this paper; these tools generally have helpful documentation available online.

1. *Gene ontology term enrichment.* Discover sets of terms describing the molecular functions, biological processes, and cellular compartments associated with mRNAs harboring RBP-binding sites. Tool is available online or for download at http://www.geneontology.org [18].

2. *Gene set/pathway enrichment.* Further characterize mRNAs harboring RBP-binding sites using comprehensive functional annotation tools such as DAVID, available through Galaxy or online at http://david.abcc.ncifcrf.gov/ [19, 20], and g:Profiler, available through Galaxy or online at http://biit.cs.ut.ee/gprofiler/ [21, 22], to identify enriched pathways, discover gene–disease associations, and identify enriched gene groups.

3. *Primary sequence motif analysis.* Identify putative functional sequence motifs using the MEME suite of sequence analysis tools, available through Galaxy, online, or for download at http://meme.nbcr.net/meme/ [23].

4. *SecondaryRNA structureanalysis.* Identify potential RNA secondary structure of interest using the ViennaRNA package of tools (e.g., RNAfold, RNAplfold) available for download at http://www.tbi.univie.ac.at/RNA/ [24].

5. *Conservation/homology analysis.* Primary sequence conservation scores can be downloaded from the UCSC genome browser [25, 26] and used to explore evolutionary conservation of calculated binding sites.

4   Notes

1. We recommend using the [--eol] parameter to ensure that the barcode is matched at the 3′-most end of the read, which is typically where barcodes are added. This ensures that matches to barcode sequences that occur randomly throughout the genome, and therefore might appear elsewhere in a read, are not mistaken for the true barcode.

2. A read length threshold of 15 nt ([-l 15]) was chosen for reads being mapped to the *S. cerevisiae* genome, but can be optimized depending on the size of the genome being used. Users should also check their sequences for 5′ "N" nucleotides, which can be trimmed before mapping.

3. Users should also remove read artifacts if they are: homopolymers, missing 3′ adapter, 5′–3′ adapter ligation products, or 5′–5′ adapter ligation products. Also note that only reads in FASTQ format may be analyzed using the Quality Filter and Quality Trimmer algorithms, as FASTA-formatted sequences do not contain quality information. The [-q 30] and [-t 30] parameters both correspond to minimum Phred quality scores to keep reads/nucleotides. A Phred quality score of 30 indicates a base call accuracy of 99.9 %. Users may choose other Phred quality score thresholds if they desire.

4. Low-frequency incorporation of 4sU into nascent mRNAs induces a mis-pairing of guanine to 4sU during reverse

transcription that manifests as T-to-C mismatches to the genome (i.e., a cytosine is sequenced where there should be a thymine). As of this publication, there are no published mapping algorithms that allow for differential treatment of different types of mismatches to the genome. Therefore, mapping algorithm parameters must be set to allow for multiple mismatches. Only reads with 0 or 1-2 T-to-C mismatches will be used in subsequent steps.

5. Some secondary analyses performed on binding sites after Gaussian curve fitting will be affected by the choice of bandwidth parameter. For example, calculating the average binding site length after Gaussian curve fitting will result in an average length close to the bandwidth parameter chosen. For most secondary analyses, however, the choice of bandwidth parameter will not have an effect on results.

6. Calculating a cross-linking score is possible because gPAR-CLIP-seq T-to-C conversions only occur when a 4sU is within a few angstroms of an amino acid side chain [27]. Because a single RBP–RNA interaction site spans many nucleotides, calculating a score for each thymine within the binding site could give insight into which thymines are the most biologically important for RBP–RNA interactions. For this calculation, we assume that rate of incorporation of the ribonucleoside analog into nascent transcripts during transcription is uniform across the length of the transcript.

Acknowledgements

This work was supported by the National Science Foundation Open Data IGERT grant 0903629 (M.A.F.), the National Institutes of Health grant GM088565 (J.K.K.), and the Pew Charitable Trusts (J.K.K.). The authors would like to thank Danny Yang, Ting Han, and James Taylor for helpful comments on the manuscript.

References

1. Darnell RB (2010) HITS-CLIP: panoramic views of protein-RNA regulation in living cells. Wiley Interdiscip Rev RNA 1(2):266–286. doi:10.1002/wrna.31

2. Ule J, Jensen KB, Ruggiu M, Mele A, Ule A, Darnell RB (2003) CLIP identifies Nova-regulated RNA networks in the brain. Science 302(5648):1212–1215. doi:10.1126/science.1090095

3. Licatalosi DD, Darnell RB (2006) Splicing regulation in neurologic disease. Neuron 52(1):93–101. doi:10.1016/j.neuron.2006.09.017

4. Jensen KB, Dredge BK, Stefani G, Zhong R, Buckanovich RJ, Okano HJ, Yang YY, Darnell RB (2000) Nova-1 regulates neuron-specific alternative splicing and is essential for neuronal viability. Neuron 25(2):359–371

5. Hafner M, Landthaler M, Burger L, Khorshid M, Hausser J, Berninger P, Rothballer A, Ascano M Jr, Jungkamp AC, Munschauer M, Ulrich A, Wardle GS, Dewell S, Zavolan M,

Tuschl T (2010) Transcriptome-wide identification of RNA-binding protein and microRNA target sites by PAR-CLIP. Cell 141(1):129–141. doi:10.1016/j.cell.2010.03.009

6. Corcoran DL, Georgiev S, Mukherjee N, Gottwein E, Skalsky RL, Keene JD, Ohler U (2011) PARalyzer: definition of RNA binding sites from PAR-CLIP short-read sequence data. Genome Biol 12(8):R79. doi:10.1186/gb-2011-12-8-r79

7. Erhard F, Dolken L, Jaskiewicz L, Zimmer R (2013) PARma: identification of microRNA target sites in AGO-PAR-CLIP data. Genome Biol 14(7):R79. doi:10.1186/gb-2013-14-7-r79

8. Chou CH, Lin FM, Chou MT, Hsu SD, Chang TH, Weng SL, Shrestha S, Hsiao CC, Hung JH, Huang HD (2013) A computational approach for identifying microRNA-target interactions using high-throughput CLIP and PAR-CLIP sequencing. BMC Genomics 14(Suppl 1):S2. doi:10.1186/1471-2164-14-S1-S2

9. Mortazavi A, Williams BA, McCue K, Schaeffer L, Wold B (2008) Mapping and quantifying mammalian transcriptomes by RNA-Seq. Nat Methods 5(7):621–628. doi:10.1038/nmeth.1226

10. Wang Z, Gerstein M, Snyder M (2009) RNA-Seq: a revolutionary tool for transcriptomics. Nat Rev Genet 10(1):57–63. doi:10.1038/nrg2484

11. Team RDC (2011) R: a language and environment for statistical computing. R Foundation for Statistical Computing, Vienna, Austria

12. Langmead B, Trapnell C, Pop M, Salzberg SL (2009) Ultrafast and memory-efficient alignment of short DNA sequences to the human genome. Genome Biol 10(3):R25. doi:10.1186/gb-2009-10-3-r25

13. Li H, Handsaker B, Wysoker A, Fennell T, Ruan J, Homer N, Marth G, Abecasis G, Durbin R, Genome Project Data Processing S (2009) The sequence alignment/map format and SAMtools. Bioinformatics 25(16):2078–2079. doi:10.1093/bioinformatics/btp352

14. DePristo MA, Banks E, Poplin R, Garimella KV, Maguire JR, Hartl C, Philippakis AA, del Angel G, Rivas MA, Hanna M, McKenna A, Fennell TJ, Kernytsky AM, Sivachenko AY, Cibulskis K, Gabriel SB, Altshuler D, Daly MJ (2011) A framework for variation discovery and genotyping using next-generation DNA sequencing data. Nat Genet 43(5):491–498. doi:10.1038/ng.806

15. McKenna A, Hanna M, Banks E, Sivachenko A, Cibulskis K, Kernytsky A, Garimella K, Altshuler D, Gabriel S, Daly M, DePristo MA (2010) The Genome Analysis Toolkit: a MapReduce framework for analyzing next-generation DNA sequencing data. Genome Res 20(9):1297–1303. doi:10.1101/gr.107524.110

16. Trapnell C, Williams BA, Pertea G, Mortazavi A, Kwan G, van Baren MJ, Salzberg SL, Wold BJ, Pachter L (2010) Transcript assembly and quantification by RNA-Seq reveals unannotated transcripts and isoform switching during cell differentiation. Nat Biotechnol 28(5):511–515. doi:10.1038/nbt.1621

17. Garber M, Grabherr MG, Guttman M, Trapnell C (2011) Computational methods for transcriptome annotation and quantification using RNA-seq. Nat Methods 8(6):469–477. doi:10.1038/nmeth.1613

18. Ashburner M, Ball CA, Blake JA, Botstein D, Butler H, Cherry JM, Davis AP, Dolinski K, Dwight SS, Eppig JT, Harris MA, Hill DP, Issel-Tarver L, Kasarskis A, Lewis S, Matese JC, Richardson JE, Ringwald M, Rubin GM, Sherlock G (2000) Gene ontology: tool for the unification of biology. The Gene Ontology Consortium. Nat Genet 25(1):25–29. doi:10.1038/75556

19. da Huang W, Sherman BT, Lempicki RA (2009) Systematic and integrative analysis of large gene lists using DAVID bioinformatics resources. Nat Protoc 4(1):44–57. doi:10.1038/nprot.2008.211

20. da Huang W, Sherman BT, Lempicki RA (2009) Bioinformatics enrichment tools: paths toward the comprehensive functional analysis of large gene lists. Nucleic Acids Res 37(1):1–13. doi:10.1093/nar/gkn923

21. Reimand J, Arak T, Vilo J (2011) g:Profiler—a web server for functional interpretation of gene lists (2011 update). Nucleic Acids Res 39(Web Server issue):W307–W315. doi:10.1093/nar/gkr378

22. Reimand J, Kull M, Peterson H, Hansen J, Vilo J (2007) g:Profiler—a web-based toolset for functional profiling of gene lists from large-scale experiments. Nucleic Acids Res 35(Web Server issue):W193–W200. doi:10.1093/nar/gkm226

23. Bailey TL, Boden M, Buske FA, Frith M, Grant CE, Clementi L, Ren J, Li WW, Noble WS (2009) MEME SUITE: tools for motif discovery and searching. Nucleic Acids Res 37(Web Server issue):W202–W208. doi:10.1093/nar/gkp335

24. Lorenz R, Bernhart SH, Honer Zu Siederdissen C, Tafer H, Flamm C, Stadler PF, Hofacker IL (2011) ViennaRNA Package 2.0. Algorithms Mol Biol 6:26. doi:10.1186/1748-7188-6-26

25. Pollard KS, Hubisz MJ, Rosenbloom KR, Siepel A (2010) Detection of nonneutral sub-

stitution rates on mammalian phylogenies. Genome Res 20(1):110–121. doi:10.1101/gr.097857.109

26. Siepel A, Bejerano G, Pedersen JS, Hinrichs AS, Hou M, Rosenbloom K, Clawson H, Spieth J, Hillier LW, Richards S, Weinstock GM, Wilson RK, Gibbs RA, Kent WJ, Miller W, Haussler D (2005) Evolutionarily con-

served elements in vertebrate, insect, worm, and yeast genomes. Genome Res 15(8): 1034–1050. doi:10.1101/gr.3715005

27. Favre A (1990) 4-Thiouridine as an intrinsic photoaffinity probe of nucleic acid structure and interactions. In: Morrison H (ed) Bioorganic photochemistry, vol 1. Wiley, New York, pp 379–425

Chapter 7

Translation Analysis at the Genome Scale by Ribosome Profiling

Agnès Baudin-Baillieu, Isabelle Hatin, Rachel Legendre, and Olivier Namy

Abstract

Ribosome profiling is an emerging approach using deep sequencing of the mRNA part protected by the ribosome to study protein synthesis at the genome scale. This approach provides new insights into gene regulation at the translational level. In this review we describe the protocol to prepare polysomes and extract ribosome protected fragments before to deep sequence them.

Key words Ribosome profiling, Ribo-seq, Translation regulation, Recoding

1 Introduction

There are a number of High throughput technics to quantify gene expression level. During the last decade microarrays and RNA-seq allowed to study in great details RNA content of cells. These approaches associated with proteomics approaches can provide a good evaluation of the gene expression level. However, limiting gene expression analysis to these approaches misses all translational regulations playing a crucial role in cell's homeostasis. Ribosome profiling fills the gap existing between data provided by transcriptomics and proteomics approaches [1, 2]. Ribosome profiling combines the observation that the nuclease digestion footprint of a ribosome on an mRNA indicates its exact position to new generation sequencing to massively sequence ribosome protected fragments (RPF) (Fig. 1). It allows to determine the amount of ribosomes on each mRNA, which will reflect the translational level of this mRNA. We can go even further with ribosome profiling to qualitatively measure translation regulation and fidelity at a given moment or in a mutant compared to a wild type cell. This gives access to the identification of new coding sequences (CDS), ribosomal A-site occupancy, upstream ORFs translational

Frédéric Devaux (ed.), *Yeast Functional Genomics: Methods and Protocols*, Methods in Molecular Biology, vol. 1361, DOI 10.1007/978-1-4939-3079-1_7, © Springer Science+Business Media New York 2016

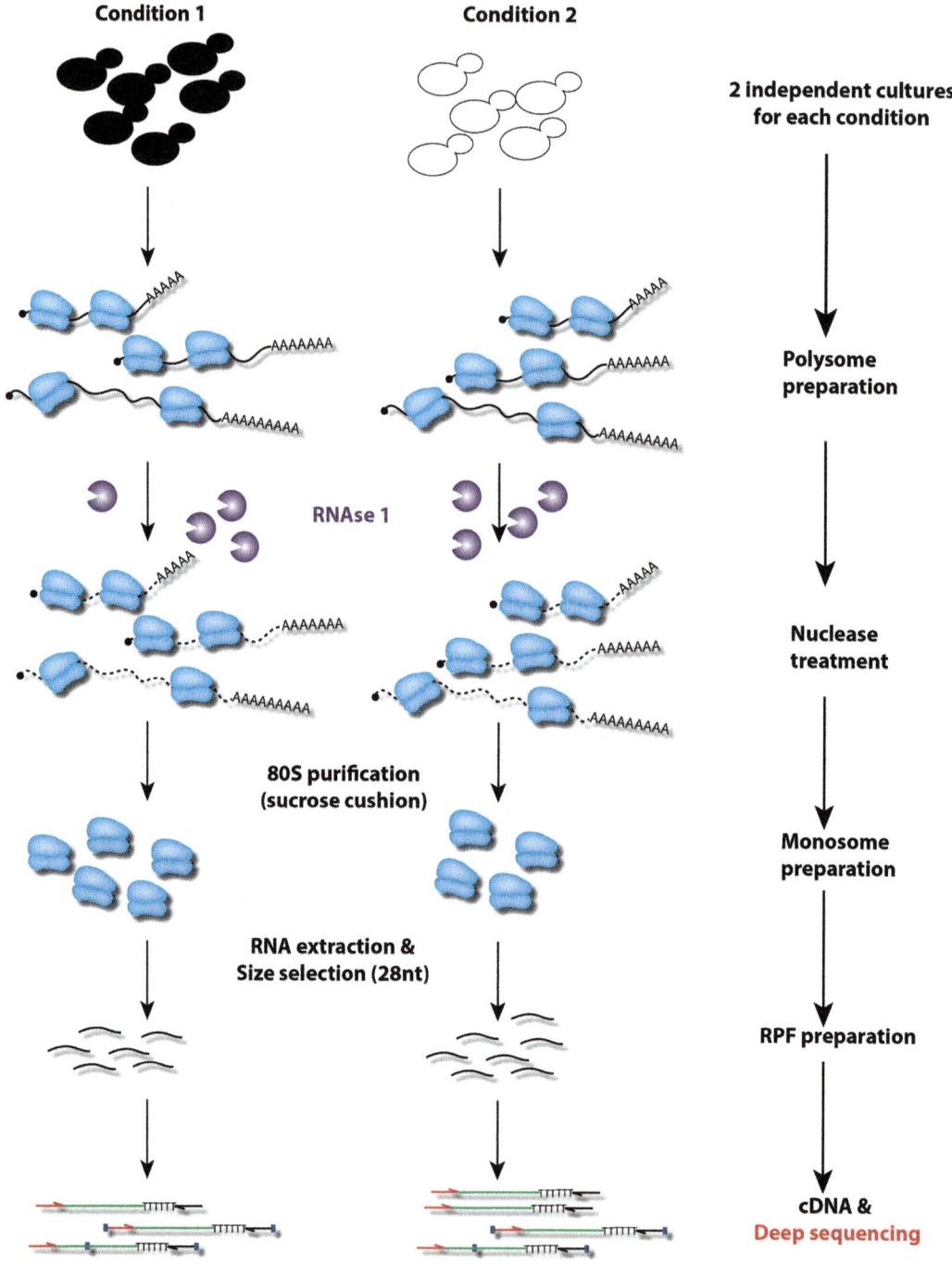

Fig. 1 Overview of ribosome profiling. The schema represents the different steps of RPF preparation from yeast cultures to the deep sequencing in two conditions

regulations, or the discovery that non coding RNA are loaded by ribosome to induce Nonsense-Mediated Decay (NMD) [1–8].

Since the initial publication by Weissman's laboratory ribosome profiling has been used in a variety of organisms to address a broad number of questions [7–11]. Despite the strong enthusiast generated by this first technics allowing genome-wide translational changes, it should be keep in mind that this is a complicated approach with many pitfalls that can generate a number of misinterpretations. Indeed small variations in growth culture, medium composition or low genome coverage can generate misinterpretations. Ribosome profiling cannot be the end of a story but instead should be the beginning of new questions. It is essential not to rely only on statistical analysis to validate data but also performing independent experiments on few genes. In this review we will describe in detail all steps needed to prepare high quality RPF and how to perform basic bioinformatics analysis to map them onto a *Saccharomyces cerevisiae* reference genome. Obviously most of the steps can be applied to other organisms since it is possible to extract polysomes.

2   Material

2.1   Media, Solutions

1. Plates of complete media YEPD or specific supplemented minimal media.

2. Solution of cycloheximide 50 mg/ml in ethanol.

3. Lysis Buffer 10×: 0.1 M Tris-HCl pH 7.4, 1 M NaCl, 0.3 M $MgCl_2$.

4. Hybridization Buffer 10×: 1.5 M NaCl, 0.5 M Tris-HcL pH 7.4, 10 mM EDTA.

5. Elution Buffer: 0.3 M sodium acetate pH 5.5, 1 mM EDTA.

6. Sucrose gradient 10–50 % (W:V) or cushion sucrose 24 % (W/V) in 50 mM Tris-acetate pH 7.6, 50 mM NH_4Cl, 12 mM $MgCl_2$, 1 mM DTT (*see* **Note 1**).

7. RNase-free distilled water.

8. Ethanol.

9. TE 1×: 10 mM tris pH 7.4, 1 mM EDTA.

10. RNase I endonuclease Ambion ref AM2295.

11. RNAse inhibitor.

12. Glycogen 20 mg/ml.

13. 3 M sodium acetate pH 5.2.

14. Ammonium persulfate 10 %.

15. TEMED (N,N,N',N'-tetramethylethylenediamine).

16. Polyacrylamide gels: 17 % 19:1 acrylamide–bis-acrylamide, 7 M urea and 1× TAE.

17. TAE 50×: 2 M tris-acetate, 50 mM EDTA pH 8.

18. 5× RNA loading dye: 50 % glycerol, 50 mM Tris pH 7.7, 5 mM EDTA pH 8, and 0.25 % bromophenol blue (BPB), aliquots are store at –20 °C.

19. Dye SYBER Gold for nucleic acid staining 10,000 concentrated with a maximum excitation wavelength at 300 nm.

2.2 Oligonucleotide Sequences

1. RNA markers of 28 and 34 nucleotides length:

   (a) oNTI199 AUGUACACGGAGUCGACCCGCAACGCGA.

   (b) oNTI34ARN AUGUACACGGAGUCGACCCGCAACG CGAUGCUAA.

2. Biotinylated RNA for subtractive hybridization:

   (a) rRNA-1 5BioTEG/TGATGCCCCCGACCGTCCCTAT TAATCATTACGACCAAGTTTGTCCAAATTCTCCG CTCTGAGA.

   (b) rRNA-2 5BioTEG/GCTAGCCTGCTATGGTTCAGCG ACGCCACAACTGATCAAATGCCCTTCCCTTTCAA CAATTTCACG.

   (c) rRNA-35BioTEG/TTCCAGCTCCGCTTCATTGAATA AGTAAAGAACTATTTTGCCGACTTCCCTTATC TACATTATTCTA.

   (d) rRNA-4 5BioTEG/ATGTCTTCAACCCGGATCAGCC CCGAAGACTTACGTCGCAGTCCTCAGTCCC AGCTGGCAGTATTCCCACAG.

   (e) rRNA-5 5BioTEG/ATTCTATTATTCCATGCTAATAT ATTCGAGCAAGCGGTTATCAGTACGACCTGG CATGAAAAC.

   (f) rRNA-6 5BioTEG/AGCTGCATTCCCAAACAACTCG ACTCTTCCCCCACTTCAGTCTTCAAAGTTCTCA TTTTTATTCTACACCCTCTATGTCTCTTCACA.

   (g) rRNA-7* 5BioTEG/GACPCCTZATTLGTETCLATC.

(* Z, P, E, L represent LNA bases.)

2.3 Ware and Accessories

1. Flasks of 100 ml and 2 l for liquid cell culture.

2. 500 ml bucket for centrifuge.

3. Conical tubes of 15 ml.

4. Microtubes of 0.5, 1.5, and 2 ml Safe-Lock.

5. Liquid nitrogen.

6. Needle of 20 gauges.

7. Large ice bucket.

8. 0.22 μm cellulose acetate filters.

9. Glass beads with a diameter 0.25–0.5 mm washed by 1 M nitric acid, then rinse with distilled water.

10. Streptavidin MagneSphere.

2.4 Apparatus

1. Thermostatic incubator with agitation at 180 rpm.

2. Water bath 25 °C, 37 °C.

3. Thermomixer for microtubes at 65 °C.

4. Heat block at 75 °C.

5. Vortex with holder for microtubes.

6. Spectrophotometer to measure cell concentration at a wavelength of 600 nm and to measure RNA concentration at a wavelength of 260 nm with quartz cuvettes or with a micro-volume UV-spectrophotometer as a NanoDrop instrument.

7. Refrigerated centrifuge for 500 ml buckets with a centrifugal force of $5000 \times g$.

8. Refrigerated centrifuge for microtubes with a centrifugal force of $16,000 \times g$.

9. −20 and −80 °C freezer.

10. Fume hood.

11. For the 10–50 % sucrose gradient fractionation of polysomes.

12. Ultracentrifuge with SW41 rotor with tubes ultra-clear 13.2 ml Beckman ref: 344059.

13. Teledyne Isco with Tris peristaltic pump ref 68-1610-010; Isco type11 optical unit with 254 nm filter ref 68-1140-005; Brandel Tube piercer used with the option cannula fractioning method ref 60-3877-060; Isco UA-6 UV-visible detector ref 68-0940-016; Retriever 500 fraction collector ref 68-3880-001 and a fraction collector Foxy R1 ref 69-2133-667.

   For sedimentation on 24 % sucrose cushion of monosomes.

14. Ultracentrifuge with TLA110 rotor with 13×56 mm polycarbonate tubes 3.2 ml.

15. Electrophoresis on denaturing 17 % Polyacrylamide-7M urea gel.

16. Vertical electrophoresis cell with central cooling core combined with outer plates of 22.3×20 cm and inner plates of 20×20 cm with spacers and comb of 1 mm.

17. Generator to apply 200 constant voltages.

18. Heat circulating water system.

19. Small RNA controlled on chip-based capillary electrophoresis machine.

20. Magnetic separation stands.

2.5 Bioinformatics	The minimal configuration needed for bioinformatics analysis is a 64-bit computer running linux, 2 CPU, 8 GB of RAM. The following software or packages are also needed:

1. FASTQC software (http://www.bioinformatics.babraham.ac.uk/projects/fastqc/).

2. Cutadapt (https://code.google.com/p/cutadapt/).

3. Bowtie software (http://bowtie-bio.sourceforge.net/index.shtml).

4. SAMtools software (http://samtools.sourceforge.net/).

5. IGV genome browser (https://www.broadinstitute.org/igv/home).

6. HTSeq-count (http://www-huber.embl.de/users/anders/HTSeq/doc/install.html#install).

7. DESeq2 (http://www.bioconductor.org/packages/release/bioc/html/DESeq2.html).

3 Methods

3.1 Cells Culture	1. Plate yeast cells on either complete (YEDP) or requested media and grow at 30 °C.

2. Pick up two colonies in order to inoculate a 20 ml YEPD pre-culture grown in a 100 ml flask for 24 h at 30 °C.

3. Prepare 2×500 ml YEPD in a 2000 ml flask and inoculate with the starter culture to an initial OD_{600} of 0.005/ml.

4. Grow cells culture at 30 °C on an orbital shaker to a final OD_{600} of 0.6 (*see* **Note 2**). This takes about 15 h for the 74-D694 strain but depends on the strain genetic background. Growth conditions regarding media composition, temperature or other considerations can vary unless the number of cell division is conserved.

3.2 Polysome Preparation	The first step consists in extracting total ribosomes. The two cultures are treated in parallel.

1. Add 500 μl of 50 mg/ml cycloheximide stock solution to each culture to a final concentration of 50 μg/ml and shake for 5 min at room temperature. The two cultures are immediately cooled in an ice bath for 15 min with occasional shaking.

2. Pellet cells by centrifugation at 4 °C, $4000 \times g$ for 10 min. Eliminate supernatant carefully and resuspend pellet in 10 ml cold Lysis Buffer containing 50 μg/ml cycloheximide.

3. Transfer to a 15 ml conical tube and centrifuge at 4 °C, $4000 \times g$ for 5 min. Eliminate supernatant.

4. Estimate the volume of the pellet and resuspend cells in two volumes of ice cold Lysis Buffer with 50 µg/ml cycloheximide (the volume of the pellet is about 1 ml for 500 ml culture).

5. Transfer to a 1.5 ml eppendorf tube for easier handling. Pulverize Cells by adding glass beads and vortexing the mixture for 10 min at 4 °C (*see* **Note 3**).

6. Remove cells debris by centrifugation at 4 °C, $5000 \times g$ for 5 min and transfer the supernatant to chilled 1.5 ml microfuge tubes on ice. The supernatant is clarified by centrifugation at 4 °C, $15,000 \times g$ for 15 min. Recover supernatant avoiding to pipet the remaining debris. At this stage, the two samples are mixed.

7. The determination of polysome concentration is done by spectrophotometric estimation, based on the fact that ribosomes are ribonucleoprotein particles. Use a 1/10 dilution in water to measure the absorbance at 260 nm. Aliquots of 30 absorbance units are flash-frozen in liquid nitrogen and stored in a −80 °C freezer. You should get a total of about 300 A_{260}.

8. An aliquot of 10 A_{260} is loaded on a 10–50 % W:V sucrose gradient and spun for 3.5 h at $188,000 \times g$, 4 °C, in an SW41 swing-out rotor. Gradient is fractionated with the ISCO gradient fractionation system to control the quality of the polysome extraction (Fig. 2a).

3.3 Nuclease Digestion and Monosome Purification on Sucrose Cushion

1. Gently thaw six samples of 30 A_{260} on ice.

2. These extracts are subjected to RNAse I digestion with 15 U of enzyme/absorbance unit, for 1 h at 25 °C (*see* **Note 4**). RNAse I digestion does not lead to complete disruption of polysomes into monosomes, rather low polysomes (mainly two ribosomes on the same RNA fragment) still persist (Fig. 2b).

3. Meanwhile, prepare the solution for 24 % sucrose and refresh it. The ratio being 3 ml sucrose cushion solution–1 ml extract, prepare two 3 ml cushions and keep on ice.

4. Once the digestion time has expired, layer three digested polysome extracts (90 A_{260}) per cushion and centrifuge at 4 °C, 100,000 rpm in a TLa110 rotor for 2 h 15 min. 24 % sucrose cushion allows to pellet 80S monosomes and the remaining undigested polysomes (Fig. 2c).

5. Each pellet is carefully washed two times with 500 µl polysome extraction buffer to eliminate sucrose and resuspended in 750 µl of polysome extraction buffer. The pellet is solubilized by pipetting up and down and transferred in a 2 ml microtube. Sample can be flash-frozen in liquid nitrogen and stored at −80 °C or subjected to RNA extraction (*see* **Note 5**).

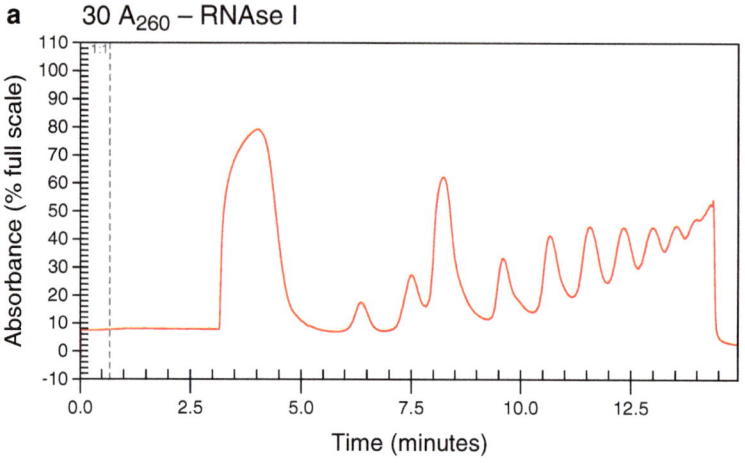

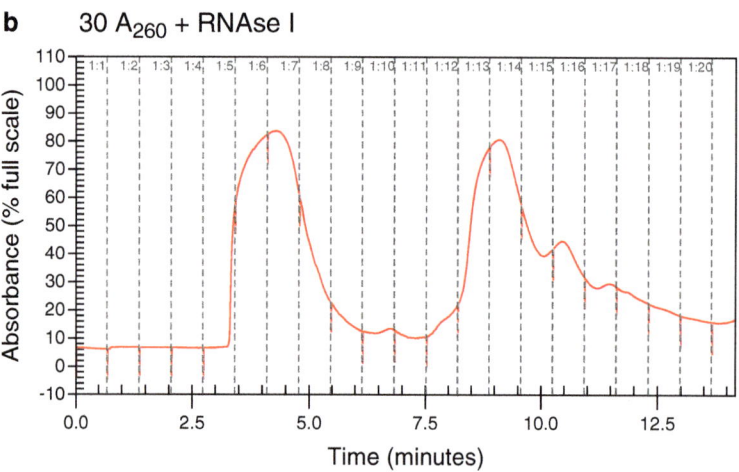

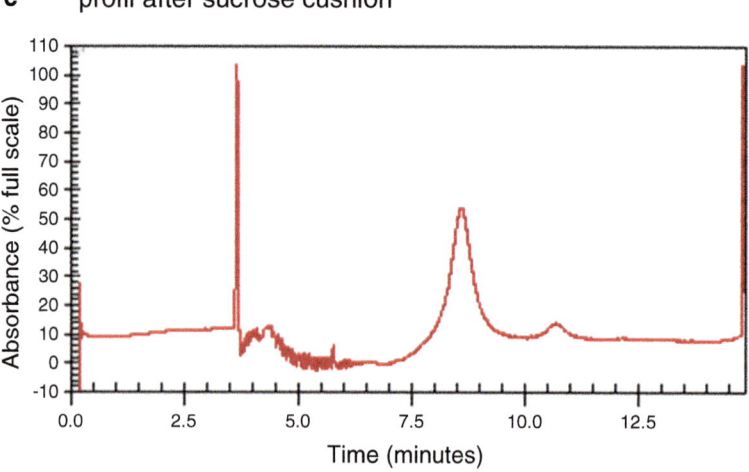

Fig. 2 Polysome profiles on sucrose gradients. (**a**) Control ribosome profile before nuclease digestion on a 10–50 % sucrose gradient. First pic at 4 min corresponds to cell debris. It is followed by the 40S and 60S free subunits (the 60S fraction should be twice the level of the 40S fraction), the monosome fraction at

3.4 RNA Extraction and Size Selection

1. An equal volume of acid phenol is added to each monosome fraction.

2. Place the mixture at 65 °C and vortex continuously using a thermomixer for 1 h in a fume hood.

3. Centrifuge at full speed for 10 min in a microfuge. Recover the aqueous phase (upper phase) and reextract with an equal volume of chloroform.

4. Vortex for 5 min and spin at full speed for 5 min in a microfuge. The aqueous phase is carefully transferred in a 1.5 ml microtube and total RNA is precipitated with 1/10 volume potassium acetate 3 M pH 5.2 and 3 volumes ethanol. Incubate samples at –20 °C overnight to enhance precipitation.

5. Spin at full speed in a microfuge for 15 min at 4 °C and eliminate as much supernatant as possible to minimize residual liquid.

6. Air-dry the pellet by leaving the tubes open for about 15 min. Dissolve each pellet in 500 μl TE + RNAse inhibitor 0.1 U/μl and mix the two samples. Measure RNA concentration at 260 nm (it should be around 2 mg/ml). RNA sample is stored at –20 °C.

7. RNA fragments are separated by electrophoresis in a polyacrylamide gel using a vertical electrophoresis cell with central cooling system. Prepare 18.5 × 20 cm gels with 1 mm thick spacer and a 15 well comb (*see* **Note 6**).

8. Prerun the gel at 150 V for 1 h with heating at 65 °C with a thermostatic circulator for obtaining high quality gel resolution and gel-to-gel reproducibility.

9. Add 5× RNA loading dye to RNA samples and load 15 μg RNA per well. The oNTI199 and oNTI34ARN RNA markers are used to demarcate the 28- to 34-nucleotide region, which is excised. A mix of 50 ng of each marker is loaded on each wells located at both extremities. A total of four gels (up to 1.2 mg of RNA) is necessary for one ribosome profiling experiment.

Fig. 2 (continued) about 8.5 min (higher pic) and the disomes, trisomes, and higher polysomes (up to ten ribosomes). (**b**) Ribosome profile after nuclease digestion on a 10–50 % sucrose gradient. Cell debris is abundant, 40S and 60S are still present unless the 60S fraction is partially masked by the 80S (monosomes) fraction that is predominant. Disomes and trisomes remain undigested although higher polysomes totally disappear. (**c**) An aliquot of the pellet after sucrose cushion is loaded on a 10–50 % sucrose gradient. It contains monosome particles (8.5 min) and disomes (10.5–11 min) exclusively

10. Run the gel at 150 V for 2 h then at 200 V, 65 °C until the blue dye reaches the gel bottom (about 6 h).

11. The gel is stained for 30 min with SYBR Gold diluted 10,000 times in 100 ml 1× TAE (100 ml is enough for four gels). This dye has a maximum fluorescence excitation when bound to RNA centered at approximately 300 nm.

12. Excise the region that corresponds to the 28 nt marker as it corresponds to the RNA region protected by a single ribosome. Store the gel slice in a tube at –20 °C (Fig. 3).

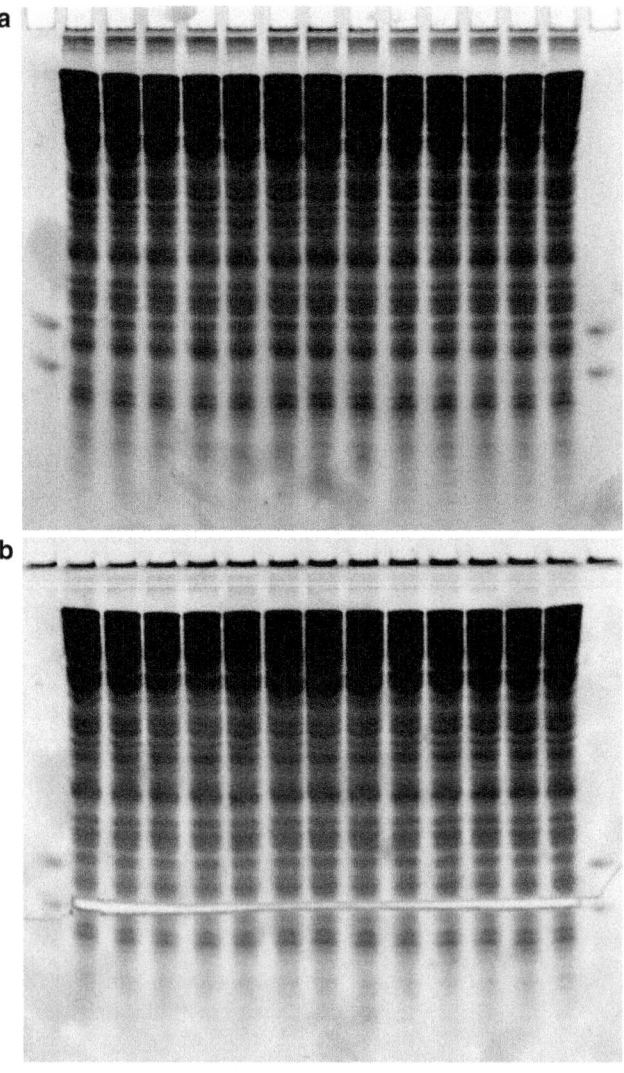

Fig. 3 Size selection of RPF. 15 µg of total monosome RNA is loaded on a 17 % acrylamide/7 M urea gel with a mix of 50 ng 28 and 34 nt marker RNA oligonucleotides on both sides. After electrophoresis (**a**), a band corresponding to the 28 nt RPF is excised (**b**) and the RNA is extracted from the gel. Note the absence of clear, definite bands in this region

13. Disrupt the gel slices by centrifugation through a needle hole in a 0.5 ml microfuge tube nested in an outer 1.5 ml collection microtube. The acrylamide fragment is introduced into the 0.5 ml tube that is capped and introduced in turn in an open 1.5 ml tube. Both are centrifuged at maximum speed the time necessary for the complete passage of the gel through the needle hole. The empty 0.5 ml tube is thrown away.

14. RNA is eluted by soaking gel debris overnight in an Elution Buffer and then recovered by filtering the eluate on a 0.22 μm cellulose acetate filter for 1 min at full speed.

15. RNA is precipitated in ethanol supplemented with 0.3 M sodium acetate and glycogen (20 μg) overnight at −20 °C. Centrifuge at maximum speed, 4 °C for 30 min. Eliminate supernatant as completely as possible and air-dry for 15 min.

16. Resuspend in 25 μl water supplemented with 0.1 U/ml RNAse inhibitor. Measure RNA concentration at 260 nm (it should be between 70 and 90 ng/μl). RNA sample is stored at −20 °C.

3.5 rRNA Depletion

Ribosome samples are subjected to subtractive hybridization with biotinylated oligonucleotides complementary to major rRNA contaminants (rRNA 1–7). These oligonucleotides are representative of the 14 main rRNA fragments recovered within the 28 nt gel slice.

1. Set a water bath or heat block to 70–75 °C.

2. To a sterile, RNase-free 1.5 ml microcentrifuge tube, add the following: 2∓g RNA, rRNA-1 to rRNA-7 (15 pmol/μl) 1 μl each, 10 μl Hybridization Buffer 10×, and water qsp 100 μl. Incubate the tube at 70–75 °C for 15 min to denature RNA.

3. Allow the sample to cool to 37 °C slowly over a period of 30 min by placing the tube in a 37 °C water bath. To promote sequence-specific hybridization, it is important to allow slow cooling. Do not cool samples quickly by placing tubes in cold water.

4. While the sample is cooling down, proceed to Beads preparation. Resuspend Magnetic Beads in its tube by thorough vortexing. Place the tube with the bead suspension on a magnetic separator for 1 min. The beads settle to the tube side that faces the magnet.

5. Gently aspirate and discard the supernatant. Add 750 μl sterile, RNAse-free water to the beads and resuspend beads by slow vortexing. Place tube on a magnetic separator for 1 min.

6. Aspirate and discard the supernatant. Repeat washing step once.

7. Resuspend beads in 750 μl Hybridization Buffer 1× and transfer 250 μl beads to a new tube and maintain the tube at 37 °C for use at a later step. Place the tube with 500 μl beads on a magnetic separator for 1 min.

8. Aspirate and discard the supernatant. Resuspend beads in 200 μl Hybridization Buffer and keep the beads at 37 °C until use.

9. Proceed to rRNA removal. After the incubation at 37 °C for 30 min of the hybridized sample (above), briefly centrifuge the tube to collect the sample at the bottom of the tube.

10. Transfer the sample to the prepared 200 μl magnetic beads. Mix well by pipetting up and down or low speed vortexing.

11. Incubate the tube at 37 °C for 15 min. During incubation, gently mix the contents occasionally. Briefly centrifuge the tube to collect the sample at the bottom of the tube.

12. Place the tube on a magnetic separator for 1 min to pellet the rRNA–probe complex. Do not discard the supernatant. The supernatant contains RNA.

13. Place the tube with 250 μl beads on a magnetic separator for 1 min. Aspirate and discard the supernatant.

14. To this tube of beads, add ~320 μl supernatant containing RNA from the other tube. Mix well by pipetting up and down or low speed vortexing. Incubate the tube at 37 °C for 15 min. During incubation, gently mix the contents occasionally.

15. Briefly centrifuge the tube to collect the sample to the bottom of the tube. Place the tube on a magnetic separator for 1 min to pellet the rRNA–probe complex. Do not discard the supernatant as the supernatant contains RNA. Transfer the supernatant (~320 μl) containing RNA to a new tube.

16. Add 1 μl glycogen, 30 μl of 3 M sodium acetate and 750 μl of 100 % ethanol. Mix well and incubate at –80 °C for a minimum of 30 min.

17. Centrifuge the tube for 15 min 12,000×g at 4 °C. Carefully discard the supernatant without disturbing the pellet. Air-dry the pellet for approximately 5 min.

18. Resuspend the RNA pellet in 20 μl water + 0.1 U/ml RNAse inhibitor. Measure RNA concentration that should be around 40 ng/μl. The efficiency of rRNA depletion is variable following the experiment but is comprised between 50 and 75 % (*see* **Note** 7).

3.6 Library Construction and HT Sequencing

1. Library from 100 ng ribosome footprint fragments is prepared. Briefly, a 3′ adapter designed to target small RNA generated from enzymatic cleavage is added to the RNA fragments. It is required for reverse transcription and corresponds to the surface bound amplification primer on the flow cell.

2. The 5′ RNA adapter ligation that serves for the amplification of the small RNA. Reverse transcription followed by PCR amplification is used to create cDNA constructs. PCR products

are then purified on acrylamide gel and the size is visualized using chip-based capillary electrophoresis machine (*see* **Note 8** and Fig. 4a, b).

3. Library is submitted to high throughput sequencing using Hiseq2000. A minimum of 10^8 reads must be achieved to perform computer analysis. Multiplex sequencing is possible unless a total of at least nine 10^7 reads per library is reached.

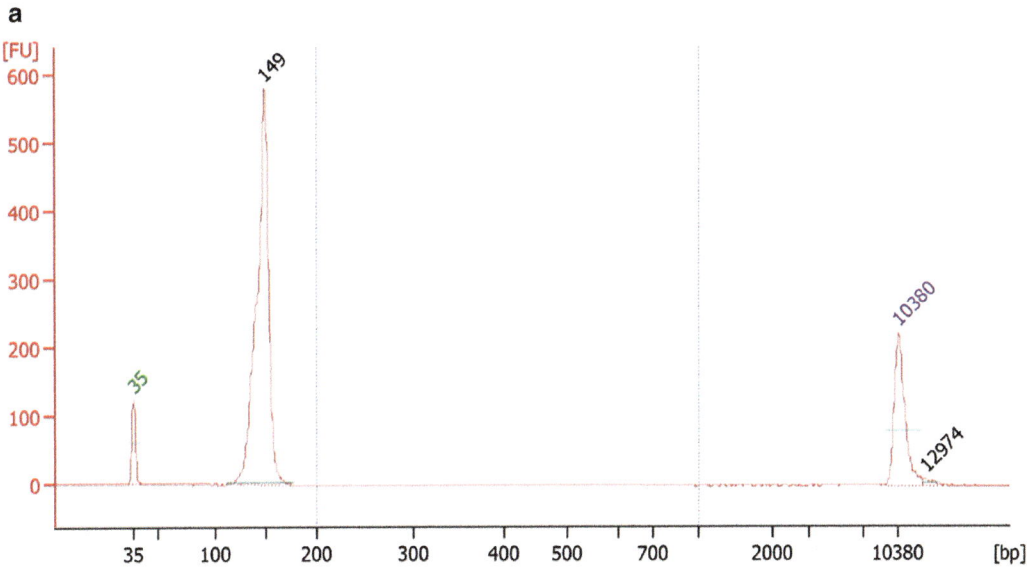

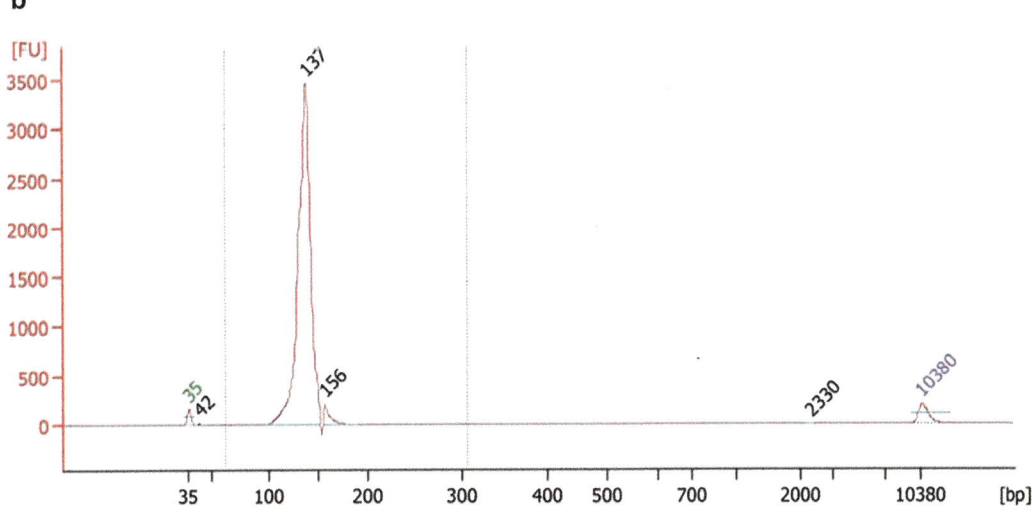

Fig. 4 NanoDrop analysis of the library. (**a**) Library prepared from 28 nt RPF excised from the acrylamide gel should be centered on 150 nt according to the Illumina v1.5 sRNA adaptors. It ensures that the reads shorter than 26 nt that are eliminated during the pre-processing raw data step (4.2) are less than 35 %. (**b**) In the case of a library mainly made of 137 nt molecules, the reads shorter than 26 nt that are eliminated during this pre-processing raw data step (4.2) represent up to 85 %, leading to a dramatic fall in the number of useful footprint

3.7 Primary Bioinformatic Analysis

Primary analysis of ribosome profiling data consists in removing adapter and rRNA contamination, and aligning footprint to yeast genome.

3.7.1 Get Sequences and Bowtie Indexes

1. Download rRNA transcripts (RDN25-1, RDN18-1, RDN58-1, and RDN5-1) from Saccharomyces Genome Database (http://yeastgenome.org) and put sequences in a single Fasta file. You can find rRNA sequences in Fasta format with other RNA genes in this URL:

   http://downloads.yeastgenome.org/sequence/S288C_ reference/rna/rna_genomic.fasta.gz

2. Build a Bowtie index for rRNA sequences:

   ```
   bowtie-build rRNA.fa rRNA
   ```

3. Download yeast reference genome in Fasta format from UCSC (http://hgdownload.cse.ucsc.edu/downloads.html#yeast). Current version is Saccer3 (*see* **Note 9**). You can download from website and regroup all chromosomes in a single Fasta or run these commands:

   ```
   wget
   http://hgdownload.cse.ucsc.edu/goldenPath/sacCer3/
       bigZips/chromFa.tar.gz
   tar xvzf chromFa.tar.gz
   cat *.fa > Saccer3.fa
   ```

4. Build a Bowtie index for yeast genome:

   ```
   bowtie-build Saccer3.fa Saccer3
   ```

3.7.2 Pre-processing Raw Data

1. After sequencing, ribosome footprints are stored in FASTQ format where each footprint is represented by a biological sequence and its corresponding quality score.

   For each sample (or fastq), you can check quality of sequencing using FASTQC software:

   ```
   fastqc -o XXX_fastqc_report XXX.fastq
   ```

2. FASTQC provides statistics about encoding, sequence quality score, GC content, overrepresented sequences, etc. The per base sequence quality is the most important things to look at Fig. 5. Generally, you have high sequence quality from the beginning of the sequence down to ¾ or even further, but as sequenced reads are longer than footprint, Illumina sequencing adapters will be removed and low quality bases too.

3. Adapters are removed by Cutadapt, a tool that discards adapter sequences from DNA sequencing reads. Because of gel slice selection, only footprints between 26 and 32 nucleotides long are kept.

   ```
   cutadapt -a "TGGAATTCTCGGGTGCCAAGGAACTCCAGTCAC" -m 26
       -M 32 XXX.fastq > XXX_trim.fastq
   ```

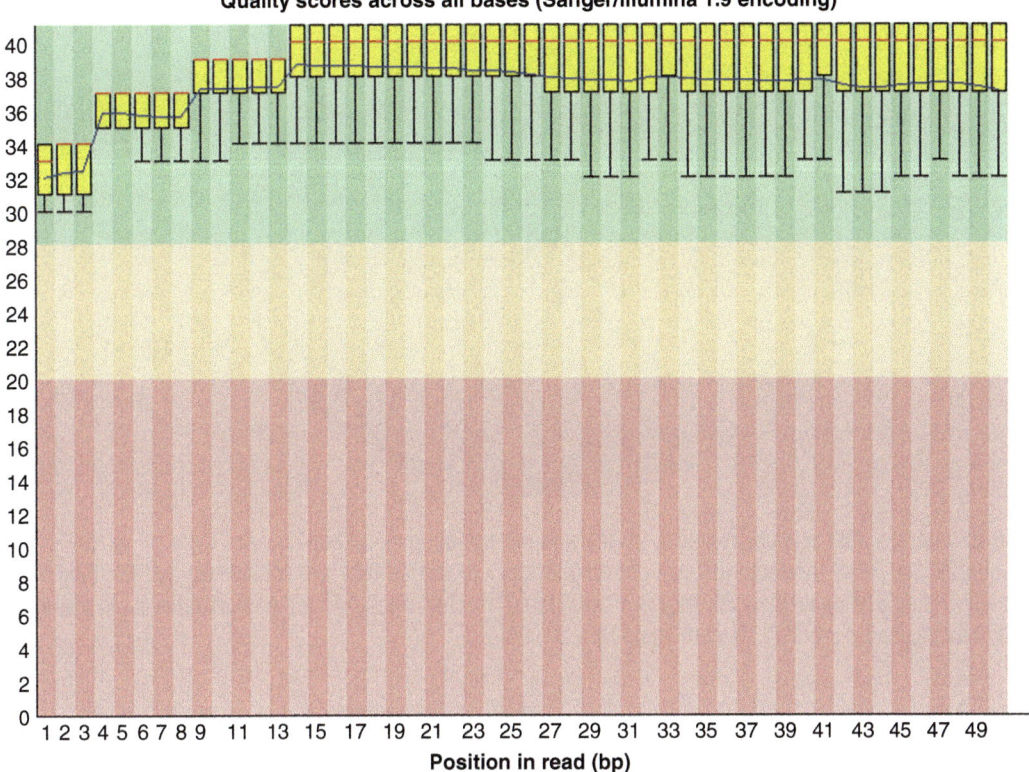

Fig. 5 Per base sequence quality. Dispersion of the quality (*yellow box*) based on position from the beginning of the sequence to the end. *Red line* is the median. *Blue line* is the mean quality score. Quality score must to stay high. The background of the graph divides the *y*-axis into three parts: very good quality calls in *green*, reasonable quality in *orange*, and poor quality in *red*. Generally, in ribosome profiling, sequence quality score stays high (in *green area*) throughout the sequence

This step is the most time-consuming because we trim a lot a reads (at least 10^8 per sample). In the end, we only keep 62 % of initial reads (average value obtained from 14 Hiseq2000 runs) (Fig. 6).

3.7.3 Alignments

1. All footprints mapping to rRNA are removed from data with Bowtie short alignment program.

    -p option refers to number of thread, if you are more than 2, adjust this parameter.

    ```
    bowtie rRNA -p 2 --un XXX_no_rRNA.fastq > /dev/null
    ```

2. No RNA footprints are mapped on yeast reference genome previously indexed, with Bowtie. We do not use a spliced read mapper because of small number of genes with introns in yeast. For enforce uniqueness of mapping, -m 1 is used.

    ```
    bowtie -m 1 -p 2 Saccer3 XXX_no_rRNA.fastq -S XXX_
        align.sam
    ```

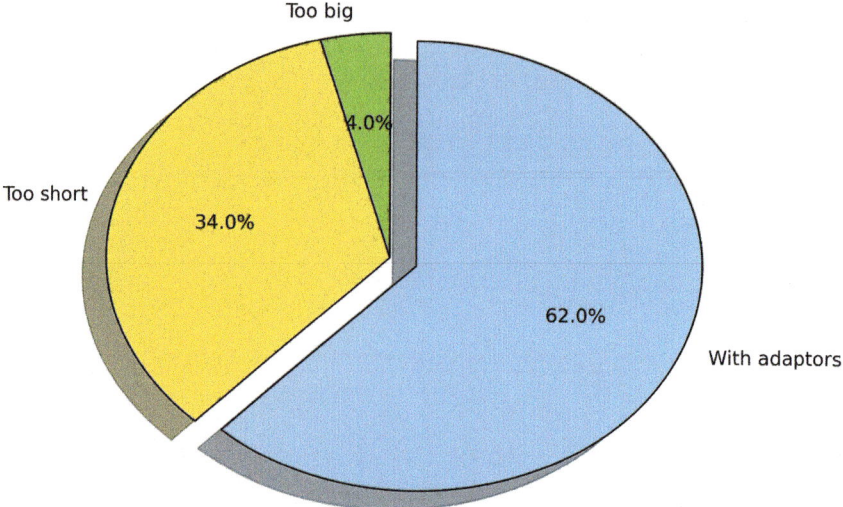

Fig. 6 Distribution after adapter removal. During adapter trimming, you keep only the footprint whose size is between 26 and 32. If the footprints are not adapters or are longer, they are considered as "Too long," and inversely if they are lower than 26, they are considered as "Too small." This pie chart represents the mean value of 14 ribosome profiling sequenced on HISEQ2000

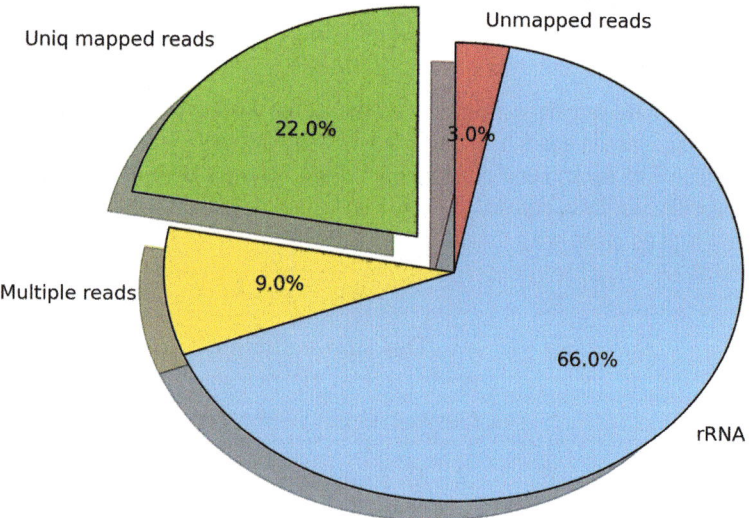

Fig. 7 Distribution of aligned reads. Aligned reads are divided into four categories: those aligned on rRNA genes, those not aligned on yeast genome, those unmapped, and those uniquely aligned. This pie chart represents the mean value of 14 ribosome profiling sequenced on HISEQ2000

3. In spite of rRNA depletion, we have a strong contamination (mean of 66 % on 14 runs). Finally, we obtain about 22 % of useful footprints that align accurately in yeast genome (Fig. 7). So the more the reads in your sample, the more the footprints you have for analysis. We estimate that 10^8 reads is enough.

4. For further analysis, we recommend to convert SAM (Sequence Alignment/Map) file, which is a generic format for storing large nucleotide sequence alignments, to BAM file.

```
samtools view -bhS XXX_align.sam | samtools sort -
    XXX_align_sort samtools index XXX_align_sort.bam
```

5. This Bam file could be visualized in a genome browser like IGV (for Integrative Genome Viewer). Start IGV browser, switch to Saccer3 genome and use File → Load from file for upload your Bam file (*see* **Notes 10** and **11**).

3.8 Differential Gene Translation Analysis

For counting how many footprints map to each gene, HTSeq-count is used. It requires a file with aligned sequencing reads (your SAM file) and a list of genomic features (a GTF file).

3.8.1 Counting Footprints in Features

You can find yeast annotation in UCSC Table browser (http://genome.ucsc.edu/cgi-bin/hgTables?org=s.+cerevisiae&db=sacCer3). Select 'Gene and gene predictions' group and 'GTF' output format and click on 'get output'. Save result as saccer3.gtf and run HTSeq-count.

```
htseq-count -m intersection-nonempty -t exon -i gene_id
    -s yes XXX_align.sam saccer3.gtf > XXX_htseq.txt
```

3.8.2 Differential Analysis of Count Data

There are two types of gene translation analysis. The first way is to determine if a gene is translated more than another in a particular condition. In this case, you must take care of large genes, in addition to library size, because a high number of footprints in a gene does not systematically indicate a high expression but could be due to a large gene. Also, you must normalize by gene length and library size and formulate expression in RPKM (Read per Kilobase per Million).

But in the context of differential expression, RPKM method is ineffective (shown by The French StatOmique Consortium [12]). Statistical methods for RNA-seq can be used, such as Poisson or negative binomial generalized linear models.

Here, we choose DESeq2 which is a version that considers complex designs, low counts, and outliers management (*see* **Note 12**).

First, create a directory with all individual results of HTSeq-count. You must to rename all count files with the name of condition. Replace Cond1 and Cond2 with your owner condition names in following code and corresponding working directory, open an R shell and run:

```
directory <- "/home/login/Ribo_count"
sampleFiles   <-   c(grep("Cond1",list.files(directory),
    value=TRUE),
grep("Cond2",list.files(directory),value=TRUE))
sampleCondition <- sub("(.*Cond1).*","\\1",sampleFiles)
sampleCondition <- sub("(.*Cond2).*","\\1",sampleFiles)
sampleTable <- data.frame(sampleName = sampleFiles,
```

```
                                        fileName = sampleFiles,
                                    condition = sampleCondition)
    dds    <-    DESeqDataSetFromHTSeqCount(sampleTable    =
        sampleTable,
                                        directory = directory,
                                        design= ~ condition)
    colData(dds)$condition <-
    factor(colData(dds)$condition,levels=c("Cond1","C
        ond2"))
    dds <- DESeq(dds)
    res <- results(dds)
    write.csv(as.data.frame(res), file="DE_results.csv")
```

The table contains information about each analyzed genes: its average expression, log2-fold change, and associated p values and adjusted p values. For a specific gene, a log2-fold change of –1 for condition 2 *vs.* condition 1, indicates a fold change level of $2^{-1} = 0.5$.

DESeq2 gives several plots to help the analysis or for illustrations in a manuscript. This is the first approach to be done for ribosome profiling analysis, and the most standard analysis. After, according to your biological question, other approaches could be considered about the qualitative analysis of the ribosome position on mRNA.

4 Notes

1. To generate 10–50 % sucrose gradient, prepare 31 % sucrose solution in 50 mM Tris-acetate pH 7.6, 50 mM NH_4Cl, 12 mM $MgCl_2$, 1 mM DTT, simply proceed to three freezing (–20 °C)–thawing cycles. The last thawing is performed immediately before using it.

2. Do not overgrow the cells (above OD_{600} of 0.6) as it modifies culture conditions and thus the pattern of expression of numerous genes.

3. Alternatively, the mixture can be subjected to ten cycles of vortexing for 15 s followed by cooling on ice for 15 s.

4. This step is crucial to generate ribosome footprints. Use a high quality, efficient RNAse I.

5. Monosome fractions can be isolated on a 10–50 % W:V sucrose gradient (50 mM Tris-acetate pH 7.6, 50 mM NH_4Cl, 12 mM $MgCl_2$, 1 mM DTT) spun for 3.5 h at 39,000 rpm, 4 °C, in an SW41 swing-out rotor. Load 30 OD of RNAse I digested polysomes per 12 ml gradient. Fractionate each gradient with the ISCO gradient fractionation system (flow rate 45 ml/h) and collect the monosome fractions which have a

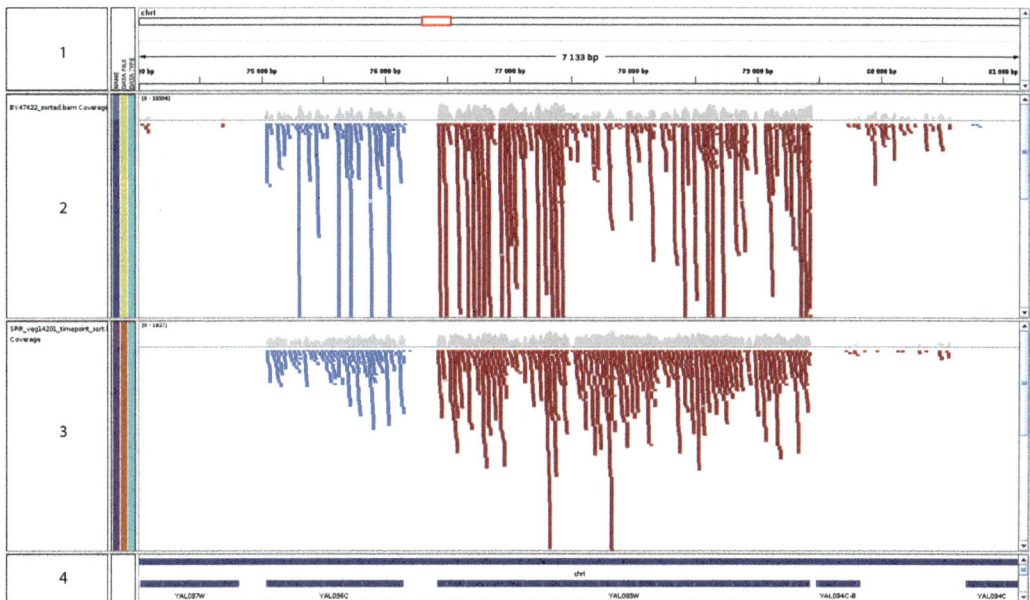

Fig. 8 IGV screenshot. Coverage representation on genome browser IGV. Screen is divided into four parts: Panel 1, localization on yeast genome. Panel 2 shows an alignment of data from our laboratory. Panel 3 shows the same alignment from Weissman's laboratory [3]. The panel 4 represents the yeast annotations. Forward footprints are in *red* and reversed footprints in *blue*

volume of about 500 μl. Pool 2 fractions and proceed to RNA extraction as described above. Be aware that the RNA containing phase corresponds to the lower phase due to high sucrose concentration.

6. A comb that delimitates a large central well dedicated to the RNA and two small wells dedicated to the ladder can be used. In that case, up to 300 μg RNA can be loaded.

7. Epicentre company sells a Ribo-Zero magnetic kit for yeast. This commercial kit is expected to remove more than 99 % of degraded rRNA. However we did not yet test the efficiency of this kit.

8. It is important to verify the size of the PCR fragments (3′ and 5′ adapters plus ribosomal footprint fragment). If it is lower than expected, this is representative of an important rRNA contamination (Fig. 4b).

9. Version 2 of yeast genome contains wrong annotations in some chromosomes, prefer use the version 3.

10. Be careful, chromosome names must be identical in genome and annotation files for visualization with IGV.

11. Our footprints coverage is not homogeneous along mRNA compared to Weissman's laboratory ribosome profiling [3], despite a similar number of footprints. This can be due to circularization during sequencing library preparation (Fig. 8).

12. We have experiments corresponding to negative binomial model but sometimes it is recommended to use another statistical model (Poisson, Bayesian, etc.). In this chapter, we demonstrate a DESeq2 analysis with default parameters but it dependent to experiment too. Do not hesitate to consult a statistician to advise you on this part.

References

1. Ingolia NT, Brar GA, Rouskin S, McGeachy AM, Weissman JS (2012) The ribosome profiling strategy for monitoring translation in vivo by deep sequencing of ribosome-protected mRNA fragments. Nat Protoc 7(8): 1534–1550

2. Ingolia NT, Ghaemmaghami S, Newman JR, Weissman JS (2009) Genome-wide analysis in vivo of translation with nucleotide resolution using ribosome profiling. Science 324(5924): 218–223

3. Brar GA, Yassour M, Friedman N, Regev A, Ingolia NT, Weissman JS (2012) High-resolution view of the yeast meiotic program revealed by ribosome profiling. Science 335(6068):552–557

4. Dunn JG, Foo CK, Belletier NG, Gavis ER, Weissman JS (2013) Ribosome profiling reveals pervasive and regulated stop codon readthrough in Drosophila melanogaster. eLife 2, e01179

5. Guydosh NR, Green R (2014) Dom34 rescues ribosomes in 3′ untranslated regions. Cell 156(5):950–962

6. Michel AM, Choudhury KR, Firth AE, Ingolia NT, Atkins JF, Baranov PV (2012) Observation of dually decoded regions of the human genome using ribosome profiling data. Genome Res 22(11):2219–2229

7. Michel AM, Fox G, M Kiran A, De Bo C, O'Connor PB, Heaphy SM, Mullan JP, Donohue CA, Higgins DG, Baranov PV (2014) GWIPS-viz: development of a ribo-seq genome browser. Nucleic Acids Res 42(Database issue):D859–D864

8. Smith JE, Alvarez-Dominguez JR, Kline N, Huynh NJ, Geisler S, Hu W, Coller J, Baker KE (2014) Translation of small open reading frames within unannotated RNA transcripts in Saccharomyces cerevisiae. Cell Rep 7(6): 1858–1866

9. Bazzini AA, Lee MT, Giraldez AJ (2012) Ribosome profiling shows that miR-430 reduces translation before causing mRNA decay in zebrafish. Science 336(6078):233–237

10. Ingolia NT (2014) Ribosome profiling: new views of translation, from single codons to genome scale. Nat Rev Genet 15(3):205–213

11. Ingolia NT, Lareau LF, Weissman JS (2011) Ribosome profiling of mouse embryonic stem cells reveals the complexity and dynamics of mammalian proteomes. Cell 147(4):789–802

12. Dillies MA, Rau A, Aubert J, Hennequet-Antier C, Jeanmougin M, Servant N, Keime C, Marot G, Castel D, Estelle J, Guernec G, Jagla B, Jouneau L, Laloe D, Le Gall C, Schaeffer B, Le Crom S, Guedj M, Jaffrezic F (2013) A comprehensive evaluation of normalization methods for Illumina high-throughput RNA sequencing data analysis. Brief Bioinform 14(6):671–683

Chapter 8

Biotin-Genomic Run-On (Bio-GRO): A High-Resolution Method for the Analysis of Nascent Transcription in Yeast

Antonio Jordán-Pla, Ana Miguel, Eva Serna, Vicent Pelechano, and José E. Pérez-Ortín

Abstract

Transcription is a highly complex biological process, with extensive layers of regulation, some of which remain to be fully unveiled and understood. To be able to discern the particular contributions of the several transcription steps it is crucial to understand RNA polymerase dynamics and regulation throughout the transcription cycle. Here we describe a new nonradioactive run-on based method that maps elongating RNA polymerases along the genome. In contrast with alternative methodologies for the measurement of nascent transcription, the BioGRO method is designed to minimize technical noise that arises from two of the most common sources that affect this type of strategies: contamination with mature RNA and amplification-based technical biasing. The method is strand-specific, compatible with commercial microarrays, and has been successfully applied to both yeasts *Saccharomyces cerevisiae* and *Candida albicans*. BioGRO profiling provides powerful insights not only into the biogenesis and regulation of canonical gene transcription but also into the noncoding and antisense transcriptomes.

Key words Nascent transcription, RNA polymerase II, RNA polymerase II I, Yeast, *Saccharomyces cerevisiae*, *Candidaalbicans*, Nascent RNA

1 Introduction

Transcription is the first step in the gene expression process. It is also believed to be the most regulated step in eukaryotes. Because of that, the study of eukaryotic transcription is one of the main topics of molecular biology. Many in vivo and in vitro procedures have been developed to study the transcription cycle of eukaryotic RNA polymerases (reviewed in 1, 2) Recently, with the advent of genomic methodologies it has been possible to study the particular features of every single gene. At the same time, doing average profiling for all genes allows to determine the real properties of a typical gene instead of extrapolating those of a particular experimental example to the whole genome [3].

Frédéric Devaux (ed.), *Yeast Functional Genomics: Methods and Protocols*, Methods in Molecular Biology, vol. 1361, DOI 10.1007/978-1-4939-3079-1_8, © Springer Science+Business Media New York 2016

To this end, some high-resolution techniques for the study of nascent transcription have been established [4–6]. Each technique has particular features that reveal different aspects of the transcription process (reviewed in 3, 7). Chromatin immunoprecipitation (ChIP) detects all RNA pol, active or not. However, it can differentiate between different RNA pol species, or carboxy-terminal (CTD) phosphorylated forms of RNA pol II, or even elongating complexes with different composition by using specific antibodies [7]. Techniques that detect nascent RNA (nRNA) only detect elongating RNA pol, allowing their mapping at high resolution [4–6]. They are, however, unable to distinguish between active RNA pol II molecules and those that are backtracked but still retaining a bound RNA molecule. Genomic run-on approaches (GRO, 8, 9), however, only detect active elongating RNA pol I, II, and III molecules.

Variants of GRO have been published by other laboratories working in yeast or higher eukaryotes [10–12]. All those methods use next generation sequencing for the analysis of purified nRNA. Purification of the very rare nRNA requires its labeling with a precursor, such as BrUTP or Biotin-UTP. Because of the small proportion of nRNA in the cell, contamination with mature RNA is an important concern. The presence of such contaminant may obscure the conclusions drawn from those methods. However, by hybridizing in vivo-biotinylated RNAs directly onto the arrays, the risk of mature RNA contamination and of any technical noise derived from amplification may be bypassed. This improvement could in turn help to draw more powerful biological insights when analyzing the results.

We have taken profit of the fact that Affymetrix arrays are based on detecting biotin labeled nucleic acids to hybridize our in vivo-biotinylated RNAs directly onto them. We call this protocol Biotin-GRO or BioGRO (Fig. 1). In this way, small amounts of contaminant mature RNA (rRNA, mRNA or any other) become unimportant because they do not fluoresce upon laser scanning. This is the same situation observed in the classic radioactive run-on protocol [9], in which a large amount of nonradioactive mature RNA neither blocks nRNA hybridization nor interferes with its detection [13]. We observed, however, that in the conditions of Affymetrix hybridizations, the presence of the much more abundant non-labeled RNA severely reduces fluorescent signal. This is probably due to the high sample concentration, >100 times higher than for macroarrays. Therefore, with the aim of reducing the amount of contaminant RNA, we treated sarkosyl-permeabilized cells with RNase A. This treatment has been previously shown to destroy most of the preexisting RNA in mammalian cells without affecting run-on efficiency because of the protection offered by elongating RNA pol to their nRNA, known as RNA pol footprinting [14]. RNase A treatment, thus, allowed to eliminate most of

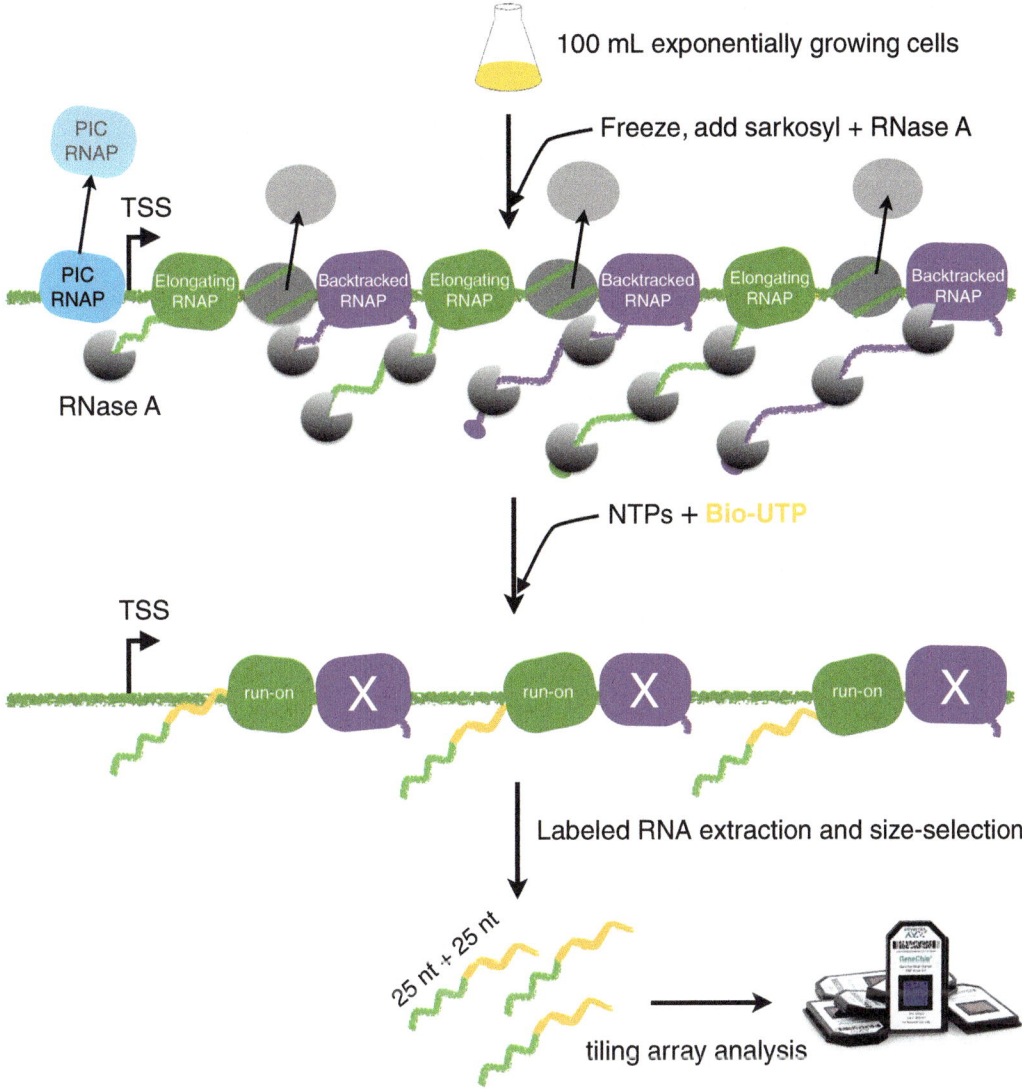

Fig. 1 Outline of the BioGRO method. The colors of the RNA polymerases (RNAP) represent different transcriptional states. Only active RNAPs (*green*) are elongation-competent during run-on. *Green* portions of nascent RNA molecules represent the footprints after RNase A digestion whereas *yellow* portions represent the run-on elongations

the mature RNA present in the cell and to trim the 5′ tail of nRNA giving a footprint of about 25 nt [14]. These molecules are then extended by around 25–30 nt during run-on, allowing the incorporation of some biotinylated uridine residues (*see* Fig. 1).

Here we describe a straightforward, strand-specific, high-resolution GRO technique for the model organisms *S. cerevisiae* and *C. albicans*, based on the use of a modified RNA precursor (biotin-UTP), and tiling microarrays. Overall, this method allows for the analysis of nRNA without any interference of mature RNA molecules for a large set of genes (*see* Figs. 2 and 3).

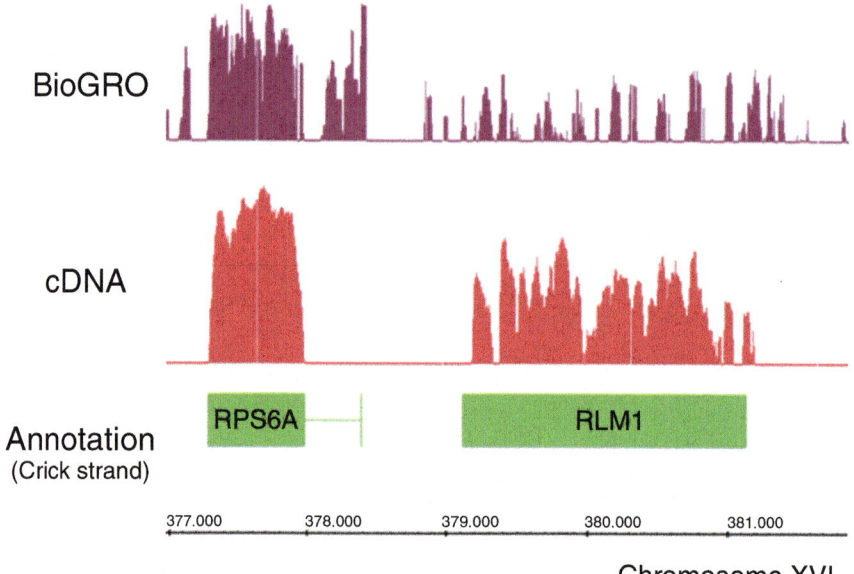

Fig. 2 Example of BioGRO signal along individual genome regions. The intronic region of gene RPS6A shows a BioGRO signal that is not seen in the same region in a mature mRNA hybridization

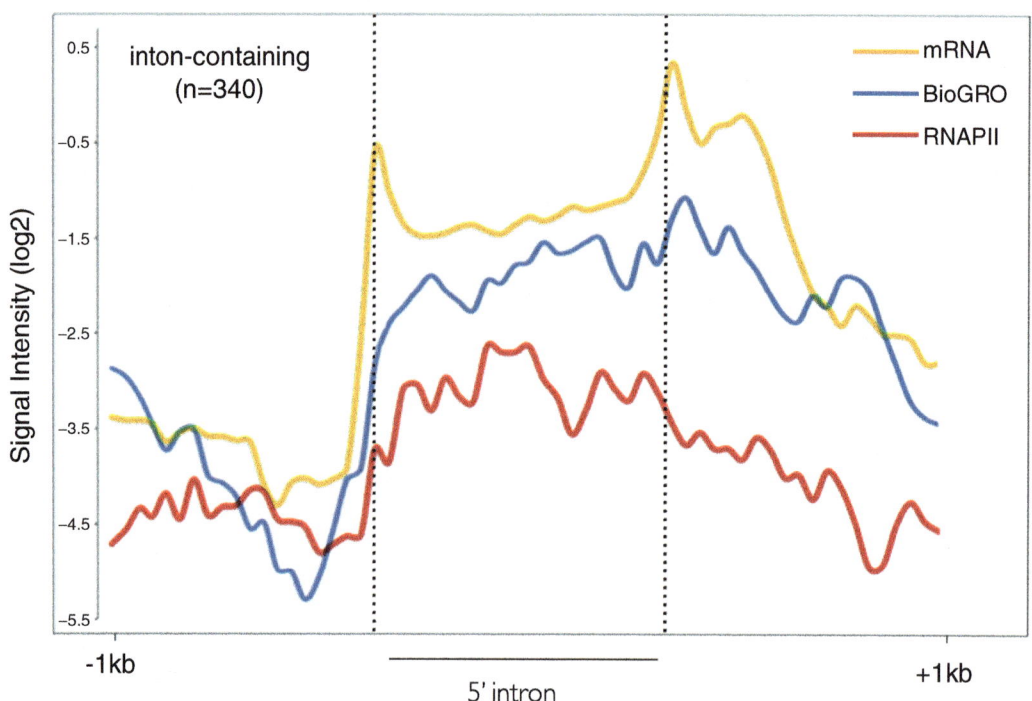

Fig. 3 BioGRO and RNA pol II ChIP measure nascent transcription, genome-wide. Average metagene profile of the 5′ intron region (in arbitrary units, normalized to 1 kb; flanking regions in real units) of 340 intron-containing genes. In contrast with the signal drop shown by the mRNA profile (*orange trace*), the generally flat density profile of both BioGRO (*blue trace*) and RNA pol II ChIP (*red trace*) data (provided by Sebastián Chávez, personal communication), argue in favor of their ability to capture elongating RNA pol II (and, thus, nRNA) and not fully processed, mature mRNAs. *Vertical dotted lines* mark exons–intron junctions

2 Materials

Precautions should be taken to minimize RNase contamination throughout all the protocol steps (*see* **Notes 1** and **2**).

2.1 Equipment

1. Low-speed table top centrifuge.
2. Refrigerated microcentrifuge.
3. Temperature-controlled orbital shaker.
4. DNA LoBind 1.5 mL Tubes (Eppendorf).
5. Eppendorf Thermomixer® Comfort Heating and Cooling Shaker.
6. NanoDrop ND1000 Spectrophotometer.
7. Affymetrix Hybridization and Wash Station and GeneChip® Scanner.
8. Thermoblock heater.
9. Savant SPD111V SpeedVac Concentrator (Thermo Scientific).
10. (Optional) GS Gene Linker UV Chamber (Bio-Rad).

2.2 Nascent RNA Biotinylation by Run-On

1. YPD medium: 1 % w/v, yeast extract, 2 % w/v, peptone, 2 % glucose. Store at room temperature (*see* **Note 3**).
2. 0.5 % w/v, l-laurylsarcosine (sarkosyl) in nuclease-free H_2O. Store at room temperature.
3. 2.5× transcription buffer: 50 mM Tris–HCl, pH 7.7, 50 mM KCl, 80 mM $MgCl_2$. Store at room temperature.
4. ACG mix (ATP, CTP, GTP, 10 mM each). Store frozen.
5. 0.1 M DTT. Store frozen.
6. Biotin-11-UTP (10 mM, Ambion). Store frozen.
7. Transcription mix: 120 μL of 2.5× Transcription buffer, 16 μL ACG mix, 6 μL 0.1 M DTT, and 20.25 μL of Bio-11-UTP. Prepare fresh.
8. RNaseOUT. Store frozen.
9. RNase A. Store at 4 °C.
10. 5 M Sodium acetate (pH 5.2). Store at room temperature.
11. 11.1 M Tris–HCl (pH 7.4). Store at room temperature.
12. Isopropanol. Store at room temperature.
13. Glycogen, for molecular biology. Store frozen.
14. DNase I, RNase-free. Store frozen.
15. Liquid nitrogen.
16. Proteinase K, recombinant, PCR grade. Store at 4 °C.
17. Ethanol, absolute.

18. Nuclease-free water, molecular biology grade.

19. MasterPure™ Yeast RNA Purification Kit (Epicentre).

2.3 Nascent RNA Size Selection

1. NucleoSpin® miRNA kit for small and large RNA species (Macherey-Nagel).

2. Ethanol, absolute.

3. Nuclease-free water, molecular biology grade.

2.4 Affymetrix Tiling Arrays

1. GeneChip® WT Terminal Labeling Kit (Affymetrix).

2. GeneChip® Hybridization, Wash, and Stain Kit (Affymetrix).

3. GeneChip® *S. cerevisiae* Tiling 1.0R Array (Affymetrix).

4. GeneChip® *S. cerevisiae* Tiling Array Custom (Affymetrix).

5. GeneChip® *Candida* Custom Array (Affymetrix).

3 Methods

The method described here has been successfully applied to both *S. cerevisiae* [15] and *C. albicans* (unpublished results). However, this protocol focuses mainly in describing the BioGRO protocol for *S. cerevisiae*. In the case of *C. albicans*, and although the procedures are very similar, there are some variations which will be described in its own subheading. The major steps of this method are outlined in Fig. 1.

3.1 Preparation of DNase-Free RNase A

1. Dissolve RNase A at a concentration of 10 mg/mL in 0.01 M sodium acetate (pH 5.2).

2. Heat to 100 °C in a thermoblock for 15 min.

3. Allow it to cool down slowly to room temperature.

4. Adjust the pH by adding 0.1 volume of 1 M Tris–HCl (pH 7.4).

5. Dispense in aliquots and store at −20 °C.

3.2 BioGRO Method

1. Allow cells to grow in YPD (*see* **Note 3**) to the desired OD_{600} (typically 0.5–0.6).

2. For each sample, an aliquot of 100 mL is needed (corresponding to 12×10^8 cells, *see* **Note 4**).

3. Collect cells by centrifugation in two 50 mL falcon tubes at $4400 \times g$ for 2.5 min. From now onwards, both tubes are processed the same, in parallel, and the extracted RNA is pooled together at the end.

4. Decant the supernatant and submerge the pellet-containing tube in liquid nitrogen for flash freezing (*see* **Note 5**).

5. Transfer the frozen pellet to −20 °C. Keep the tube in the freezer for at least 3 h. This is a safe stopping point, as cells can be stored for longer periods (*see* **Note 6**).

6. Slowly thaw cells on ice and add 10 mL of a 0.5 % sarkosyl solution. Mix by inversion.

7. Pellet cells by centrifugation as in **step 3**, and discard the supernatant.

8. Resuspend cells in 3.2 mL of 0.5 % sarkosyl and add 32 µL of 10 mg/mL DNase-free RNase A. Mix by pipetting up and down several times.

9. Incubate cells with RNase A for 10 min at 30 °C in an orbital shaker to avoid sedimentation of cells at the bottom of the tube (*see* **Note 7**).

10. After 10 min, bring the volume up to 45 mL with sarkosyl 0.5 %. Mix vigorously by inversion to wash the cells and eliminate RNase A.

11. Recover cells by centrifugation as in **step 3**. Discard the supernatant.

12. Resuspend cells in 45 mL of 0.5 % sarkosyl. Shake vigorously by inversion and pellet the cells again.

13. Repeat previous step for a third and final wash.

14. Resuspend the pellet in 1 mL of 0.5 % sarkosyl and transfer cells to an Eppendorf tube (1.5 mL).

15. Recover cells by centrifugation at $5400 \times g$ for 1 min in a microcentrifuge. Carefully remove the supernatant by pipetting and centrifuge again, if necessary, to eliminate any remaining sarkosyl.

16. Resuspend cells in 113.5 µL of nuclease-free water.

17. Add 5 µL of RNase inhibitor (RNaseOUT) to protect the integrity of nascent RNAs from any residual RNase that might be present after the washes. Mix by pipetting up and down several times. Keep cells on ice until needed.

18. Prepare the transcription mix: 120 µL of 2.5× transcription buffer, 6 µL 0.1 M DTT, 16 µL of ACG mix, and 20.25 µL of 10 mM Biotin-11-UTP (*see* **Notes 8** and **9**).

19. Pre-warm both cells and transcription mix at 30 °C for 5 min.

20. Add the transcription mix (162.25 µL) to the cell suspension and mix by pipetting.

21. Perform the run-on reaction by incubating the mix for 5 min at 30 °C in a thermomixer, with 550 rpm agitation (*see* **Note 10**).

22. Stop the reaction by adding 1 mL of ice-cold nuclease-free water to the tube. Snap cool and maintain on ice for 5 min.

23. Harvest cells by centrifugation for 1 min at $11,000 \times g$ at 4 °C. Remove the supernatant (containing unincorporated nucleotides).

3.3 RNA Extraction and DNA Removal

RNA extraction was done with the MasterPure™ Yeast RNA Purification Kit, with some major modifications. Thus, and for clarity purposes, this section describes a continuous protocol that integrates our modifications with the kit manufacturer's instructions (*see* **Notes 11** and **12**).

1. Dilute 2.78 µL (50 µg) of Proteinase K into 300 µL of Extraction Reagent for RNA.

2. Add the mixture to the cell pellet from **step 23** of Subheading 3.2 and resuspend by pipetting up and down several times.

3. Incubate at 70 °C for 15 min in a thermomixer, with constant 600 rpm shaking. Additionally, vortex mix every 5 min to avoid cell deposition at the bottom of the tube.

4. Place the samples on ice for 3–5 min and add 175 µL of MPC Protein Precipitation Reagent. Vortex for 10 s.

5. Pellet the debris by centrifugation for 10 min at 4 °C at $12,000 \times g$.

6. Instead of a normal microcentrifuge tube, transfer the supernatant to a clean DNA LoBind Tube (*see* **Note 13**). Discard the pellet.

7. Add 500 µL of isopropanol and 10 µg of glycogen to the recovered supernatant (*see* **Note 14**). Mix by inversion 5–10 times.

8. Precipitate RNA overnight at −20 °C.

9. Pellet the RNA by centrifugation at 4 °C for 20 min at $12,000 \times g$.

10. Carefully pour off the isopropanol without dislodging the RNA pellet.

11. Add 500 µL of 70 % ethanol and centrifuge for 5 more minutes.

12. Pour off the ethanol and dry the pellet by incubating the open tube for 10 min at 45 °C in a thermomixer.

13. Resuspend in 32 µL of nuclease-free water (*see* **Note 15**).

14. Use 2 µL for spectrophotometric quantitation with a NanoDrop system.

15. Bring the sample volume up to 87.5 µL with nuclease-free water and add 10 µL of 10× DNase I Reaction Buffer, 0.5 µL of RNaseOUT, and 20 U (2 µL) of RNase-free DNase I. Final reaction volume should be 100 µL.

16. Incubate mix for 30 min at 37 °C.

17. Add 200 µL of 2× T and C Lysis Solution. Vortex mix for 5 s.

18. Add 200 µL of MPC Protein Precipitation Reagent. Vortex mix for 10 s and then place on ice for 3–5 min.

19. Repeat **steps 5–13** of this protocol. Be careful not to carry portions of the white pellet when transferring the supernatant to a new tube in **step 6**. To avoid it, centrifuge for five more minutes if needed.

20. Pool together resuspended RNA from tubes 1 and 2. The final volume should be 60 µL and total RNA amount obtained should be around 30 µg.

3.4 Nascent RNA Size-Selection

1. For the isolation of RNA fragments shorter than 200 bases, follow the instructions in section 6.4 of the NucleoSpin® miRNA kit for small and large RNA species manual. *See* **Note 16**.

2. Discard the blue column (containing the large RNA fraction), and elute the small RNA fraction from the green column with 30 µL of nuclease-free water.

3. Use 2 µL to quantitate RNA. The expected yield lies in a range of 2–5 µg of RNA.

4. Bring the sample volume down to 45 µL with a SpeedVac system, or similar. *See* **Notes 17** and **18**.

3.5 Total RNA Extraction for Conventional Transcriptomic Analysis

The sample collection for total RNA extraction can be performed in parallel with the sample collection for the BioGRO.

1. Allow cells to grow to the desired OD_{600} (typically 0.5–0.6).

2. For each sample, an aliquot of 50 mL cells is needed (corresponding to 6×10^8 cells).

3. Collect cells by centrifugation in a falcon tube at $4400 \times g$ for 2.5 min.

4. Discard the supernatant and submerge the pellet-containing tube in liquid nitrogen for flash freezing.

5. Transfer the frozen pellet to –20 °C. Keep the tube in the freezer until needed.

6. Slowly thaw cells on ice and proceed with the same RNA extraction method described in Subheading 3.3.

7. Due to the higher amount of RNA extracted, compared to the BioGRO method, resuspend RNA in 200 µL of nuclease-free water.

3.6 Tiling Array Direct Hybridization of BioGRO Samples

1. Follow the instructions of the *GeneChip®Whole Transcript (WT) Sense Target Labeling Assay Manual*, starting from *Chapter 5*: *Hybridization*. Use the GeneChip® Hybridization, Wash and Stain Kit (*see* **Note 19**).

2. Perform the staining and washing of the array as described in *Chapter 6* of the manual.

3. Repeat the Fluidics Station 450 protocol sequence twice consecutively to increase the signal of the biotinylated nascent RNAs (*see* Fig. 4).

4. At the end of the second Fluidics protocol, scan the array as described in *Chapter 7: Scanning*.

3.7 Tiling Array Hybridization of Total RNA Sample

1. For the preparation of total RNA/T7-(N)$_6$ Primers/Poly-A RNA Controls, follow the instructions of the *GeneChip® Whole Transcript (WT) Sense Target Labeling Assay Manual*, starting from *Chapter 4: 100 ng Total RNA Labeling Protocol*.

2. For the next steps: First-Cycle, First-Strand cDNA Synthesis, First-Cycle, Second-Strand cDNA Synthesis, First-Cycle,

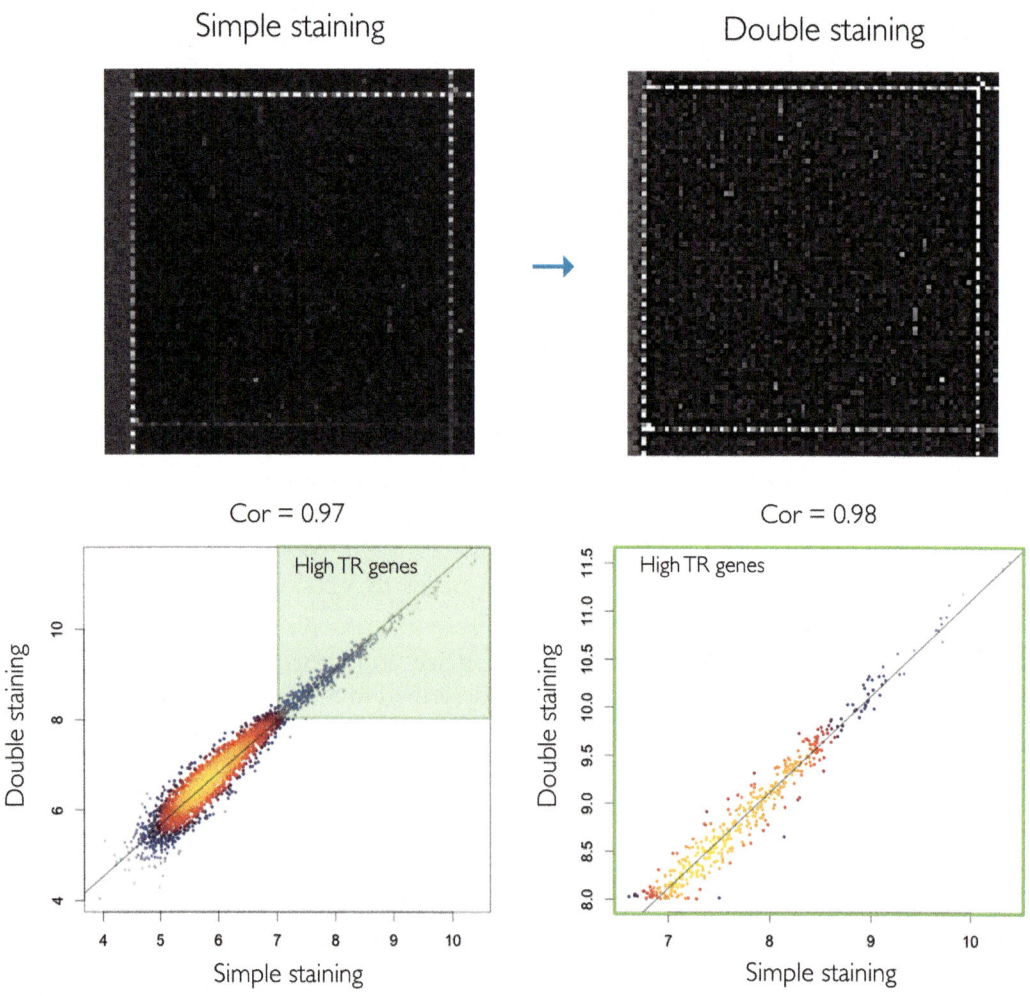

Fig. 4 Uniform re-staining of the arrays. The *top half* images show the detailed view of a tiling array quadrant region after simple and double staining. *Bottom half* graphs show how the re-staining strategy results in a uniform 2× increase in the signal

cRNA Synthesis and Cleanup, Second-Cycle, First-Strand cDNA Synthesis, Hydrolysis of cRNA and Cleanup of Single-Stranded DNA, Fragmentation of Single-Stranded DNA, and Labeling of Fragmented Single-Stranded DNA, follow instructions in Chapter 3, Procedures B–H.

3. Follow the instructions on Chapters 5 and 6 for hybridization, washing, staining and scanning of the arrays.

3.8 Adaptation of the BioGRO Protocol to C. albicans cells (See Note 20)

1. Allow cells to grow in YPD (*see* **Note 3**) at 37 °C to the desired OD_{600} (typically 0.5–0.6). *See* **Note 21**.

2. Collect cells by centrifugation in two 50 mL falcon tubes at $4400 \times g$ for 3 min.

3. Slowly thaw cells on ice and add 10 mL of a 0.05 % sarkosyl solution (*see* **Note 22**). Mix by inversion.

4. Resuspend cells in 3.2 mL of 0.05 % sarkosyl and add 32 µL of 1 mg/mL DNase-free RNase A. Mix by pipetting.

5. Perform the following three washes using 0.05 % sarkosyl instead of 0.5 %.

6. Pre-warm both cells and transcription mix at 37 °C for 5 min.

7. Perform the run-on reaction by incubating the mix for 5 min at 37 °C (*see* **Note 23**) in a thermomixer, with 550 rpm agitation.

8. From this step on, follow the same RNA extraction, size-selection and hybridization procedures described for *S. cerevisiae*.

4 Notes

1. General precautions of working with RNA should be taken. Always use RNase-free water, and prepare all reagents with it. Whenever possible, work with nuclease-free materials, such as filter tips, do not touch anything that is going to be in contact with RNA without gloves, and keep your workbenches always clean. You will also minimize potential RNA degradation if you store nuclease-free materials and reagents in separate compartments inside your laboratory.

2. All homemade buffers and most solutions are autoclaved at 2 kg/cm^2 for 1 h to inactivate DNases and RNases.

3. Although YPD is the most common culture medium, other complete or synthetic media may also be used.

4. The cell number in each GRO experiment should be very similar between samples to avoid differences in labeling during the run-on. We estimate the real number of cells used from the amount of RNA obtained after purification. If the amount of

RNA per cell is known (this can be obtained from a series of independent RNA purifications from the known amount of cells), the number of cells is derived from it.

5. We have observed that the slow freezing of sarkosyl-treated cells causes some RNA degradation. It is recommended to freeze cell pellets immediately in liquid nitrogen or dry ice before storing them at the freezer.

6. Cell pellets can be stored more than 3 h, even for months without any negative impact on the run-on performance. We recommend storing at −80 °C for long-term periods (>1 month).

7. RNase A digestion prior to the run-on reaction is a very variable step in terms of final extracted RNA yield. We calculated that using 0.1 μg/μL RNAse A was the appropriate concentration to obtain the correct final RNA yield. However, even when using 0.1 μg/μL of RNase A, sometimes the yield is either lower or higher than expected. This variability might be due to two main things: incubation temperature, and RNase A activity. To try to optimize the first aspect we always do the incubation at 30 °C instead of at room temperature. For the second aspect, we recommend doing a trial BioGRO experiment (using unmodified UTP instead of Biotin-UTP), just before the real experiment, to test different RNase A concentrations (for example 1, 5, and 10 μg/μL) and see which one yields the expected final RNA amount (around 30 μg prior to size selection; around 2–5 μg after size selection).

8. For multiple reactions, prepare a 1/10th excess of the master mixes (transcription run-on mix and DNA removal mix).

9. Due to its photoreactive potential, it is better to avoid extended direct light exposure to Biotin-11-UTP.

10. We have checked that longer incubation times do not increase labeling. Probably, the run-on reaction is completed in only a few minutes.

11. RNA extraction was done using the "MasterPure Yeast RNA Purification Kit". This kit is designed to extract large amounts of intact RNA from yeast cells. Due to the fact that biotinylated RNAs of interest are short (50 nt on average) and present in low proportion compared to non-labeled mRNAs, some modifications to the general protocol were implemented in order to optimize the biotinylated RNA recovery.

12. Alternative RNA extraction strategies based in organic phase separation, such as acid phenol and TRIzol (Ambion), are also possible, but in our hands we recovered less biotinylated RNA than with the MasterPure kit.

13. In order to minimize loss of RNA material in the final steps of the BioGRO protocol, we used low nucleic acid retention plastic tubes (such as DNA LoBind Tubes, Eppendorf).

14. Glycogen is an inert polysaccharide that we used as carrier for the precipitation of RNA. Adding a few micrograms of glycogen significantly increased the recovery of the RNA in isopropanol precipitations. During centrifugation, it forms a visible pellet, which greatly facilitated handling of the precipitated RNA.

15. Over-drying the pellet results in a difficulty to dissolve RNA. For a complete dissolution keep the RNA pellet with water in a bench-top shaker at 45 °C for about 10–15 min. Lower temperatures and longer times may also be used. Check the dissolution by carefully inspecting while pipetting.

16. For the size selection of biotinylated RNAs, we tried many different approaches, including concentration with Amicon Ultra 0.5 centrifugal filters (Millipore), extraction from normal or low melting point agarose gels with β-Agarase, electro elution cartridges, and others, but the only one that recovered enough RNA material was the miRNA kit. This kit is designed to separately isolate both large (>200 bases) and small (<200 bases) RNA molecules from a mixed population of fragments. The 200 base-cut-off is enough to significantly enrich the run-on sample in biotinylated nascent RNAs (average size <100 bases, *see* Fig. 1).

17. We typically hybridize 5 μg of nascent RNA to the arrays. This applies for the three types of Affymetrix tiling arrays we have used (*see* Subheading 2.4).

18. *C. albicans* is a human commensal organism, with opportunistic pathogenic behavior, so general safety precautions should be taken when handling *C. albicans* cells.

19. *C. albicans*, as a human commensal organism, has an optimal growth temperature around 37 °C.

20. In our hands, *C. albicans* cells are more fragile than *S. cerevisiae* to 0.5 % sarkosyl treatments. When exposed to 0.5 % sarkosyl solutions, *C. albicans* cells break and are more difficult to recover by centrifugation, resulting in a far lower number of cells available for the run-on reaction. To avoid that, we use 0.05 % sarkosyl for initial permeabilization and post-RNase A digestion washes. It has been reported that 0.05 % sarkosyl is enough to stop ongoing transcription in the cells and prevent PIC RNA polymerases from starting new rounds of transcription [16].

21. Performing the run-on at higher temperatures (37 °C for example) resulted in better biotin-UTP incorporations. In any case we used 30 °C for *S. cerevisiae* and 37 °C for *C. albicans* in order to keep during run-on the usual culture conditions for each yeast species.

22. If you want to check/optimize for the biotinylation efficiency and RNA trimming parameters, you can run an electrophore-

sis gel (native, 2 % agarose) and then transfer the RNA to a nylon membrane using a standard Northern protocol. Once transferred, crosslink the RNA to the membrane with 50 mJ of UV radiation with a GS Gene Linker (Bio-Rad), or similar, and detect it by streptavidin-horseradish peroxidase (Streptavidin-HRP, Pierce).

23. Purity of your biotinylated RNA (proportion of labeled vs. non-labeled) can be estimated by means of a dot-blot hybridization assay, comparing your final extracted RNA against a synthetic, biotinylated RNA. Briefly: equal starting amounts of each RNA (we start with 20 ng) are placed on a nylon membrane as 1 μL dots, followed by a number of serial ½ dilutions (typically 5–8). Once dots are deposited, the membrane is air-dried for 20 min, cross-linked, and biotin signal detected as in **Note 17**.

Acknowledgments

The authors are grateful to Jesica Portero for her help with Affymetrix arrays, to Lola de Miguel and Sebastián Chávez from the Universidad de Sevilla for providing RNA pol II ChIP-chip data, and to the members of the laboratories in Valencia and Heidelberg for discussion and support. The work in the Valencia laboratory is supported by grants from the Spanish MINECO and the European Union funds (FEDER) (BFU2010-21975-C03-01 to J.E. P.-O. and PIM2010EPA-00658 to E. Herrero), and from the Regional Valencian Government (Generalitat Valenciana—PROMETEO 2011/088).

References

1. Coulon A, Chow CC, Singer RH, Larson DR (2013) Eukaryotic transcriptional dynamics: from single molecules to cell populations. Nat Rev Genet 14(8):572–584

2. Dangkulwanich M, Ishibashi T, Bintu L, Bustamante CJ (2014) Molecular mechanisms of transcription through single-molecule experiments. Chem Rev 114(6):3203–3223

3. Pérez-Ortín JE, de Miguel-Jiménez L, Chávez S (2012) Genome-wide studies of mRNA synthesis and degradation in eukaryotes. Biochim Biophys Acta 1819(6):604–615

4. Ameur A, Zaghlool A, Halvardson J, Wetterbom A, Gyllensten U, Cavelier L, Feuk L (2011) Total RNA sequencing reveals nascent transcription and widespread co-transcriptional splicing in the human brain. Nat Struct Mol Biol 18(12):1435–1440

5. Churchman LS, Weissman JS (2011) Nascent transcript sequencing visualizes transcription at nucleotide resolution. Nature 469(7330): 368–373

6. Carrillo-Oesterreich F, Preibisch S, Neugebauer KM (2010) Global analysis of nascent RNA reveals transcriptional pausing in terminal exons. Mol Cell 40(4):571–581

7. Guo J, Price DH (2013) RNA polymerase II transcription elongation control. Chem Rev 113(11):8583–8603

8. García-Martínez J, Aranda A, Pérez-Ortín JE (2004) Genomic run-on evaluates transcription rates for all yeast genes and identifies gene

regulatory mechanisms. Mol Cell 15(2): 303–313

9. García-Martínez J, Pelechano V, Pérez-Ortín JE (2011) Genomic-wide methods to evaluate transcription rates in yeast. Yeast Genetic Networks: Methods and Protocols. Methods Mol Biol 734:25–44

10. Core LJ, Waterfall JJ, Lis JT (2008) Nascent RNA sequencing reveals widespread pausing and divergent initiation at human promoters. Science 322(5909):1845–1848

11. Kwak H, Fuda NJ, Core LJ, Lis JT (2013) Precise maps of RNA polymerase reveal how promoters direct initiation and pausing. Science 339(6122):950–953

12. McKinlay A, Araya CL, Fields S (2011) Genome-wide analysis of nascent transcription in Saccharomyces cerevisiae. Genes Genomes Genetics 1(7):549–558

13. Hirayoshi K, Lis JT (1999) Nuclear run-on assays: assessing transcription by measuring density of engaged RNA polymerases. Methods Enzymol 304:351–362

14. Jackson DA, Iborra FJ, Manders EMM, Cook PR (1998) Numbers and organization of RNA polymerases, nascent transcripts, and transcription units in HeLa nuclei. Mol Biol Cell 9:1523–1536

15. Jordán-Pla A, Gupta I, de Miguel-Jiménez L, Steinmetz LM, Chávez S, Pelechano V, Pérez-Ortín JE (2015) Chromatin-dependent regulation of RNA polymerases II and III activity throughout the transcription cycle. Nucleic Acids Res 43(2):787–802

16. Szentirmay MN, Sawadogo M (1994) Sarkosyl block of transcription reinitiation by RNA polymerase II as visualized by the colliding polymerases reinitiation assay. Nucleic Acids Res 22(24):5341–5346

Chapter 9

Genome-Wide Probing of RNA Structures In Vitro Using Nucleases and Deep Sequencing

Yue Wan, Kun Qu, Zhengqing Ouyang, and Howard Y. Chang

Abstract

RNA structure probing is an important technique that studies the secondary and tertiary conformations of an RNA. While it was traditionally performed on one RNA at a time, recent advances in deep sequencing has enabled the secondary structure mapping of thousands of RNAs simultaneously. Here, we describe the method Parallel Analysis for RNA Structures (PARS), which couples double and single strand specific nuclease probing to high throughput sequencing. Upon cloning of the cleavage sites into a cDNA library, deep sequencing and mapping of reads to the transcriptome, the position of paired and unpaired bases along cellular RNAs can be identified. PARS can be performed under diverse solution conditions and on different organismal RNAs to provide genome-wide RNA structural information. This information can also be further used to constrain computational predictions to provide better RNA structure models under different conditions.

Key words RNA, Structure, Biochemistry, Genomics, High-throughput sequencing

1 Introduction

RNA structure plays important roles in almost every step of the RNA lifecycle [1]. As such, studying how an RNA folds can provide valuable information into how an RNA functions and/or is regulated during different cellular processes. Traditionally, the secondary structure of an RNA can be probed in solution using chemicals or enzymes, which modify or cleave at double stranded or single stranded regions [1–3]. Upon structure probing, the cleavage sites can then be read out by size fractionation using gel electrophoresis, while the modification sites are detected by reverse transcription (RT) stoppages that are resolved by running a sequencing gel or performing capillary sequencing. The resolution of a sequencing gel and capillary electrophoresis is typically

Frédéric Devaux (ed.), *Yeast Functional Genomics: Methods and Protocols*, Methods in Molecular Biology, vol. 1361, DOI 10.1007/978-1-4939-3079-1_9, © Springer Science+Business Media New York 2016

around 100 and 600 bases respectively. Although structure probing coupled to capillary sequencing can be multiplexed, structure probing of long RNAs and of many RNAs at one time is still slow and tedious [4]. Furthermore, as the sequence content of an RNA needs to be available to enable its cloning or the design of RT primers for reverse transcription, the de novo discovery of structural information of genes with unknown sequence is not possible. To enable the large scale probing of RNA structures from known genes or genes that are discovered de novo, RNA structure probing can be coupled to high throughput sequencing to provide structure information for thousands of RNAs simultaneously [5–7].

One high throughput approach, known as Parallel Analysis of RNA Structures (PARS), utilizes structure specific nucleases and high throughput sequencing to enable structure probing of RNA structures globally in vitro [7]. PARS has been applied to yeast and human transcriptomes, as well as under different conditions, to obtain structural information for thousands of genes [8, 9]. Here, we describe the procedures to obtain large scale yeast secondary structure information starting from yeast total RNA isolation (Fig. 1). Briefly, isolated cellular RNAs are structure probed using two nucleases, RNase V1 and S1 nuclease, which cleaves at double and single stranded regions respectively. The cleaved RNAs are then fragmented to around 200 bases to enable cloning into a cDNA library. As RNase V1 and S1 nuclease cleaves leaving behind 5′-phosphate (5′ P) and 3′ hydroxyl (3′ OH), the nuclease cleavage sites can be ligated to 5′ and 3′ RNA adapters. The ligated fragments are then reverse transcribed and amplified using polymerase chain reaction (PCR) into a cDNA library. Upon deep sequencing and mapping to the yeast transcriptome, the positions at which the nuclease cleavages had occurred can be identified. As these two nucleases cut at the 3′-end of either paired or unpaired bases, the exact cleavage site is one nucleotide in front of the base that the sequencing reads are mapped to. The number of double or single stranded reads that initiate at a particular base indicate the intensity of cleavages that occurred at that base. To determine the propensity of a base to be double or single stranded, we calculate a PARS score per base by taking the ratio of double and single stranded reads at that base. A large positive PARS score indicates that a base is likely to exist in a paired conformation while a large negative PARS score indicates that the base is likely to exist in an unpaired conformation. PARS data can also serve as constraints to generate more accurate RNA secondary structure models of cellular RNAs [10]. In total, the experimental procedure take around 5 days, the sequencing takes about 4 days and the mapping of the sequencing reads takes about 2 days to complete.

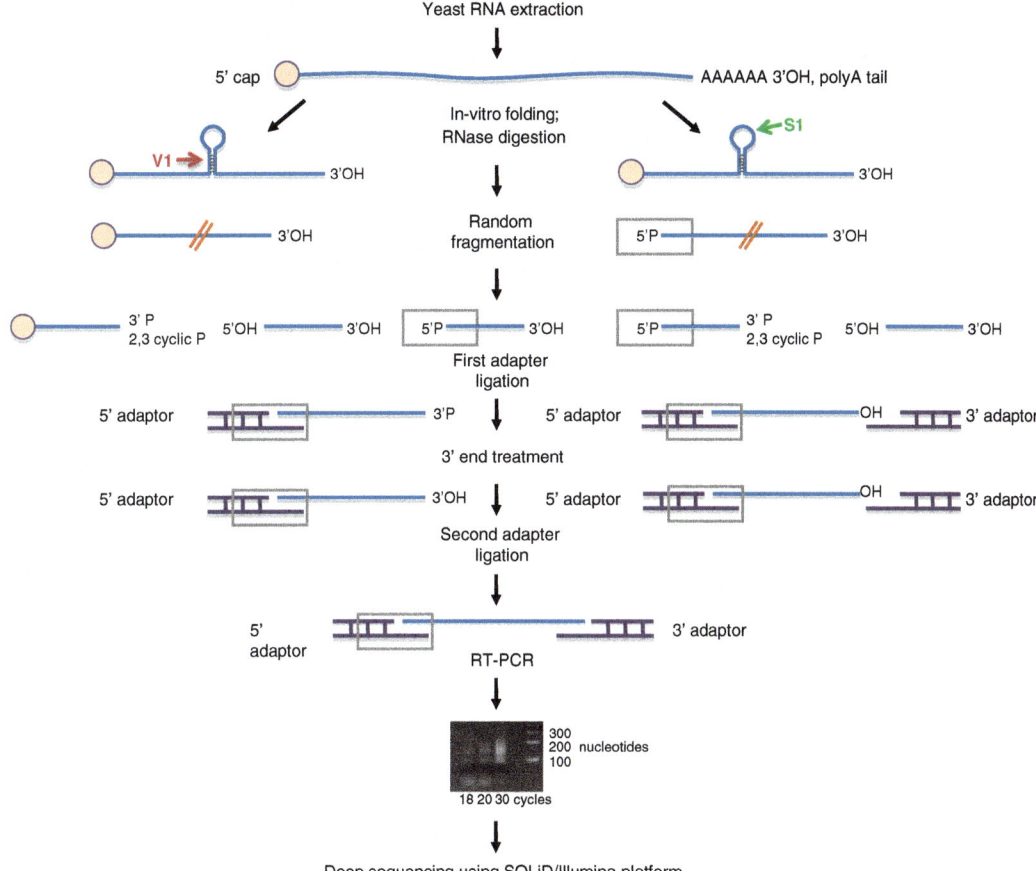

Fig. 1 Outline of the PARS procedure. Isolated cellular RNAs are structure probed using two nucleases, RNase V1 and S1 nuclease, which cleaves at double and single stranded regions respectively. The cleaved RNAs are then fragmented to around 200 bases to enable cloning into a cDNA library. As RNase V1 and S1 nuclease cleaves leaving behind 5′ phosphate (5′ P) and 3′ hydroxyl (3′ OH), the nuclease cleavage sites can be ligated to 5′- and 3′-RNA adapters. The ligated fragments are then reverse transcribed and amplified using polymerase chain reaction (PCR) into a cDNA library. Upon deep sequencing and mapping to the yeast transcriptome, the positions at which the nuclease cleavages had occurred can be identified

2   Materials

Prepare all solutions in nuclease-free water. All reagents need to be nuclease free. Use the highest quality reagents whenever possible.

2.1   Yeast Lysis and PolyA Selection

1. Corning bottle top vacuum filters.
2. 50 ml Falcon polypropylene conical tube.
3. Acid phenol solution (pH 4.3).
4. 3 M Na acetate.
5. 5 mg/ml glycogen.

6. PolyA selection kit (Poly(A)Purist MAG kit or other alternative polyA selection kits).

7. Yeast lysis buffer: 10 mM EDTA pH 8, 0.5 % SDS, 10 mM Tris–HCl pH 7.5.

2.2 RNA Structure Probing and Library Construction

1. 10× RNA structure buffer (100 mM MgCl$_2$, 1.5 M NaCl, 500 mM Tris–HCl pH 7.4).

2. Phenol–chloroform–isoamyl alcohol (25:24:1).

3. RNase V1.

4. S1 nuclease.

5. RiboMinus concentration module.

6. Ambion RNA-Seq library construction kit (Life Technologies, cat#4454073) for Illumina sequencing or SOLiD® Total RNA-Seq Kit (Life Technologies, cat#4445374) for SOLiD sequencing.

7. T4 PNK.

8. Antarctic phosphatase.

9. Superasin RNase Inhibitor.

10. Qiagen MinElute PCR purification kit.

11. Qiagen MinElute Gel extraction kit.

12. Phusion Master Mix with HF buffer (NEB, cat#F-531S).

13. NuSieve GTG agarose.

14. 6 % TBE-Urea Gels 1.0 mm, ten wells.

15. Zero Blunt TOPO PCR Cloning Kit with One Shot TOP10 Chemically Competent *E. coli* (Life Technologies, cat#K2800-20).

16. Costar® Spin-X® centrifuge tube filters (Sigma-Aldrich, cat#CLS8162-96EA).

17. Gel elution buffer (5 mM Tris–HCl pH 8, 0.5 mM EDTA, 2.5 M ammonium acetate).
Sequence of the positive control Tetrahymena p4p6 domain:

ggaauugcgggaaaggggucaacagccguucaguaccaagucu-
caggggaaacuuugagauggccuugcaaagggguaugguaau-
aagcugacggacaugguccuuaaccacgcagccaaguccuaagucaa-
cagaucuucuguugauauggaugcaguuc

2.3 Mapping of Sequencing Data

1. FASTQC software (http://www.bioinformatics.babraham.ac.uk/projects/fastqc/).

2. PERL software (http://www.activestate.com/activeperl).

3. Bowtie software (http://bowtie-bio.sourceforge.net/index.shtml/).

4. SeqFold software (http://www.stanford.edu/~zouyang/seqfold/).

5. Sfold software (http://sfold.wadsworth.org/cgi-bin/index.pl).

6. Treeview software (http://jtreeview.sourceforge.net/).

7. VARNA software (http://varna.lri.fr/).

8. Sample data and perl scripts for data analysis: https://s3.amazonaws.com/changbackup/ywan/PARS_Nature_Protocols/sample_data.tar.gz. This link contains the following files: (a) V1.csfasta: sample SOLiD sequencing data for RNase V1 library; (b) sam2tab.pl: Script to calculate the total V1 and S1 reads per base after mapping to the yeast transcriptome; (c) normalize.pl: Script for normalizing mapped reads across libraries; (d) calculate_PARS.pl: Script to calculate PARS score from mapped V1 and S1 reads.

---

3 Methods

3.1 Yeast Total RNA Extraction and Poly(A)+ Selection

1. Filter 500 ml of log phase growing yeast cells using a 0.45 μm, 250 ml, Corning bottle-top vacuum filters to collect the yeast (see **Note 1**).

2. Remove the yeast-containing filter paper from the rest of the bottle top filter. Snap-freeze the yeast-containing filter paper by putting in it a 50 ml Falcon polypropylene conical tube that is filled with liquid nitrogen. Store yeast cells at −80 °C until they are ready to be lysed for RNA extraction.

3. Loosen the cap of a bottle of acid phenol solution, pH 4.3, and warm up the phenol at 65 °C for 20 min. Add 10 ml of yeast lysis buffer and 10 ml of the phenol solution to the frozen yeast cells in the 50 ml Falcon tube and vortex vigorously. Lyse the cells at 65 °C for 1 h, with vigorous vortexing at every 20 min.

4. Spin the 50 ml Falcon polypropylene tube at $10,000 \times g$ for 15 min. Transfer the upper aqueous layer to a new 50 ml Falcon polypropylene tube without disturbing the white intermediate layer. In the new tube, add 10 ml of acid phenol solution to the aqueous solution and mix well by vortexing vigorously.

5. Repeat **step 4** for a total of two phenol extractions (see **Note 2**).

6. Spin the 50 ml Falcon polypropylene tube at $10,000 \times g$ for 15 min. Transfer the upper aqueous layer to a new 50 ml Falcon polypropylene tube and add 10 ml of chloroform to the aqueous solution in the new tube. Vortex vigorously.

7. Spin the 50 ml Falcon polypropylene tube at $10,000 \times g$ for 15 min.

8. Transfer the upper aqueous layer to a new 50 ml Falcon polypropylene tube. Add 1 ml of 3 M sodium acetate and 30 ml of

100 % ethanol to precipitate the RNA. Mix well by inverting the tubes a few times and precipitate the RNA at −20 °C overnight or at −80 °C for 1 h. The RNA can be stored at −80 °C indefinitely.

9. Spin down the RNA in the 50 ml Falcon polypropylene tube at 10,000 ×g for 30 min at 4 °C. Decant the supernatant from the tube.

10. Add 40 ml of 70 % (vol/vol) ethanol to the 50 ml Falcon tube to wash the RNA. Spin down the RNA at 10,000 ×g for 30 min at 4 °C. Decant the supernatant from the tube.

11. Spin at 5000 ×g for 1 min to collect residual ethanol from the walls of the Falcon tube. Remove excess ethanol using a p200 pipette.

12. Add 10 ml of nuclease-free water to the RNA. Rotate the Falcon tube at room temperature (23 °C) for 15 min to completely dissolve the RNA. Measure RNA concentration using a NanoDrop spectrophotometer.

13. Perform poly(A)+ selection by following manufacturer's instructions according to the Poly(A)Purist MAG Kit. Add 40 μl of 5 M ammonium acetate, 1 μl of glycogen, and 1.1 ml of 100 % ethanol to each poly(A)+ selection reaction in the provided 2 ml microfuge tube. Mix by inverting the tubes several times. Precipitate the RNA at −20 °C overnight or at −80 °C for 1 h (*see* **Note 3**).

14. Spin the microfuge tube containing the RNA at 13,000 ×g for 30 min, at 4 °C, to pellet the RNA precipitate.

15. Add 1 ml of 70 % (vol/vol) ethanol to wash the RNA. Mix thoroughly by vortexing. Spin the microfuge tube at 13,000 ×g for 15 min, at 4 °C, to re-pellet the RNA. Carefully remove the supernatant from the RNA pellet.

16. Spin the microfuge tube briefly to collect the residual ethanol at the bottom of the tube. Remove the excess ethanol carefully using a 10 μl pipette tip. Add 50–100 μl of nuclease-free water to the RNA to resuspend it. Measure the concentration of the RNA using a NanoDrop spectrophotometer. The RNA can be stored at −80 °C indefinitely (*see* **Note 4**).

3.2 RNA Structure Probing

1. Add 1 μg of poly(A)+ enriched RNA in 80 μl of nuclease-free water to each of a 200 μl thin wall PCR tube for two reactions (one for RNase V1 digestion and another for S1 nuclease digestion) (*see* **Note 5**).

2. Heat the RNA at 90 °C for 2 min in a thermal cycler, with heated-lid on, then immediately place the tubes on ice for 2 min.

3. Add 10 μl of 10× RNA structure buffer (ice-cold) to the RNA; pipette up and down the mixture several times. Transfer the

tubes from ice to the thermal cycler. Program the thermal cycler so that the temperature slowly increases from 4 to 23 °C over 20 min.

4. Add 10 μl of S1 nuclease (diluted ten-fold in nuclease-free water) to one tube and 10 μl of RNase V1 (diluted 100-fold in nuclease-free water) to the second tube; mix by pipetting. Incubate the samples at 23 °C for 15 min (see **Note 6**).

5. Transfer the two reaction mixtures from *step 4* to two 1.5 ml microfuge tubes containing 100 μl of phenol–chloroform–isoamyl alcohol each. Vortex the tubes vigorously. Spin the tubes in a microcentrifuge at 4 °C, 13,000 ×g, for 10 min.

6. Remove the top aqueous layers carefully and transfer them to two new 1.5 ml microfuge tubes. Add 10 μl of 3 M sodium acetate, 1 μl of glycogen, and 300 μl of 100 % cold ethanol to the aqueous solution. Mix well by inverting the tubes several times. Precipitate the RNA by incubating at −80 °C for 1 h or at −20 °C overnight. The RNA can be stored at −80 °C indefinitely.

7. Spin at 13,000 ×g, 4 °C, for 30 min in a centrifuge to pellet the RNA. Remove the supernatant; add 1 ml of 70 % (vol/vol) ethanol to the RNA pellets.

8. Spin at 13,000 ×g, 4 °C, for 15 min in a centrifuge to re-pellet the RNA. Remove the supernatants; spin the tubes briefly to collect the residual ethanol at the bottom of the tube. Remove the residual ethanol using a p10 pipette. Resuspend the RNA pellets in 4 μl of nuclease-free water and transfer the RNA to clean PCR tubes. Keep the tubes on ice.

3.3 PARS Library Preparation

3.3.1 RNA Fragmentation and Cleanup

1. Add 40 μl of 1× alkaline hydrolysis buffer to a clean PCR tube and place it in a thermo cycler that is set to 95 °C. After 45 s, transfer the two PCR tubes containing 4 μl of RNA each from ice to the thermal cycler set at 95 °C; heat for 15 s.

2. Add 16 μl of the heated 1× alkaline hydrolysis buffer to each of the heated RNA solution; pipette up and down several times. Incubate at 95 °C for 3.5 min, then immediately place the RNA on ice. Add 2 μl of 3 M sodium acetate to stop the fragmentation reaction (see **Note 7**).

3. Add 6 ml of 100 % ethanol to 1.5 ml of Wash Buffer (W5) from the RiboMinus Concentration Module (see **Note 8**).

4. Add 78 μl of 10 mM Tris pH 7.0 to each of the two fragmentation RNA reaction mixtures. Pipette up and down several times to mix.

5. Add 100 μl of Binding Buffer (L3) from the RiboMinus Concentration Module to the 100 μl RNA fragmentation reaction mixtures, followed by adding 250 μl of 100 % ethanol. Mix by pipetting up and down.

6. Place a spin column in a new 1.5 ml wash tube and transfer 450 μl of the RNA sample onto the spin column. Spin at $12,000 \times g$ for 1 min and discard the flow through. Put the spin column back to the empty wash tube.

7. Add 500 μl of wash buffer (W5, with ethanol added), to the spin column. Spin at $12,000 \times g$ for 1 min and discard the flow through. Place the spin column back to the empty wash tube and spin at maximum speed for 2 min to dry the column.

8. Discard the wash tube that contains the flow through and place the spin column into a clean 1.5 ml recovery tube. Add 12 μl of nuclease-free water to the center of the spin column. Wait for 1 min at room temperature. Spin the column at $13,000 \times g$ for 1 min to elude the RNA (*see* **Note 9**).

9. Dry the fragmented RNA in each of the two tubes in a vacuum centrifuge at low heat (<40 °C) for approximately 15 min until the volume is reduced to less than 3 μl (*see* **Note 10**).

3.3.2   First Adapter Ligation

The library preparation steps below are modifications made to the Ambion RNA-Seq Library Construction Kit, for sequencing on the Illumina platform, and SOLiD® Total RNA-Seq Kit, for sequencing on the SOLiD platform. For simplicity, the components used in the library preparation process are from the Ambion RNA-Seq Library construction kit, unless specified otherwise (*see* **Note 11**).

1. Add nuclease-free water to the RNA in 1.5 ml microfuge tube to a final volume of 3 μl. Mix by pipetting.

2. Add 2 μl of adapter mix A and 3 μl of hybridization buffer to the RNA; mix by pipetting.

3. Heat the tube to 65 °C in a thermal cycler for 10 min then incubate it at 16 °C for 5 min.

4. Add 10 μl of 2× ligation buffer slowly to the mixture, pipette up and down gently to mix. Add 2 μl of the ligation enzyme mix. Mix well by flicking the tube such that the sample looks homogenous. Briefly spin the tubes to collect the sample at the bottom of the tubes. Incubate the sample in a thermal cycler at 16 °C overnight, with heated lid off.

3.3.3   Treatment of 3′ ends

1. Add 20 μl of nuclease-free water to the sample and mix by pipetting. Add 5 μl of 10× Antarctic phosphatase buffer to the sample, followed by adding 2.5 μl of Superasin RNase inhibitor and 2.5 μl of Antarctic phosphatase enzyme. Mix by flicking the tube. Briefly spin the tube to collect the contents at the bottom of the tube and incubate sample at 37 °C in a thermo cycler for 1 h.

2. Transfer the reaction from the PCR tube to a 1.5 ml microfuge tube. Add 150 μl of nuclease-free water, then add 100 μl of 100 mM Tris, pH 8. Pipette up and down to mix.

3. Add 300 μl of phenol–chloroform–isoamyl alcohol to the RNA and vortex to mix. Spin at $13,000 \times g$ for 10 min at 4 °C.

4. Transfer the top aqueous layer to a new 1.5 ml tube and add 30 μl of 3 M sodium acetate, followed by 3 μl of glycogen and 900 μl of 100 % ethanol. Pipette up and down to mix. Incubate at −20 °C overnight or at −80 °C for 1 h to precipitate the RNA. The RNA can be stored at −80 °C indefinitely.

5. Spin at $13,000 \times g$, for 30 min, at 4 °C. Remove the supernatant; then add 1 ml of 70 % (vol/vol) ethanol to wash the pellet.

6. Spin at $13,000 \times g$, for 15 min, at 4 °C. Remove the supernatant; spin briefly to collect the excess 70 % (vol/vol) ethanol at the bottom of the tube. Remove the residual ethanol using a 10 μl pipette. Dissolve the RNA pellet in 3 μl of nuclease-free water.

3.3.4 Second Adapter Ligation

1. Add 2 μl of adapter mix A and 3 μl of hybridization buffer to the RNA in a 1.5 ml microfuge tube. Mix by pipetting and transfer the 8 μl of RNA mix to a new 200 μl PCR tube.

2. Heat the PCR tube to 65 °C in a thermal cycler for 10 min, then incubate it at 16 °C for 5 min.

3. Add 10 μl of 2× ligation buffer slowly to the RNA mix, pipette up and down gently to mix. Add 2 μl of the ligation enzyme mix. Mix well by flicking the tube until the sample looks homogenous. Briefly spin the tubes to collect the sample at the bottom of the tubes. Incubate the sample in a thermal cycler at 16 °C overnight, with heated lid off.

3.3.5 Reverse Transcription

1. Prepare the reverse transcription (RT) master mix on ice:
   Add 9 μl of nuclease-free water to the ligated RNA sample, followed by adding 4 μl of RT buffer, 2 μl of dNTP mix, and 4 μl of RT primer. Pipette up and down to mix.

2. Heat the RNA in the PCR tube in a thermal cycler at 70 °C, with heated lid on, for 5 min. Snap-cool the reaction on ice for 2 min.

3. Add 1 μl of ArrayScript RT enzyme to each sample. Flick the tubes several times to mix the sample and spin briefly to collect the sample at the bottom of the tube. Heat the PCR tubes at 42 °C for 30 min in a thermal cycler, with heated lid on. The resulting cDNA can be stored at −20 °C for months.

4. Add 60 μl of nuclease-free water to the 40 μl cDNA and transfer the 100 μl sample to a clean 1.5 ml microfuge tube. Add

500 µl of Buffer PB from Qiagen MinElute PCR purification kit to the sample, pipette up and down, and transfer the mixture to a spin column.

5. Spin at $13,000 \times g$ for 1 min and discard the flow through. Place the column back into the tube and add 750 µl of buffer PE (with ethanol added) to the spin column. Spin for 1 min at $13,000 \times g$. Discard the flow through.

6. Spin at $13,000 \times g$ for 1 min to completely dry the column.

7. Transfer the column to a clean 1.5 ml microfuge tube. Add 10 µl of nuclease-free water to the center of the column. Wait for 1 min before spinning at $13,000 \times g$ for 1 min.

3.3.6 cDNA Size Selection

1. Add 24 µl of nuclease-free water to 1 µl of 50 bp DNA ladder in a microfuge tube. Transfer 5 µl of the diluted DNA into a new 1.5 ml microfuge tube and add 5 µl of 2× Novex TBE-Urea Sample buffer. Mix by pipetting.

2. Add 5 µl (out of 10 µl) of cDNA to a clean 1.5 ml microfuge tube. Add 5 µl of 2× Novex TBE-Urea Sample buffer to the cDNA. Mix by pipetting.

3. Heat the tubes containing 10 µl of DNA ladder and 10 µl of cDNA from **steps 1** and **2** at 95 °C for 3 min. Snap-cool on ice.

4. Load each sample into a well in a 6 % Novex TBE-Urea PAGE gel (1 mm). Separate the samples from each other by at least two lanes. Run the PAGE gel in 1× TBE at 180 V until the leading blue dye is about 1 cm below the center of the gel. This takes approximately 25 min.

5. Add 1 µl of SYBR-GOLD to 10 ml of 1× TBE; mix and incubate the solution with the gel for 5 min in the dark. Wrap the gel between two pieces of plastic wrap. Visualize the size distribution of the ladder and the sample by either using UV or blue-light.

6. Puncture two holes at the bottom of a clean 0.6 ml microfuge tube using a clean 18.5 G needle and place the 0.6 ml tube inside a 1.5 ml microfuge tube.

7. Excise the gel slice that contains cDNAs between 100 and 300 bases using a sterile scalpel. Transfer the gel slice to the punctured 0.6 ml tube. Spin at $12,000 \times g$ for 1 min. Make sure that all the gel slices are at the bottom of the 1.5 ml tube and discard the 0.6 ml tube.

8. Add 700 µl of gel elution buffer to the shredded gel; incubate at room temperature overnight on a rotator (*see* **Note 12**).

9. Cut the end of a 1 ml pipette tip to increase surface area and use it to transfer the elution buffer containing the gel pieces

into a Costar® Spin-X® centrifuge tube filter. Spin at $10,000 \times g$ for 2 min in a centrifuge.

10. Transfer 350 μl of the elution buffer from **step 9** into each of the two clean 1.5 ml microfuge tubes. Add 3.5 μl of glycogen and 1 ml of ethanol to the elution buffer in each tube, pipette up and down, and incubate the tubes at –20 °C overnight or –80 °C for 1 h.

11. Spin the tubes at $13,000 \times g$ for 30 min to pellet the cDNA. Remove the supernatant and add 1 ml of 70 % (vol/vol) ethanol to wash the pellet. Remove the supernatant and add 10 μl of water to resuspend the cDNA pellet. Transfer 1 μl of cDNA to a clean 0.2 ml PCR tube to perform a small-scale PCR.

3.3.7 Small Scale PCR Amplification

1. Prepare the PCR master mix on ice. Add 22 μl of nuclease-free water, 1 μl of Ambion 5′ PCR primer, 1 μl of Ambion 3′ PCR primer, and 25 μl of 2× Phusion High Fidelity PCR Master Mix to the cDNA. Mix by pipetting up and down.

2. Use the lowest number of PCR cycles that is sufficient to amplify the sample. Perform the PCR reaction using the cycling conditions below. After 15 cycles, pause the program at 72 °C, transfer 10 μl of the PCR reaction to a clean 1.5 ml tube and place it on ice. Un-pause the program and repeat the same sample collection step after 20, 25, and 30 cycles. At the end of the PCR, the researcher should have four tubes, each with a sample that has undergone 15, 20, 25, or 30 PCR cycles (*see* **Note 13**, Table 1).

3. Load 500 ng of 100 bp ladder and 50 bp ladder into two separate wells in a 2 % agarose gel made with 1× TBE.

4. Add 1 μl of 10× BlueJuice Gel Loading Buffer to each of the 10 μl PCR product at 15, 20, 25, and 30 cycles; then load each sample into a well in the 2 % agarose gel. Skip at least one lane between the sample and the ladders.

Table 1
Small scale PCR conditions

Cycle number	Denature	Anneal	Extend
1	98 °C, 2 min		
2–31	98 °C, 30 s	65 °C, 30 s	72 °C, 30 s
32			72 °C, 5 min

PCR cycle conditions for determining the minimum number of PCR amplification needed for library generation

Table 2
Large scale PCR conditions

Cycle number	Denature	Anneal	Extend
1	98 °C, 2 min		
X	98 °C, 30 s	65 °C, 30 s	72 °C, 30 s
X+1			72 °C, 5 min

PCR cycle conditions for large scale generation of sequencing library

5. Run the gel at 120 V in 1× TBE until the running dye for the ladder is near the bottom of the gel.

6. Visualize the gel under UV light. The amplified samples should have a smear from 250 to 300 bases. Choose the lowest number of PCR cycles that show the presence of such a smear, to amplify the samples in the large scale PCR amplification.

3.3.8 Large-Scale PCR Amplification

1. Prepare the PCR reaction on ice, in a new 1.5 ml microfuge tube. The PCR is set up to have a total volume of 200 µl, to be split into two 0.2 ml PCR tubes. Add 4 µl of cDNA library, 88 µl of nuclease-free water, 4 µl of Ambion 5′ PCR primer, 4 µl of Ambion 3′ PCR primer, and 100 µl of 2× Phusion High Fidelity PCR Master Mix to the microfuge tube. Mix by pipetting up and down.

2. Perform the PCR using the cycling conditions described in the table below. *X* is the lowest number of PCR cycles needed to amplify the sample, as determined by the small scale PCR reaction in Table 2.

3. Pool the replicate large-scale PCR reactions from the same sample together. Add 1 ml (5 volumes) of buffer PB from MinElute Gel Extraction Kit to the sample; mix by pipetting up and down.

4. Add 700 µl of the sample to a MinElute column. Spin at 13,000×*g*, for 1 min and discard the flow through. Load the remaining 500 µl of the sample into the column. Spin at 13,000×*g*, for 1 min and discard the flow through.

5. Add 750 µl of buffer PE (add ethanol prior to using PE) to the column. Spin at 13,000×*g* for 1 min and discard the flow through.

6. Spin at 13,000×*g* for 1 min to dry the column. Transfer the column to a clean 1.5 ml microfuge tube.

7. Add 20 µl of nuclease-free water to the center of the column, wait for 1 min and spin at 13,000×*g* for 1 min. Repeat this elution step using another 20 µl of nuclease-free water.

3.3.9 Size Selection of PCR Products by Gel Electrophoresis

1. Add 5 μl of 10× BlueJuice Gel Loading Buffer to 40 μl of the PCR product from **the previous step.**.

2. Load 500 ng of 100 and 50 bp ladder into two separate wells in a large 3 % NuSieve GTG agarose gel (made with 1× TBE) (*see* **Note 14**).

3. Skip at least two lanes between the sample and the ladders and load 45 μl of the PCR product into one well in the agarose gel. Skip two lanes between each sample. Run the agarose gel at 100 V for 2–3 h, until the leading dye front from the ladder is at the end of the gel (*see* **Note 15**).

4. Visualize the gel in blue-light (preferred but has lower sensitivity) or under UV light. Cut out a gel slice containing the PCR product between 150 and 300 bases using a clean scalpel. Transfer the gel slice to a 2 ml microfuge tube.

5. Add 1 ml of QG buffer from MinElute Gel Extraction Kit. Incubate the tube containing the gel slice and QG buffer at room temperature in a rocking shaker until the gel slice is dissolved (*see* **Note 16**).

6. Add 300 μl of isopropanol to the dissolved gel slice. Mix by pipetting up and down.

7. Transfer 700 μl of the dissolved gel slice to a clean MinElute column. Spin the MinElute column at $13,000 \times g$ for 1 min and discard the flow through. Repeat this step until all of the dissolved gel slice has been loaded onto the column.

8. Add 750 μl of PE buffer (with ethanol added) to the column. Spin the column at $13,000 \times g$ for 1 min and discard the flow through. Replace the column back into the empty microfuge tube and spin the column at $13,000 \times g$ for 1 min to dry the column.

9. Transfer the column into a clean 1.5 ml microfuge tube. Add 15 μl of nuclease-free water to the center of the column and let the column sit at room temperature for 1 min. Spin the column at $13,000 \times g$ for 1 min.

10. Measure the concentration of the eluted PARS cDNA library using the Qubit Fluorometer.

11. Measure the size and concentration of the PARS cDNA library using Agilent Bioanalyzer.

12. Clone the cDNA products using Zero Blunt TOPO PCR Cloning Kit and pick 20 colonies for capillary sequencing. Blast the reads against the yeast genome and PCR adapter sequences (*see* **Note 17**).

13. Sequence the libraries (from **step 9**) using Illumina's Hi-seq machine, according to manufacturer's protocol.

3.4 Mapping and Analysis of PARS Data

All of the executable commands below are prefixed with a "$" character, to be used in the UNIX shell (e.g., bash or csh). Most of the commands can be run using UNIX shell prompt, and are described to run from the example working directory.

1. Download the raw sequencing reads, fastq or csfasta files from Illumina or SOLiD sequencers, respectively, from the sequencers.

2. For Illumina sequencing, the quality of the sequencing reads in the fastq files can be determined using the FastQC program. The low quality bases from the 5′ and 3′ ends of each read are trimmed to enable accurate mapping of the reads.

3. Create a transcriptome index file of yeast RNA sequences using the following commands below for the Illumina platform (option A, if Ambion RNA-Seq library construction kit is used) or the SOLiD platform (option B, if the SOLiD® Total RNA-Seq Kit is used).

4. Type the following command line to create a transcriptome index file for mapping fastq sequences:

```
$ bowtie-build sce_genes.fa sce_genes
```

5. Type the following command line to create a transcriptome index file for mapping csfasta sequences:

```
$ bowtie-build -C sce_genes.fa sce_genes_c
```

6. Map the sequencing reads to the yeast transcriptome by aligning the raw reads to the transcriptome indexes using the Bowtie software. This creates a SAM format file that indicates the positions along the transcriptome where nuclease cleavages have occurred. Map the sequencing reads from (1) Illumina single-end sequencing (Option A), (2) Illumina paired-end sequencing (Option B), (3) SOLiD single-end sequencing (Option C) or (4) SOLiD paired-end sequencing (Option D), to the transcriptome using the command lines below. The meanings of the parameters, and their suggested initial values, are clarified in the Table 3 at the end of this step.

7. (a) To trim and map the single-end reads from Illumina sequencing, type the following command lines:

```
$ bowtie -5 NumA -3 NumB -p NumC -S sce_
    genes S1.fastq S1.sam
$ bowtie -5 NumA -3 NumB -p NumC -S sce_
    genes V1.fastq V1.sam
```

(b) To trim and map the paired-end reads from Illumina sequencing, type the following command lines:

```
$ bowtie -5 NumA -3 NumB -p NumC -S sce_
    genes -1 S1_R1.fastq -2 S1_R2.fastq
    S1.sam
```

Table 3
Parameters for mapping Illumina and Solid sequencing reads using Bowtie

Parameter	Description
NumA	The number of low quality bases that are trimmed from the 5′-end of a raw read. If no trimming is desired, the suggested initial value is 0
NumB	The number of low quality bases that are trimmed from the 3′-end of the raw read. If no trimming is desired, the suggested initial value is 0
NumC	The number of alignment threads to launch, e.g., 8 for an 8-core processor

```
$ bowtie -5 NumA -3 NumB -p NumC -S sce_
  genes -1 V1_R1.fastq -2 V1_R2.fastq
  V1.sam
```

(c) To trim and map the single-end reads from SOLiD sequencing, type the following command lines:

```
$ bowtie -5 NumA -3 NumB -p NumC -S -C -f
  sce_genes_c S1.csfasta S1.sam
```

```
$ bowtie -5 NumB -3 NumB -p NumC -S -C -f
  sce_genes_c V1.csfasta V1.sam
```

(d) To trim and map the paired-end reads from SOLiD sequencing, type the following command lines:

```
$ bowtie -5 NumA -3 NumB -p NumC -S -C -f
  sce_genes_c -1 S1_R1.csfasta -2 S1_
  R2.csfasta S1.sam
```

```
$ bowtie -5 NumA -3 NumB -p NumC -S -C -f
  sce_genes_c -1 V1_R1.csfasta -2 V1_
  R2.csfasta V1.sam
```

8. Calculate the total number of V1 or S1 reads at each base, for all the transcripts, using the "sam2tab.pl" script. This script generates a table whereby each row consists of the name and structural information, in the form of the number of RNase V1 or S1 nuclease reads mapped to each base, for that transcript. The name and the data are tab-delimited, while the number of reads at each base is semicolon-delimited.

Use the following command line to calculate the number of RNase V1 cleavages at each base from the SAM file. The parameters x, y, and z are defined in the Table 4 below.

```
$ perl sam2tab.pl x y z V1.sam V1.tab
```

Table 4
Parameters to calculate the number of RNase V1 cleavages at each base from SAM file

Parameter	Input	Description
x	0	The read is mapped onto the forward strand of the transcriptome
y	1	The number of mapped reads, only from the first read of a paired-end sequence or single-end sequence, is counted
z	NumA + 1	NumA is the number of bases that were trimmed from the 5′-end of a read. $z = $ NumA + 1, which is the base in front of the first base of the read prior to any trimming. By correcting for z, we identify the accurate position at which the cleavage has occurred. For example, when base 10 along a RNA of 100 bases is cut by a nuclease, bases 11–100 are sequenced as a read. If we trim two low-quality bases from the 5′-end (NumA = 2), the read will be mapped to base 13 on the RNA. However, if 2 + 1 bases are subtracted from the mapped position, the read will be mapped back to base 10, which is the exact base that the nuclease cleavage occurred

Use the following command line to calculate the number of S1 nuclease cleavages at each base from the SAM file.

```
$ perl sam2tab.pl x y z S1.sam S1.tab
```

9. Normalize for sequencing depth across different PARS samples using the total number of mapped reads for the V1 and S1 libraries. Type the following command line:

```
$ perl normalize.pl Sample1_S1.tab Sample1_
V1.tab   Sample2_S1.tab   Sample2_V1.tab   …
SampleN_S1.tab SampleN_V1.tab
```

10. Calculate the PARS score at each base by taking the log ratio of V1 over S1 reads at every base using the script "calculate_PARS.pl". Type the following command line:

```
$ perl calculate_PARS.pl norm.S1.tab norm.
V1.tab sample.pars.txt
```

RNA secondary structure modeling using PARS data and Seqfold program.

11. Generate structure preference profiles using the following command line:

```
$ python $HOME/opt/seqfold/pars2spp.py
sce_S1.tab sce_V1.tab sce
```

The resulting file sce.spp contains the structure preference profiles with one transcript per row in a tab-delimited format: transcript name (column 1), structure preferences (column 2, the structure preference for each base along the transcript is semicolon-delimited).

12. Create sample structures and clusters for each transcript using the following command line:

```
$ perl $HOME/opt/seqfold/sfold_wrapper.pl
sfold_executable_file sce_genes.fa sfold_out-
put_directory
```

The sfold_executable_file is the path to the executable file of Sfold. The sfold_output_directory is the directory whereby each transcript contains a folder with structure sampling results. By default, 1000 sample structures are generated and clustered into distinct groups for each transcript (*see* **Note 18**).

13. Generate RNA secondary structure predictions and base-level accessibilities using the following command line. Choose the optional parameters below:

-d: SeqFold output directory. Default:

-o: Prefix of output summary files. Default: out

-f: Cut-off used to filter transcripts with the fraction of sites having experimental data<= cutoff_frac. Default: 0

```
$ python $HOME/opt/seqfold/seqfold.py
sfold_output_directory sce.spp
```

Two file sets, A and B, are generated under the SeqFold output directory. (A) *.seqfold.ct: Each ct file contains the predicted secondary structure for a transcript whereby * represents the name of the transcript. The structures are in CT format. (B) out.acc: Each acc file contains the estimated accessibility of each base in a transcript with one transcript per row in a tab-delimited format: transcript name (column 1), accessibilities (column 2, the accessibility of each base is semicolon-delimited).

14. Visualize the predicted RNA secondary structures, in the *.seqfold.ct file, for each transcript using the program VARNA.

---

4 Notes

1. The cells can also be collected by spinning the cells down in a 50 ml Falcon tube at $3000 \times g$ for 3 min at room temperature. However there might be some gene expression changes in the yeast as the cells are being spun down.

2. The intermediate phase should be a very thin layer after two phenol extractions. If the intermediate phase is still a white and thick layer after two phenol extractions, perform a third phenol extraction.

3. To reduce rRNA populations to the minimum in the RNA pool, we perform poly(A)+ enrichment twice. The same amount of beads was used in round 1 and round 2 to enrich for the poly(A)+ transcripts.

4. To determine that the poly(A)+ enrichment is successful and that the cellular RNAs are not degraded during the poly(A)+ selection process, run 60 ng of total RNA, 1× poly(A)+-enriched RNA, and 2× poly(A)+-enriched RNA on the Agilent Bioanalyzer. Alternatively, 200 ng of total RNA, 1× poly(A)+ enriched RNA, and 2× poly(A)+ enriched RNA can also be resolved on a 1 % agarose gel. The yeast rRNA bands, which are two dominant bright bands at 2 and 3.8 kb, should be progressively fainter with poly(A)+ selection. There should also be an increasing smear indicating the presence of other RNAs in the cell with poly(A)+ enrichment.

5. A positive control RNA with a known RNA secondary structure, such as the Tetrahymena ribozyme, should be doped into the poly(A)+ RNA pool to a concentration of 1 % (in terms of moles) of the total RNAs present [11]. After sequencing and mapping of the reads to the Tetrahymena sequence, the double and single stranded regions determined by sequencing should resemble the known secondary structure of the Tetrahymena ribozyme. This ensures that the structure probing, library generation, sequencing and mapping processes to be accurate.

6. The amount of nucleases used may need to be titrated to ensure single hit kinetics during structure probing. In general, a 1:10 dilution of S1 nuclease and 1:100 dilution of RNase V1 is a good starting point for structure probing at 23 °C. However, we always test each batch of nucleases to ensure that their reaction rates are the same by performing traditional RNA structure probing on P32 labeled Tetrahymena ribozyme. We use the nuclease concentration that provides structural information while leaving most of the transcripts intact (>80 % of the transcripts remain full length). If the structure probing is performed at a different temperature, such as at 37 °C, enzyme concentrations need to be titrated as less enzyme will needed for structure probing, due to faster reactivity, at the higher temperatures.

7. The size of the RNA fragments is critical to the success of library generation. Depending on the population size of the starting pool of RNAs, the RNA fragmentation time may need to be titrated to obtain an optimum size of around 200 bases. To determine the optimal fragmentation time, RNAs are fragmented to different times, such as 3, 4, 5, 6, and 7 min, and run on the Agilent Bioanalyzer. The fragmentation time that results in an average population size 200 bases will be used for the library preparation.

8. The fragmented RNAs can also be purified using Qiagen's MinElute RNA Cleanup kit using a modified protocol that retains all fragments >=18 bases. Add 78 µl of 10 mM Tris pH

7.0 to each of the two fragmentation RNA reaction mixtures, then add 350 μl of RLT lysis buffer to the mixture. Pipette up and down to mix. Add 900 μl of 100 % ethanol to the mixture and pipette up and down to mix. Follow the manufacturer's instructions to load the column and spin the columns. Wash the columns three times, first by adding 700 μl of RWT buffer to the column, then proceed with two washes with 500 μl of RPE buffer. Elute the RNA in a clean microfuge tube using 12 μl of nuclease-free water.

9. At this point, 1 μl of the fragmented RNA can be run on the Agilent Bioanalyzer to determine that the fragmentation sizes are correct before proceeding with the rest of the library preparation protocol.

10. The timing of this step is important as over-drying of RNA can cause the RNA to be very difficult to resuspend.

11. Other small RNA cloning kits, such as the NEBNext® Multiplex Small RNA Library Prep Set for Illumina can also be modified for PARS library preparation. The protocol should be modified such that the 5′ adapter ligation occurs first, followed by 3′ adapter ligation, reverse transcription, and PCR amplification.

12. The cDNAs exit out of the gel pieces into the elution buffer by passive diffusion. A second elution can be performed to ensure that all of the cDNA is in solution.

13. Ideally, the number of PCR cycles required to amplify the samples should be less than 18 cycles. The higher the number of PCR cycles, the greater the possibility of signal distortion as certain regions along the transcriptome might be more easily amplified than other regions. We usually do not continue the library preparation or sequence process if the amplification cycles required is greater than 25 cycles.

14. NuSieve GTG agarose gel is much better at resolving fragments between 100 and 300 bases than regular agarose gels and hence can separate adapter dimers from the actual PCR amplified products cleanly.

15. Running the gel at higher voltage may result in poorer resolution of the cDNA fragments and higher percentage of adapter dimers in the sequencing library.

16. Dissolve the gel slices at room temperature, and not at higher temperatures such as 50 °C. Incubating the gel slices at higher temperatures causes local melting of AT-rich cDNA. These melted, single stranded cDNAs are preferentially lost through column purification, resulting in a GC rich sequencing library [12].

17. cDNA products should contain different yeast mRNAs and a low percentage of ribosomal RNA and adapter dimers. If the same yeast mRNA fragment is cloned repeatedly, this suggests

that bottlenecking has occurred during the library preparation process. Typically this is caused by either very low amounts of starting material or over-amplification of the PCR products. If there is a large fraction of ribosomal RNA reads (e.g., >50–60 %) in the cloned fragments, this suggests that poly(A) + selection is not efficient.

18. In a parallel computing environment, the runtime for Seqfold can be increased by changing the value of $para in sfold_wrapper.pl. For example:

```
$para = "bsub -M 3072000 -W 6:00";
```

or

```
$para = "qsub -cwd -V -l h_vmem=3G -l h_rt=6:00:00 -m ea -w e -b y";
```

Under these settings, each transcript will be processed by one CPU core independently and hundreds to thousands of transcripts can be processed in parallel.

Acknowledgements

This work is supported by NIH R01-HG004361 (H.Y.C.) and Agency for Science, Technology and Research of Singapore (Y.W.).

References

1. Wan Y, Kertesz M, Spitale RC, Segal E, Chang HY (2011) Understanding the transcriptome through RNA structure. Nat Rev Genet 12:641–655

2. Weeks KM (2010) Advances in RNA structure analysis by chemical probing. Curr Opin Struct Biol 20:295–304

3. Ehresmann C et al (1987) Probing the structure of RNAs in solution. Nucleic Acids Res 15:9109–9128

4. Mitra S, Shcherbakova IV, Altman RB, Brenowitz M, Laederach A (2008) High-throughput single-nucleotide structural mapping by capillary automated footprinting analysis. Nucleic Acids Res 36, e63

5. Lucks JB et al (2011) Multiplexed RNA structure characterization with selective 2′-hydroxyl acylation analyzed by primer extension sequencing (SHAPE-Seq). Proc Natl Acad Sci U S A 108:11063–11068

6. Underwood JG et al (2010) FragSeq: transcriptome-wide RNA structure probing using high-throughput sequencing. Nat Methods 7:995–1001

7. Kertesz M et al (2010) Genome-wide measurement of RNA secondary structure in yeast. Nature 467:103–107

8. Wan Y et al (2012) Genome-wide measurement of RNA folding energies. Mol Cell 48:169–181

9. Wan Y et al (2014) Landscape and variation of RNA secondary structure across the human transcriptome. Nature 505:706–709

10. Ouyang Z, Snyder MP, Chang H (2013) SeqFold: genome-scale reconstruction of RNA secondary structure integrating high-throughput sequencing data. Genome Res 23(2):377–387

11. Guo F, Gooding AR, Cech TR (2004) Structure of the Tetrahymena ribozyme: base triple sandwich and metal ion at the active site. Mol Cell 16:351–362

12. Quail MA et al (2008) A large genome center's improvements to the Illumina sequencing system. Nat Methods 5:1005–1010

Chapter 10

Genome-Wide Chromatin Immunoprecipitation in *Candida albicans* and Other Yeasts

Matthew B. Lohse, Pisiwat Kongsomboonvech, Maria Madrigal, Aaron D. Hernday, and Clarissa J. Nobile

Abstract

Chromatin immunoprecipitation experiments are critical to investigating the interactions between DNA and a wide range of nuclear proteins within a cell or biological sample. In this chapter we outline an optimized protocol for genome-wide chromatin immunoprecipitation that has been used successfully for several distinct morphological forms of numerous yeast species, and include an optimized method for amplification of chromatin immunoprecipitated DNA samples and hybridization to a high-density oligonucleotide tiling microarray. We also provide detailed suggestions on how to analyze the complex data obtained from these experiments.

Key words Chromatin immunoprecipitation, *Candidaalbicans*, Yeast, ChIP-chip, ChIP-seq

1 Introduction

Chromatin immunoprecipitation (ChIP) is a method that allows for the investigation of the interaction between a protein of interest and DNA in a cell or biological sample. In order to perform ChIP, live cells are cross-linked then lysed, the chromatin is sheared, specific DNA fragments associated with the protein of interest are immunoprecipitated from the lysate using antibodies, and the bound DNA fragments are purified away from the protein upon reversal of the cross-links (Fig. 1). The overall goal of the ChIP procedure is to determine the specific genomic binding sites of the DNA-associated protein of interest. In general, genome-wide ChIP procedures with *Candidaalbicans* are similar to those performed in the model yeast *Saccharomyces cerevisiae* [1]. There are, however, several *Candida*-specific modifications in terms of cell lysis and DNA shearing that we highlight in this chapter that are critical for successful *Candida* genome-wide ChIP experiments. The protocol described below has been used successfully for

Frédéric Devaux (ed.), *Yeast Functional Genomics: Methods and Protocols*, Methods in Molecular Biology, vol. 1361,
DOI 10.1007/978-1-4939-3079-1_10, © Springer Science+Business Media New York 2016

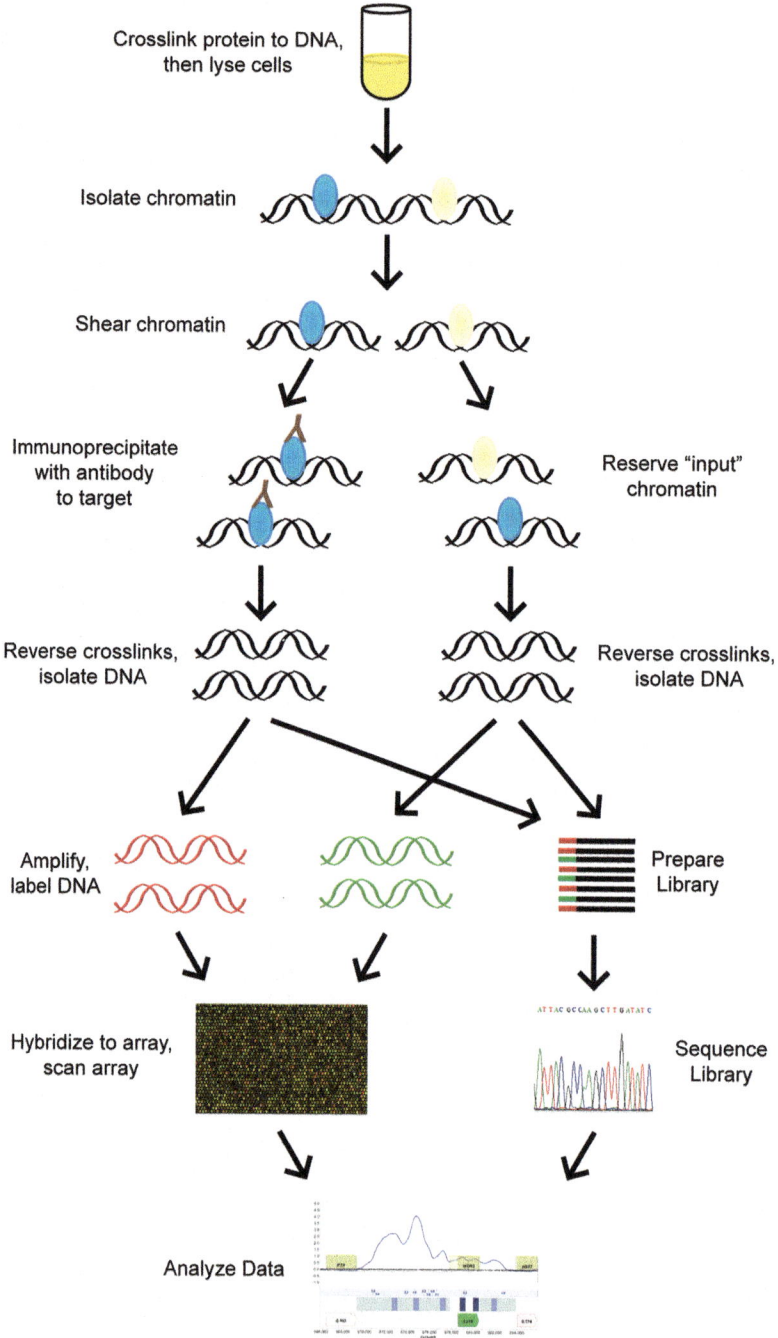

Fig. 1 Overview of the ChIP-chip and ChIP-seq experimental workflows. In brief, DNA is cross-linked to proteins, isolated from lysed cells, and then sheared into fragments. At this point, a fraction of the sample is separated to process independently as the "input" sample. The protein of interest is then immunoprecipitated from the experimental sample with an antibody against that protein. The cross-links are then reversed for both samples and the DNA isolated. For ChIP-chip, the DNA is amplified and labeled in preparation for hybridization to a high-density oligonucleotide tiling microarray. If performing ChIP-seq, the DNA would be used for library generation in preparation for sequencing. Data generated from both approaches are the starting points for further analysis

several distinct morphological forms of numerous yeast species, including *C. albicans*, *S. cerevisiae*, *Kluyveromyces lactis*, and *Histoplasma capsulatum* [2–9]. The detailed methods described in this chapter include an optimized method for amplification of ChIP DNA samples and hybridization to a high-density oligonucleotide tiling microarray (ChIP-chip) (*see* also ref.[10]). We also include a section on how to analyze the data obtained from genome-wide ChIP experiments. Although the protocols described here are focused on ChIP-chip, much of what we outline also applies to genome-wide ChIP-seq methods, which combine ChIP with high-resolution next-generation sequencing.

2 Materials

2.1 Chromatin Immunoprecipitation Buffers (See Note 1)

1. TBS: 20 mM Tris–HCl (pH 7.5), 150 mM NaCl.
2. Lysis buffer: 50 mM HEPES–KOH (pH 7.5), 140 mM NaCl, 1 mM EDTA, 1 % Triton X-100, 0.1 % Na-deoxycholate.
3. Lysis buffer with 500 mM NaCl: 50 mM HEPES/KOH (pH 7.5), 500 mM NaCl, 1 mM EDTA, 1 % Triton X-100, 0.1 % Na-deoxycholate.
4. Wash buffer: 10 mM Tris–HCl (pH 8.0), 250 mM LiCl, 0.5 % NP-40, 0.5 % Na-deoxycholate, 1 mM EDTA.
5. Elution buffer: 50 mM Tris–HCl (pH 8.0), 10 mM EDTA, 1 % SDS.
6. TE/0.67 % SDS: 10 mM Tris–HCl pH 8.0, 1 mM EDTA, 0.67 % SDS.
7. TE/1 % SDS: 10 mM Tris–HCl pH 8.0, 1 mM EDTA, 1 % SDS.
8. 4 M LiCl.
9. 2.5 M glycine (prepared fresh) in ddH_2O.
10. 10 mg/mL proteinase K in TE (prepared fresh).
11. 10 mg/mL glycogen (in TE).

2.2 Culture Growth and Cross-Linking

1. 37 % formaldehyde solution (use freshly opened bottles).
2. 2.5 M glycine (make fresh in ddH_2O).
3. Ice-cold TBS.
4. Liquid nitrogen.

2.3 Cell Lysis and Immunoprecipitation

1. Ice-cold lysis buffer.
2. Complete protease Inhibitor cocktail EDTA-free.
3. 0.5 mm glass beads.
4. Clamped horizontal shaking vortex adaptor.

5. 70 % ethanol.

6. 18-G needles.

7. 26-G needles.

8. Diagenode Bioruptor™ (preferred) or Microtip sonicator (alternative).

9. TE/1 % SDS.

10. 5 µg of affinity-purified polyclonal antibody or 2–10 µg of monoclonal antibody.

11. 50 % slurry of protein A or protein G Sepharose beads.

12. TBS.

2.4 Recovery of Immunoprecipitated DNA

1. 18-G needles.

2. Lysis buffer.

3. Lysis buffer with 500 mM NaCl.

4. Wash buffer.

5. TE.

6. Elution buffer.

7. TE/0.67 % SDS.

2.5 Cross-Link Reversal and DNA Cleanup

1. Proteinase K mix: 238 µL TE, 1 µL 10 mg/mL glycogen, 10 µL 10 mg/mL proteinase K (per sample).

2. TE.

3. 5 mg/mL glycogen.

4. 10 mg/mL proteinase K.

5. 4 M LiCl.

6. Phenol–chloroform–isoamyl alcohol (25:24:1), pH 8.0.

7. Ice-cold 100 % ethanol.

8. Ice-cold 70 % ethanol.

9. TE with 100 µg/mL RNaseA.

2.6 Strand Displacement Amplification

1. ddH$_2$O.

2. 2.5× SDA buffer: 125 mM Tris–HCl (pH 7.0), 12.5 mL MgCl$_2$, 25 mM βME, 750 µg/mL random DNA nonamers (dN9) (make fresh or store aliquots without βME at −20 °C and add βME immediately prior to use).

3. dNTP mix (1.25 mM each nucleotide).

4. 50 U/µL exo-Klenow.

5. 0.5 M EDTA.

6. DNA Clean and Concentrator™ Columns (Zymo Research).

7. DNA binding buffer (Zymo Research).

8. DNA wash buffer (Zymo Research).

9. 10× aminoallyl-dNTP stock solution (12.5 mM dATP, 12.5 dCTP, 12.5 mM dGTP, 5 mM dTTP, 7.5 mM aa-dUTP).

2.7 Dye Coupling

1. ddH_2O.

2. Fresh 1 M sodium bicarbonate, pH 9.0.

3. Cy3 and Cy5 monoreactive dye (Amersham).

4. DMSO.

5. DNA binding buffer (Zymo Research).

6. DNA wash buffer (Zymo Research).

2.8 ChIP-Chip Hybridization

1. ddH_2O.

2. 1 mg/mL Human Cot-1 DNA (Invitrogen).

3. 10× CGH/CoC blocking agent (Agilent).

4. 2× Hi-RPM hybridization buffer (Agilent).

5. Oligo aCGH/ChIP wash buffer 1 (Agilent).

6. Oligo aCGH/ChIP wash buffer 2 (Agilent).

7. Acetonitrile.

8. Drying and stabilization solution (Agilent).

3 Methods

3.1 Culture Growth and Cross-Linking

1. Grow 200–400 mL of planktonic cells to an OD_{600} of 0.4 (*see* **Note 2**).

2. Add a final concentration of 1 % fresh formaldehyde (stock is at 37 %) and cross-link for 15 min at room temperature on a platform shaker.

3. Quench cross-linking with freshly made 2.5 M glycine to a final concentration of 125 mM, and incubate for 5 min at room temperature on a platform shaker.

4. Collect cells by centrifugation for 10 min at $1,000 \times g$ in a fixed angle centrifuge rotor.

5. Decant and resuspend pellets in 10 mL ice-cold TBS.

6. Transfer cell suspension to 15 mL Falcon tubes, pellet, decant and repeat the wash once more.

7. Resuspend pellet in 2 mL ice-cold TBS, and separate cell suspension to two 2 mL Sarstaedt tubes (for 400 mL cell volume) (*see* **Note 3**).

8. Pellet and decant before proceeding to lysis step, or freeze the decanted pellets in liquid nitrogen and store at –80 °C.

3.2 Cell Lysis and Immunoprecipitation

1. Thaw cell pellets on ice, and resuspend in 700 μL ice-cold lysis buffer containing complete protease inhibitor cocktail (EDTA-free) (Roche).

2. Transfer cell suspension to a clean 1.75 mL microfuge tube preloaded to the 500 μL mark with 0.5 mm glass beads.

3. Place on a clamped horizontal tube adaptor on a vortex mixer at 4 °C, and lyse cells for 30 min to 2 h (*see* **Note 4**).

4. Observe cell lysis under a microscope; if more than 90 % of the cells are lysed continue to next step (*see* **Note 5**).

5. Recover the lysate by inverting the microfuge tubes containing the lysate/bead mixture, wipe the bottom of the tube with 70 % ethanol, allow the tube to dry, and then pierce the bottom of the tube with a 26-G needle. Open the microfuge tube, place it into a 5 mL falcon tube (right side up), and pierce the falcon tube (above the level of the bottom of the microfuge tube) using an 18-G needle attached to a vacuum line. This will cause the lysate to flow through to the bottom of the falcon tube (*see* **Note 6**). Recover the lysate, and transfer 300 μL to each of two new 1.75 mL microfuge tubes for Bioruptor shearing (*see* **Note 7**).

6. Shear chromatin by sonication in a Diagenode Bioruptor™ with the following settings: 15 min, high setting, 30 s on, 1 min off (*see* **Note 8**).

7. Pellet the cell debris for 5 min at full speed at 4 °C and transfer the supernatant containing the lysate to a new tube.

8. Remove 50 μL of the lysate and transfer to a new tube containing 200 μL TE/1 % SDS. This is the "input DNA" sample; store at −20 °C until it will be processed along with the immunoprecipitated DNA.

9. Aliquot and dilute the sheared lysate according to the number of IPs that will be performed. Use 50–500 μL of crude lysate in 500 μL (final volume) lysis buffer (with fresh protease inhibitors) for each IP.

10. Add appropriate antibody (e.g., 5 μg of custom affinity-purified polyclonal antibody, 2 μg of monoclonal anti-c-myc antibody, or 10 μg of anti-FLAG M2 monoclonal antibody), and incubate at 4 °C overnight on a nutator agitator.

11. The following day, wash bead slurry as follows. For 20 reactions, you will need ~1 mL of a 50 % slurry of appropriate type of Sepharose beads (protein A or protein G; ~500 μL bed volume). Use wide-bore P1000 tips for bead dispensing. Spin 4000×*g*, 10 s (double bed volume for washes since you lose ~30–40 % beads during washes). Wash two times with TBS and three times with lysis buffer. Resuspend in lysis buffer to 50 % slurry.

12. Add 50 µL of washed 50 % slurry of protein A or protein G Sepharose beads to each immunoprecipitation sample, using wide-bore P200 tips to dispense the bead slurry. Incubate at 4 °C for 2 h on a nutator agitator.

3.3 Recovery of Immunoprecipitated DNA (See Note 9)

1. Wash beads as follows (see **Note 10**): Pellet for 1 min at $1000 \times g$ at room temperature.

2. Draw off the supernatant with an 18-G needle on a vacuum line.

3. Wash with buffers indicated below for 5 min each while mixing on a nutator:

   – Two times with 1 mL lysis buffer.
   – Two times with 1 mL lysis buffer with 500 mM final NaCl.
   – Two times with 1 mL wash buffer.
   – One time with 1 mL TE.

4. After final wash, draw off TE and add 110 µL elution buffer, vortex on gentle setting, and incubate for 10 min at 65 °C, mixing every 2 min by gentle vortex.

5. Pellet for 30 s at full speed at room temperature, and remove 100 µL of supernatant to a new tube.

6. Add 150 µL of TE with 0.67 % SDS to remaining bead bed and vortex vigorously. Pellet by spinning full speed, 1 min, remove 150 µL of supernatant, and pool together with previous eluate (250 µL total).

3.4 Cross-Link Reversal and DNA Cleanup

1. Incubate the IP samples and "input DNA" samples (from previous steps) at 65 °C for 16 h.

2. Add 250 µL of proteinase K mix (for each sample: 238 µL TE, 1 µL 10 mg/mL glycogen, 10 µL 10 mg/mL proteinase K). Incubate at 37 °C for 2 h (see **Note 11**).

3. Add 55 µL of 4 M LiCl and 555 µL of cold phenol–chloroform–isoamyl alcohol (25:24:1), pH 8.0. Mix by vortexing briefly, spin at $10,000 \times g$ for 2 min, and remove 500 µL of the top aqueous layer to a fresh tube.

4. Add 1 mL of ice-cold 100 % ethanol to the collected aqueous layer and incubate at –20 °C overnight or at –80 °C for at least 1 h.

5. Centrifuge at $10,000 \times g$ at 4 °C for 30 min. Decant carefully using a 1 mL pipette.

6. Wash the pellet with 950 µL ice-cold 70 % ethanol (wash by gentle inversion), spin for 10 min at $10,000 \times g$ at 4 °C, decant, spin briefly, and remove any residual ethanol.

7. Air-dry the pellets and resuspend as follows: use 25 µL of TE for the IP samples and 100 µL of TE with 100 µg/mL RNaseA for the "input DNA" samples (*see* **Note 12**).

8. Incubate the input samples at 37 °C for 1 h to degrade RNA, and the IPs at 37 °C for 10 min to ensure pellet is dissolved, and store all at –20 °C (*see* **Note 13**).

3.5 Strand Displacement Amplification of Chromatin IP Samples (See Note 14)

The following amplification protocol uses high concentration exo-Klenow with random DNA nonamers (dN9s) in order to perform strand displacement amplifications of the IP and input DNA samples for ChIP experiments. Prior to amplification, input and IP DNA concentrations are normalized by dilution of the input DNA for each corresponding IP based on qPCR values for a non-enriched locus (e.g., *ADE2* locus). Input and IP samples are amplified separately, in parallel, and should yield similar amounts of product after each round of amplification (typically three rounds). Round B amplification can be omitted if the IP DNA concentration is sufficient.

3.5.1 Round A Primary Amplification

1. Mix 12 µL of IP sample or 12 µL of diluted input (diluted in TE) (*see* **Note 15**), 12 µL of ddH$_2$O, and 20 µL of 2.5× SDA buffer.

2. Incubate at 95 °C for 5 min and then immediately transfer samples to an ice water bath for 5 min.

3. Add 5 µL of dNTP mix (1.25 mM each nucleotide).

4. Add 1 µL of 50 U/µL exo-Klenow, and mix by pipetting.

5. Incubate at 37 °C for 2 h with a heated lid thermal cycler (*see* **Note 16**).

6. Purify product using Zymo[25] columns by adding 10 volumes of binding buffer (450–50 µL sample), bind to column, wash 1× with 200 µL binding buffer, 2× with 200 µL wash buffer, spin for 1 min at 10,000×g to dry, and elute with 30 µL H$_2$O into a new tube.

7. Check 1.5 µL of the sample on a NanoDrop spectrophotometer. If the total yield is ≥400 ng, then skip to Round C. If not, continue to Round B.

3.5.2 Round B Secondary Amplification

1. Mix 24 µL of Round A DNA, 20 µL of 2.5× SDA buffer.

2. Repeat **steps 2–7** of Round A Primary Amplification, but elute with 50 µL of H$_2$O after purifying product.

3.5.3 Round C Aminoallyl-dUTP Incorporation and Final Amplification (See Note 17)

For this final amplification, perform 100 µL reactions with 1–2 µg total Round B DNA for each sample. This will yield ~2.5- to 3-fold amplification.

1. Mix 1–2 µg of Round B DNA in H$_2$O to 48 µL total volume, then add 40 µL of 2.5× SDA.

2. Incubate for 5 min at 95 °C and immediately transfer the samples to an ice water bath for 5 min.

3. Add 10 μL of 1.25 mM aminoallyl-dNTP mix (1:10 dilution of stock solution).

4. Add 2 μL of 50 U/μL exo-Klenow.

5. Incubate at 37 °C for 2 h with heated lid in a thermal cycler (*see* **Note 16**).

6. Purify the Round C product using Zymo[25] columns by adding 10 volumes of binding buffer (900–100 μL sample), bind to column, wash 1× with 200 μL binding buffer, 2× with 200 μL wash buffer, spin for 1 min at 10,000×g to dry, and elute with 50 μL of H_2O into a new tube.

7. Check 1.5 μL of the sample on a NanoDrop; the yield should be about 5 μg of total DNA per reaction.

3.6   Dye Coupling

1. Speed-vac the amplified input and IP reactions from Round C to ≤9 μL volume, or until dry.

2. Resuspend with H_2O to 9 μL final volume and add 1 μL of fresh 1 M sodium bicarbonate, pH of 9.0 (*see* **Note 18**).

3. Add 1.25 μL of Cy3 for the input sample or Cy5 for the IP sample (*see* **Note 19**).

4. Incubate the labeling reactions at room temperature for 1 h in the dark.

5. Purify the dye-coupled DNA with Zymo[25] columns by adding 800 μL of Zymo DNA binding buffer to each of the samples and load onto a Zymo column. Wash once with 200 μL binding buffer, wash twice with 200 μL wash buffer, and spin at 10,000×g for 1 min to dry. Elute using 50 μL of H_2O into a new tube. Check concentration on a NanoDrop spectrophotometer using the "microarray" setting to quantitate the total yield and dye-coupling efficiency. A minimum of 2–4 pmol/μL Cy3 or Cy5 is ideal.

6. Equalize the input and IP samples to 5 μg each in order to hybridize to a 1× 244 K format Agilent microarray (*see* **Note 20**).

3.7   ChIP-Chip Hybridization (Adapted from the Agilent Oligo aCGH/ChIP-On-Chip Hybridization Kit)

1. Mix 5 μg each of the input and IP samples, and bring volume to 150 μL in ddH₂O (*see* **Note 21**).

2. Add 50 μL of 1 mg/mL Human Cot-1 DNA.

3. Add 50 μL of 10× CGH/CoC blocking agent.

4. Add 250 μL of 2× Hi-RPM hybridization buffer.

5. Mix and quick spin to collect sample.

6. Incubate at 95 °C for 3 min and transfer to 37 °C for 30 min.

7. Spin at full speed in a microcentrifuge for 1 min, carefully remove 490 μL, and load the sample onto a gasket slide. Cover with the array slide and assemble the hybridization chamber.

 – *See* demo video at: http://agilent.cnpg.com/video/flatfiles/189/

8. Hybridize at 65 °C for 40 h in an Agilent microarray hybridization oven at a rotation speed of 20 rpm.

9. Disassemble the array and wash using Agilent wash buffers with mixing (using a magnetic stir bar) as follows (*see* **Note 22**):

 – Agilent oligo aCGH/ChIP-on-Chip wash buffer 1 for 5 min at 25 °C.

 – Agilent oligo aCGH/ChIP-on-Chip wash buffer 2 for 5 min at 32 °C.

 – Acetonitrile for 1 min at 25 °C.

 – Agilent drying and stabilization solution (*see* **Note 23**) for 30 s at 25 °C (*see* **Note 24**).

10. Scan slides in GenePix 4000B scanner, and grid array.

3.8 Preliminary Data Analysis (See Note 25)

The data analysis sections described below provide an overview of many of the analyses we have commonly performed for sequence-specific DNA binding proteins. Not all of the analyses described are relevant to every situation and our analyses may not include analyses relevant to your situation. These analyses may also be performed with data generated from ChIP-seq datasets.

1. For all of the downstream ChIP-chip data analysis described below, we use MochiView [11], however there are alternative programs available. MochiView is freely available at http://johnsonlab.ucsf.edu/mochi.html (*see* **Note 26**). An example plot of ChIP-chip data visualized using MochiView is shown in Fig. 2.

2. Normalize the enrichment values for every probe on the array by LOWESS normalization using Agilent Chip Analytics software or other normalization software, and import the data into MochiView.

3. Raw ChIP-chip data should be uploaded to a publicly available database, such the Gene Expression Omnibus (www.ncbi.nlm.nih.gov/geo), and an accession number should be obtained (*see* **Note 27**).

3.9 Peak Calling

1. Identify binding events by smoothing the data using MochiView. This utility applies a smoothing function to the Chip Analytics $\log_2$ enrichment values, followed by the application of a peak detection algorithm, where all binding peaks are assigned a *P* value using permutation testing.

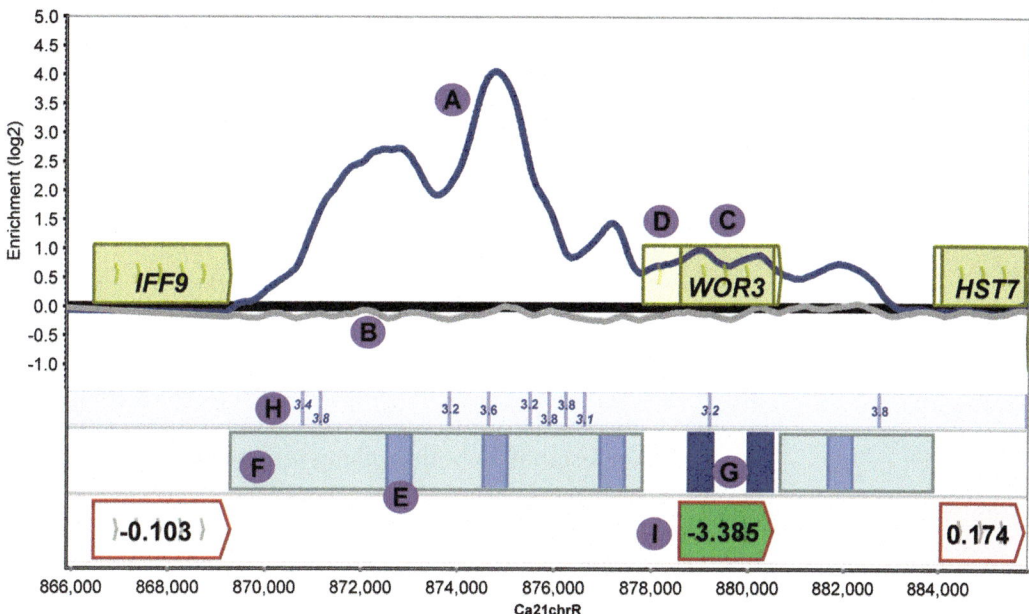

Fig. 2 Sample of a MochiView screenshot illustrating concepts relating to the analysis of ChIP-chip data. Data in this figure is from a ChIP-chip experiment of *C. albicans* Efg1, a regulator of white-opaque switching, in the opaque cell type. ChIP-chip data for Efg1 (*blue, A*) and an Efg1 delete control (*grey, B*) are shown. Open reading frames are represented by *yellow boxes* (*C*), *lighter yellow* represents the untranslated region (*D*), genes above the *bold line* are transcribed in the sense direction and genes below the *bold line* are transcribed in the anti-sense direction. The *x*-axis represents ORF chromosomal locations and the *y*-axis represents the ChIP-chip enrichment value (log₂). A lower track illustrates 500 bp peaks of Efg1 binding (*blue boxes, E*) and intergenic regions bound by Efg1 (*light blue boxes, F*) as well as 500 bp Efg1 peaks that do not fall in an intergenic region and would normally be excluded from further analysis (*dark blue boxes, G*). High-scoring instances of the Efg1 DNA-binding motif (maximum possible score 4.17) is indicated (*H*). The plot also contains data from a microarray analysis of an opaque *efg1* deletion strain versus wildtype opaque cells; values are on a log₂ scale (*I*) with downregulated genes in green and upregulated genes in *red*, color intensity represents differences from wild type (with darker colors having greater differences). The data for Chip-chip, Efg1 motif, and Efg1 delete microarray analysis was taken from Hernday et al. [9]. Plots were constructed using MochiView 1.46, *see* Hornman et al. [11]

2. Use peak-finding significance thresholds at the default settings, $P \leq 0.001$ for the experimental IPs, and $P \leq 0.05$ for the control IPs (*see* **Note 28**).

3. User-defined cut-offs for the minimum value for peak inclusion post-smoothing should be determined using the distribution of log-ratios for each experimental IP, and should be set at two standard deviations from the mean of log₂-transformed fold enrichments. User-defined cut-offs for the minimum value for peak inclusion post-smoothing ranging from 0.27 to 0.36 (1.5 standard deviations from the mean of log₂-transformed fold enrichments) should be used for the untagged or delete IP control data sets (*see* **Note 29**).

3.10 Peak Curation and Assignment (See Note 30)

1. Eliminate any peaks that overlap with a "red-flagged" location set (*see* **Note 27**).

2. Map the cleaned list of peaks to intergenic regions and to specific genes using defined criteria, such as those described in the following steps (*see* **Note 31**).

3. If a peak falls entirely over an intergenic region map it to that intergenic region.

4. If a peak partially overlaps with both an intergenic region and an open reading frame (ORF), assign it to that intergenic region.

5. If a peak is positioned over a short open reading frame such that it also overlaps with the intergenic regions on each side of the ORF, assign it to both flanking intergenic regions.

6. If a peak is positioned entirely over an open reading frame such that it does not overlap any intergenic region, omit it from further analysis.

7. If multiple peaks fall within a single intergenic region, assign the maximum enrichment value of the individual peaks to that intergenic region (and any associated genes) for the purpose of further analysis.

8. Consider any gene whose start codon is located immediately downstream of a given intergenic region (that is whose start codon is on the same side of the ORF as a given intergenic region rather than on the opposite end of the ORF) as potentially regulated if that intergenic region is bound by the factor in question. As such, a given intergenic region may be associated with two (in the case of divergent ORFs), one (in the typical case), or zero (in the case of convergent ORFs) genes.

9. In the case of divergent ORFs (two genes controlled by a given intergenic region), a distinction is not made between the regulation of the two genes based on the distance from the binding site/s. Both genes are considered to be potential targets of the binding site.

3.11 Binding Site and Regulon Analysis

1. Once lists of the intergenic regions and genes that are potential targets for the regulator of interest are obtained, begin the following analysis (*see* **Note 32**).

2. If transcription data is available for deletion or overexpression of the regulator in question or for conditions related to the regulator, determine if the target genes are preferentially regulated in a regulator-dependent manner relative to the genome as a whole.

3. Perform Gene Ontology (GO term) mapping or other similar analysis on the target set of genes in order to determine if there are enriched classes of genes in the regulon.

4. Determine whether the binding events occur at a constant distance from a common feature, such as a start codon, transcription start site, or other known regulator target sites.

5. Use multiple motif search algorithms, such as SCOPE [12], Bioprospector [13], and MEME [14], to identify DNA sequences whose occurrence correlates with the binding sites (*see* **Note 33**).

6. Having identified a list of potential DNA-binding motifs, perform the following analyses to evaluate their predictive power (*see* **Note 34**).

7. Create a control location set randomly selected from unbound intergenic regions. The average size of members of this control location set should be the same as the average size of the experimental peaks. If whole intergenic regions are used, the control set should contain unbound intergenic regions whose average size is equivalent to the experimental set (*see* **Note 35**).

8. Once the control set has been created, determine how many sites in the experimental and control location sets have one or more instances of the motif for a series of stringency criteria. These can then be plotted on a graph, such as an ROC (Receiver Operating Characteristic) plot, with true positives (percent of experimental locations passing) and false positives (percent of control locations passing) on the two axes.

9. Search for DNA-binding motifs using only a fraction of the binding sites (i.e., the top 50 % of peaks or a randomly selected subset of the peaks) and subject the resulting motifs to **steps 7** and **8** of Subheading 3.11 (*see* **Note 36**).

10. Once you have identified one or more motifs with good predictive value, ideally for full as well as partial datasets, proceed to the following analyses.

11. If the factor in question has a published DNA-binding motif or one or more homologs with a published motif, compare your motif to the existing motifs. Comparisons between two or more motifs can be performed using programs like MochiView and motifs can be compared against libraries of motifs using online databases, such as YeastTract (http://www.yeastract. com) [15], MOTIF (http://www.genome.jp/tools/motif/), and TOMTOM (http://meme.nbcr.net/meme/cgi-bin/ tomtom.cgi) [16]. Using these same tools, determine whether part or your entire motif matches that of a known regulator (*see* **Note 37**).

12. Determine whether the motif/s occurs at a constant distance from a common feature, such as a start codon, a transcription start site, or the motif for another transcription factor (*see* **Note 38**).

13. Assess the individual occurrences of your motif at your binding sites. Determine whether your motif/s can be subdivided into distinct sub-motifs or whether any pairs of your motifs consistently occur with fixed spacing between them.

3.12 Analysis of Multiple Datasets (See Note 39)

1. If possible, produce equivalently sized and formatted location sets for each of the datasets to be compared (*see* **Note 40**). There are three general approaches for comparing the overlap of binding datasets.

2. Compare the lists of genes downstream of the binding sites for each factor (*see* **Note 41**).

3. Compare the overlap of the intergenic regions bound by the various factors rather than the genes (*see* **Note 42**).

4. Compare the overlap of the binding peaks (*see* **Note 43**).

5. For whichever comparison/s is made, subdivide the overlap into all the various possible combinations (i.e., "A and B" and "A and B and C"), as well as broader categories (i.e., "bound by *n* or more factors"). Perform the following analysis on the different subgroups (*see* **Note 44**).

6. Perform Gene Ontology (GO) term analysis for the genes regulated by specific categories of regulator binding events.

7. If transcriptional profiling is available for deletions or overexpressions of various regulators or for conditions related to the regulators, determine if the genes regulated by specific categories of regulators are preferentially regulated (*see* **Note 45**).

8. If available, determine whether the DNA-binding motifs for the factors in question are better at predicting certain types of binding events.

9. Determine whether certain types of binding events correspond with higher or lower levels of binding enrichment for the factors in question.

10. Determine whether the overlap of binding events for any combination of factors occurs more frequently than would be expected by chance. There are several possible metrics (described below) for making such a calculation and we recommend using all of them to account for possible biases. These metrics all work on a similar principle, comparing the number of observed binding events to the number of predicted events for a given combination (*see* **Note 46**).

11. Make a comparison based on the fraction of intergenic regions bound by each regulator, treating all intergenic regions as equally probable targets for regulatory binding. For each regulator determine the fraction of possible intergenic regions bound; divide the number of intergenic regions the regulator binds by the total number of intergenic regions in the genome. Proceed to **step 14** of Subheading 3.12.

12. Make a comparison based on the fraction of intergenic regions bound by each regulator, correcting for the difference in intergenic length. To do this, determine a length correction factor for each regulator by dividing the mean length of the intergenic regions bound by each regulator by the mean length of all intergenic regions in the genome. Multiply the fraction of possible intergenic regions bound by this length correction factor, before proceeding with the calculations in step 14 of Subheading 3.12.

13. Make a comparison that accounts for the fact that short intergenic regions provide a relatively smaller evolutionary target, and thus are less likely by chance than longer intergenic regions to acquire active binding motifs. For each intergenic region, calculate the region's weighted length as the product of the region's length and the inverse of the mean length of all intergenic regions. For each regulator, calculate a length-corrected fraction of bound intergenic regions by taking the sum of weighted lengths for all intergenic regions bound by that regulator and dividing by the sum of weighted lengths for all intergenic regions in the genome. Use this corrected value for the calculations in step 14 of Subheading 3.12.

14. Determine the fraction of regulators predicted to be bound by any given combination of regulators by multiplying the fraction of possible intergenic regions for the relevant regulators. Substitute in the relevant length corrected factors from **step 12** or **13** of Subheading 3.12 if performing either of those analyses. Multiply this predicted fraction of binding sites with a particular combination by the total number of intergenic regions in the genome in order to get the predicted number of regulators bound by a given combination of regulators. Compare the observed number of binding events to the predicted number of events.

15. If data is available for the binding of a homolog of your target/s in another species, compare the targets of a given factor/s between those species. Mapping gene orthologs between species is beyond the scope of this chapter, however we recommend mapping orthologs in both directions and using multiple orthology lists if possible (*see* **Note 47**).

4   Notes

1. Use autoclaved ddH_2O and baked glassware when making buffers to avoid DNA contamination. This is especially important for the final wash buffers and post-elution steps.

2. 200 mL of planktonic cells at an OD_{600} of 0.4 is sufficient for a batch of lysate which equates to about 10 individual ChIPs.

For biofilm cells, use 1×6-well plate/strain (4 mL/well; 24 mL total volume).

3. For 200 mL planktonic cell volume and for 24 mL biofilm cell volume, resuspend in 1 mL ice-cold TBS and store in one 2 mL Sarstaedt tube.

4. Lysis times may vary depending upon the cell type and growth conditions.

5. Cells should appear as a mixture of dead cell "ghosts" and fragmented cell debris by phase contrast microscopy.

6. As an alternate method, recover the lysate by centrifugation into a larger tube.

7. Alternatively, if using a microtip sonicator for shearing, transfer the entire lysate to one new 1.75 mL microfuge tube.

8. If using a microtip sonicator, use the following settings: 5×20 s at level 2, 100 % duty cycle, with 1 min on ice between each pulse. In our experience, using the Bioruptor to shear results in yields of smaller fragment sizes, tighter shear distribution, and greater consistency than the microtip sonication method.

9. For optimal recovery of immunoprecipitated DNA during the washes, keep samples at room temperature and buffers ice cold.

10. Wash buffer temperatures, incubation temperatures, and incubation times can be optimized for each antibody; however, we have found that ice-cold buffers and 5 min incubations at room temperature work best for most antibodies.

11. Make a fresh proteinase K solution from lyophilized powder.

12. RNase A stock is prepared at 10 mg/mL in ddH_2O. The RNase A stock solution should be boiled for 10 min before adding to TE to remove DNase activity.

13. An optional DNA clean up step could be performed on the "input DNA" using a commercial DNA clean up kit, however this adds an additional variable (relative to the IP DNA), and could produce spikes in the ChIP-chip data. Therefore, it is recommended to leave the RNase A in the "input DNA" sample and avoid the cleanup steps prior to amplification. We also recommend monitoring the sheer distribution of the input DNA sample prior to proceeding with subsequent analysis of ChIP samples. Test the sheer distribution by running ~200–500 ng of purified input DNA on a 2 % agarose gel at ~5 V/cm. Average sheer size from the Bioruptor is typically ~200 bp, with most fragments distributed between 100 and 400 bp.

14. This is a nonspecific amplification, and any contaminating DNA will be amplified. Therefore, perform all amplification rounds with gloves, filter tips, autoclaved ddH_2O, and reagents free of any potential DNA contamination.

15. Equalize the input and IP samples based on the qPCR values for a non-enriched locus (e.g., the *ADE2* locus).

16. The reactions may sit up to ~2 h at 10 °C following amplification or add 5 μL 0.5 M EDTA at pH 8.0 and store at −20 °C.

17. If Round B yields less than 1 μg total DNA, follow this alternate Round C approach:

    Set up two 100 μL Round C reactions for each sample, using 200–400 ng of Round B DNA per tube. Perform amplification and cleanup as described in the standard Round C approach, but pool the two independent reactions prior to Zymo[25] column purification.

18. Prepare the sodium bicarbonate fresh in ddH$_2$O on the day of labeling and pH using a pH meter.

19. We use Amersham monoreactive dye packs (Cat. # PA 23001 and PA25001). Each tube contains enough dye for eight labeling reactions. Resuspend the dye in 10 μL DMSO and use 1.25 μL of dye per labeling reaction. If fewer than eight labeling reactions are to be performed, any unused dye can be desiccated and stored at 4 °C in the dark.

20. Agilent custom oligonucleotide arrays, hybridization buffers, and wash buffers consistently yield high-quality data.

21. Although 5 μg is optimal, a minimum of 1 μg each of input and IP samples is sufficient for hybridization without any significant decrease in data quality. Be sure that equal amounts of the input and IP sample are used for hybridization.

22. For disassembly, hold the microarray/gasket slide submerged in wash buffer 1 while gently gripping sides of the microarray slide. Gently pry the gasket slide off of the array by inserting the tip of a plastic forceps between the outer edge of the two slides and lightly twist the forceps. The gasket slide will fall away, while the array should remain in your hands. Be sure to avoid any contact with the printed array surface.

23. If Agilent drying and stabilization solution contains precipitate, place bottle in a 37 °C water bath prior to use. Depending on the amount of precipitate, it may be necessary to dissolve the precipitate overnight in the water bath; cool to RT before use.

24. In order to ensure even drying of the array, remove the slide holder from the Agilent drying and stabilization slowly; minimize water droplets on the arrays.

25. At least two independent biological replicates for each strain should be used for data analysis.

26. Some alternatives to MochiView include UCSC's Genome Browser, freely available at http://genome.ucsc.edu/ [17] and CisGenome, freely available at http://www.biostat.jhsph.edu/~hji/cisgenome/ [18].

27. In addition to submitting datasets to a curated database, several additional steps will help readers to make use of your data. Downloading and interpreting entire datasets is a barrier to many readers, so we suggest including certain supplemental files with any manuscript containing genome-wide ChIP data. If you have a list of "red flagged" locations used to remove spurious peaks, a sheet with said locations should be provided. To allow for quick examination of the data, we include an Excel file with a list of genes and the maximum enrichment value for any peak upstream of that gene. Ideally, this list should include all genes but at a minimum it should contain all genes with upstream binding sites. There should be columns for each genome-wide ChIP dataset as well as other datasets being used. *See* supplementary files in Hernday et al. [9] and Nobile et al. [2] for examples of such files. Also important is an Excel or text file with peak locations, enrichment values, and the adjacent genes for each peak can also be of use. *See* supplementary files from Cain et al. [4], Nobile et al. [2], and Hernday et al. [9] for examples. In addition, a file with plots centered on each peak overlaid with the experimental and control data traces are useful. We generate these plots in MochiView, *see* supplementary files from Hernday et al. [9] and Nobile et al. [2] for examples. Include also tables for any higher order analysis such as binding site overlap comparisons. If any DNA binding motifs are reported, include an Excel or text file with the data used to make the Position Specific Weight Matrix as such information cannot readily be determined based on figures alone. *See* supplementary files from Hernday et al. [9] for an example of such a file. Describe the predictive value of all motifs, ideally by the relevant ROC plots. *See* supplemental Fig. 1 from Cain et al. [4] for an example of such a figure. The methods section of the paper reporting the ChIP data should include information for any antibodies used, such as sequences polyclonal antibodies were raised against or suppliers of commercially available monoclonal antibodies. Methods sections should also explain the criteria used for peak calling as well as the details of any analyses performed.

28. Peak-finding significance thresholds may be adjusted accordingly to assess the quality of the data. For greater confidence, the amount of sampling can be increased tenfold from the default setting to 100,000 (number of random samples to compare against each peak), and 100 (maximum number of random samples passing for inclusion of peak).

29. Adjustments to peak inclusion cutoffs should not alter the majority of called peaks in a good dataset.

30. The peak curation steps described stem in part from our observation of a number of likely artifactual peaks near tRNA and

over ribosomal genes, based on the fact that these loci showed variable but substantial enrichment in the majority of deletion control ChIP-chip experiments that were performed with antibodies against a deleted target [9, 19]. Recent reports have indicated that ChIP-chip data often results in spurious areas of enrichment at or near highly transcribed regions, such as tRNA [20]. Such spurious peaks often escape culling during the peak calling process depending on the nature of a given control; as such we take the precaution of automatically eliminating anything overlapping our "red flagged" dataset. That said, entirely different sets of assumptions may apply with other types of proteins, such as histone deacetylases, where certain categories of peaks, including peaks covering ORFs, may be physiologically relevant [21]. Peaks near tRNA or over ribosomal genes are not spurious a priori, so it may not be possible to eliminate them unless they have appeared in multiple control datasets. Care should be taken when dealing with such peaks and in making any conclusions based on their presence or absence.

31. We perform the peak curation and assignment analysis using MochiView but other software packages could be used. We use the criteria described in the methods section for assigning peaks to intergenic regions and genes, but other criteria may be more appropriate in a given situation. Normally we ignore small features such as tRNAs or spurious ORFs when making gene proximity assignments. Depending on the specific circumstances, different assumptions may be appropriate. Whatever criteria are chosen, it is important to apply them consistently across the dataset and to list them when reporting the dataset.

32. In general, the results and significance of a given analysis will improve as the number of locations (or genes) being used increases and the average size of the locations being used decreases. Although searches can be performed using entire intergenic regions, it is preferable to use discrete peaks and to have these peaks be as narrowly defined as possible.

33. When dealing with online servers for DNA-binding motif identification, it is important to remember that the different servers have their own strengths and weaknesses (for review, *see* [22]) and that the server used as well as the settings chosen for a given server may limit the type of motifs that can be found. The use of multiple tools or servers as well as different settings within a server may provide a better motif, while also allowing for the identification of a broader range of motif types.

34. It is important to remember that the top motif hit (or even top several hits) may not necessarily be the best at explaining the location set. Rather, the output of these tools should be seen

as a series of candidates that need to be further examined in detail. Likewise, the statistical value (i.e., *P* or *E* value) for a given putative motif does not in and of itself indicate that the motif is valid. Depending on the overall makeup of the genome (i.e., AT or GC rich), it is possible to get spurious but statistically significant motifs that consist of repeats of a given nucleotide or a repeated sequence of 2–3 nucleotides that occur throughout the genome. For example, spurious motif hits that are rich in "A" and "T" nucleotides are not uncommon given that promoters generally tend to be "AT" rich. Thus, as a general rule, the motifs produced should be considered as a set of candidates for further analyses.

35. There are several important considerations to take into account in order to avoid unintentionally skewing of the results or potentially misinterpreting data. For almost every analysis or calculation, it is important to consider how the specific intergenic length in question compares to the average intergenic length of the genome as a whole. It is critical to make sure that any control location sets are equivalently sized to the relevant experimental set as most analyses will tend to favor the longer location set if the two sets are not equivalently sized.

36. Using only the full location set for motif analysis contains a level of bias, even when coupled with the follow-up tests we describe, as it is evaluating a motif against the dataset used to create the motif. To reduce this bias, we recommend performing parallel analyses on motifs developed using only portions of the binding sets (i.e., the full target list, the top half of the target list based on enrichment values, and the top third or quarter of the list based on enrichment values). An example of the multiple rounds of motif analysis described in the methods section can be found in Cain et al. [4]. We consider this analysis to be critical in the motif identification process given the frequency with which our searches returned spurious motifs as the best hit for a given location set.

37. Although changes in preferred DNA-binding sites for homologs of a given regulator have been reported [23], it is more common that they remain similar or even identical [3]. A match between your motif and a previously reported one for a homolog of your target is another piece of evidence in support of the motif you have developed. If there is a large difference between your motif and a previously reported motif, it may be useful to verify the new motif through in vivo or in vitro experiments.

38. If your target has one or more cofactors, motif searches may return the motifs for the cofactors in addition to or in lieu of your target's motif. Depending on the motif search settings and the spacing of the DNA-binding sequences, these may

occur together in one longer motif or as two independent motifs. As such, you should look to see whether your motif/s can be subdivided into multiple motifs (one of which might match a previously known motif) or whether any pair of your motifs occurs with a constant spacing between them. If the analysis suggests that your target does have a cofactor, it can be informative to rerun motif finding analysis on the subsets of sites with or without the cofactor in order to improve the relevant motifs in each case and to determine if the DNA-binding preferences of your target differ between the two cases [9, 24].

39. Depending on what datasets are available in your case, a wide range of analyses can potentially be performed. When dealing with more complex regulons, you may have produced or have access to previously reported binding data for a number of other factors. If such datasets exist, it can be informative to examine the degree of overlap between the different sets of binding sites. Although we have traditionally performed this analysis in MochiView, it can be performed with a wide variety of available software.

40. As with motif identification, it is important to avoid unintentionally skewing of the results or potentially misinterpreting data. As a rule of thumb, more importance should be attached to the presence of a peak rather than the absence of a peak. For almost every analysis or calculation, it is important to consider how the specific intergenic length in question compares to the average intergenic length of the genome as a whole. It is critical to make sure that any control location sets are equivalently sized to the relevant experimental set as most analyses will tend to favor the longer location set if the two sets are not equivalently sized. It is also often fruitful to perform each analysis in parallel with different target lists (i.e., the full target list, the top half of the target list based on enrichment values, and the top third or quarter of the list based on enrichment values). Likewise, it may be worthwhile to evaluate whether various metrics or analyses are linked in some way to binding enrichment values.

41. The target gene based overlap analysis, which can be quickly conducted in Microsoft Excel, will provide a basic understanding of the overlap (if any) between the factors. Although quick, it is of limited use compared to some of the other analyses we describe.

42. The intergenic regions overlap analysis is conceptually similar to the gene based analysis, but remains useful for many of the follow-up analyses. The most involved version of this analysis looks for overlap in binding peaks themselves. Since intergenic regions will sometimes contain multiple binding peaks for a given factor and since different factors will often be positioned

slightly askew to each other, this analysis requires careful curation of the various types of binding events. Some events may require manual judgment calls as overlapping patterns may not necessarily fit into a simple category.

43. The overlap between binding sites or intergenic regions can also be examined using a visualization program such as Cytoscape (http://www.cytoscape.org/) [25]. Regardless of the approach taken, when dealing with high degrees of overlap between two or more factors in ChIP datasets, it is important to remember that the absence of evidence for a peak for one factor at a location should not be taken as evidence of certain absence as there are a number of experimental reasons, such as epitope masking, as to why binding could be occluded or missed.

44. There are a variety of approaches to take in regard to this analysis, each with their own benefits and drawbacks. In all cases, it is worth remembering that the number of possible combinations for "n" factors equals $(2n) - 1$ so the effort involved will greatly increase as more factors are included.

45. When looking for overrepresentation or underrepresentation of some condition in regard to an experimental set relative to the genome as a whole, the hypergeometric distribution test is often the most relevant statistical test. In such cases, the chi-squared test may also be valid if there are a large number of targets. It is worth remembering that a lack of regulation of targets for a given regulator in a specific condition does not necessarily mean that binding events have no function. The Gal4 binding sites upstream of the galactose metabolic genes in *S. cerevisiae* are a good example of this; Gal4 binds under almost all conditions, but is only functional in response to a specific metabolic cue ([26], for review *see* ref. [27]). At the same time, remember that not all binding sites are functional.

46. We have discussed three possible metrics for determining whether binding site overlap occurs more frequently than would be expected by chance. We recommend using all three of them to account for possible biases in each method. These are described in detail in the Supplemental Methods from Hernday et al. [9].

47. Depending on the data available, it may be informative to compare the targets of a given factor/s across multiple species. Although many of the analyses described above can be performed in this case, additional considerations come into play. The method described assumes a number of genome rearrangements between the species in question, making it necessary to map orthologs of targets between the species. When comparing between closely related species with few to

no genome rearrangements, it may be possible to directly compare equivalent intergenic regions and even specific instances of DNA-binding motifs. We have not provided a detailed method for mapping gene orthologs between species as it is beyond the scope of this chapter. The exact details of the mapping process will depend on the resources available for the species in question, our methods are general recommendations based on our experience. If multiple independent orthology calls are available, the different mappings will often have at least subtle differences. Therefore, we recommend performing this analysis with at least two sets of orthology calls if multiple sets are available. Map orthology in both directions (i.e., from species A to B as well as for from B to A) and perform the analysis for both gene lists. Limiting mapping to clear cases of one to one homology is simplest, but often results in little useable information. As such, we recommend considering more ambiguous cases like one to two, two to two, or even many to many. An example of this approach can be found in Cain et al. [4].

Acknowledgments

This work was supported by National Institutes of Health (NIH) grants R00AI100896 and R01AI049187.

References

1. Lee TI, Johnstone SE, Young RA (2006) Chromatin immunoprecipitation and microarray-based analysis of protein location. Nat Protoc 1:729–748

2. Nobile CJ et al (2012) A recently evolved transcriptional network controls biofilm development in Candida albicans. Cell 148:126–138

3. Tuch BB et al (2008) The evolution of combinatorial gene regulation in fungi. PLoS Biol 6, e38

4. Cain CW et al (2012) A conserved transcriptional regulator governs fungal morphology in widely diverged species. Genetics 190:511–521

5. Baker CR et al (2012) Protein modularity, cooperative binding, and hybrid regulatory States underlie transcriptional network diversification. Cell 151:80–95

6. Nguyen VQ, Sil A (2008) Temperature-induced switch to the pathogenic yeast form of Histoplasma capsulatum requires Ryp1, a conserved transcriptional regulator. Proc Natl Acad Sci U S A 105:4880–4885

7. Beyhan S et al (2013) A temperature-responsive network links cell shape and virulence traits in a primary fungal pathogen. PLoS Biol 11, e1001614

8. Pérez JC, Johnson AD (2013) Regulatory circuits that enable proliferation of the fungus Candida albicans in a mammalian host. PLoS Pathog 9, e1003780

9. Hernday AD et al (2013) Structure of the transcriptional network controlling white-opaque switching in Candida albicans. Mol Microbiol 90:22–35

10. Hernday AD et al (2010) Genetics and molecular biology in Candida albicans. Methods Enzymol 470:737–758

11. Homann OR, Johnson AD (2010) MochiView: versatile software for genome browsing and DNA motif analysis. BMC Biol 8:49

12. Chakravarty A et al (2007) A novel ensemble learning method for de novo computational identification of DNA binding sites. BMC Bioinformatics 8:249

13. Liu X, Brutlag DL, Liu JS (2001) BioProspector: discovering conserved DNA motifs in upstream regulatory regions of co-expressed genes. Pac Symp Biocomput 127–138

14. Bailey TL et al (2009) MEME SUITE: tools for motif discovery and searching. Nucleic Acids Res 37:W202–W208

15. Teixeira MC et al (2014) The YEASTRACT database: an upgraded information system for the analysis of gene and genomic transcription regulation in Saccharomyces cerevisiae. Nucleic Acids Res 42:D161–D166

16. Gupta S et al (2007) Quantifying similarity between motifs. Genome Biol 8:R24

17. Kent WJ et al (2002) The human genome browser at UCSC. Genome Res 12:996–1006

18. Ji H et al (2008) An integrated software system for analyzing ChIP-chip and ChIP-seq data. Nat Biotechnol 26:1293–1300

19. Lohse MB et al (2013) Identification and characterization of a previously undescribed family of sequence-specific DNA-binding domains. Proc Natl Acad Sci U S A 110: 7660–7665

20. Teytelman L et al (2013) Highly expressed loci are vulnerable to misleading ChIP localization of multiple unrelated proteins. Proc Natl Acad Sci U S A 110:18602–18607

21. Hnisz D et al (2013) A histone deacetylase adjusts transcription kinetics at coding sequences during Candida albicans morphogenesis. PLoS Genet 8, e1003118

22. Tompa M et al (2005) Assessing computational tools for the discovery of transcription factor binding sites. Nat Biotechnol 23:137–144

23. Baker CR, Tuch BB, Johnson AD (2011) Extensive DNA-binding specificity divergence of a conserved transcription regulator. Proc Natl Acad Sci U S A 108:7493–7498

24. Askew C et al (2011) The zinc cluster transcription factor Ahr1p directs Mcm1p regulation of Candida albicans adhesion. Mol Microbiol 79:940–953

25. Shannon P et al (2003) Cytoscape: a software environment for integrated models of biomolecular interaction networks. Genome Res 13:2498–2504

26. Ren B et al (2000) Genome-wide location and function of DNA binding proteins. Science 290:2306–2309

27. Traven A, Jelicic B, Sopta M (2006) Yeast Gal4: a transcriptional paradigm revisited. EMBO Rep 7:496–499

Chapter 11

ChIPseq in Yeast Species: From Chromatin Immunoprecipitation to High-Throughput Sequencing and Bioinformatics Data Analyses

Gaëlle Lelandais, Corinne Blugeon, and Jawad Merhej

Abstract

Chromatin immunoprecipitation (ChIP) followed by high-throughput sequencing (ChIPseq) is a powerful technique for the genome-wide location of protein DNA-binding sites. The ChIP experiment consists in treating living cells with a cross-linking agent to bind proteins to their DNA substrates. After fragmentation of DNA, specific fractions associated with a particular protein of interest are purified by immunoaffinity. They are next sequenced and identified on the reference genome using dedicated bioinformatics programs. Several technical aspects are important to obtain high-quality ChIPseq results. This includes the quality of antibodies, the sequencing protocols, the use of accurate controls and the careful choice of bioinformatics tools. We present here a general protocol to perform ChIPseq analyses in yeast species. This protocol has been optimized to identify target genes of specific transcription factors but can be used for any other DNA binding proteins.

Key words Chromatin immunoprecipitation, High-throughput sequencing, DNA binding sites of proteins, Yeasts, Bioinformatics

1 Introduction

Genome wide discovery of protein–DNA interactions is a prerequisite to better understand transcriptional regulations [1], epigenetic modifications [2, 3], or chromatin organisation [4]. Cross-linked chromatin immunoprecipitation technique (ChIP) is the most powerful method to determine in vivo, whether a specific protein (such as transcription factors or other chromatin-associated proteins) interacts with a specific genomic region [5]. The ChIP experimental procedure can be divided into five main steps (summarized Fig. 1, steps 1–5): (1) treating of living cells with a reversible cross-linking agent (typically formaldehyde or UV), (2) cell lysis and DNA shearing (typically by sonication or DNA restriction enzyme digestion), (3) immunoprecipitation using antibody targeting the protein of interest, (4) cross-link reversal, and finally

Frédéric Devaux (ed.), *Yeast Functional Genomics: Methods and Protocols*, Methods in Molecular Biology, vol. 1361,
DOI 10.1007/978-1-4939-3079-1_11, © Springer Science+Business Media New York 2016

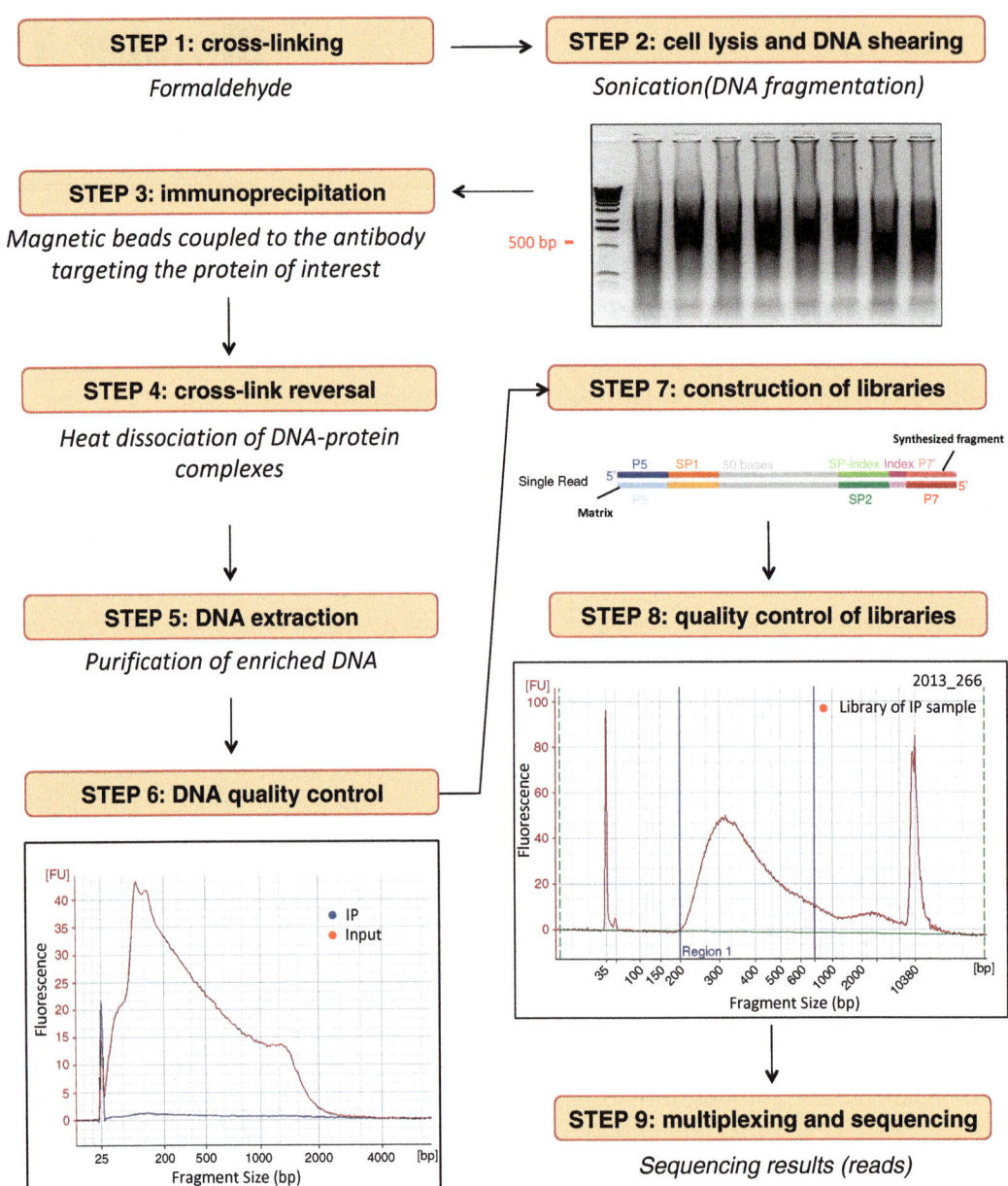

Fig. 1 Procedure for chromatin immunoprecipitation followed by high-throughput sequencing (ChIPseq). ChIP technique is summarized in five main steps. *Step 1*: Cross-linking of the living cells to temporary fix the proteins to their DNA binding sites. *Step 2*: Cells are lysed and DNA is sheared by sonication into ~500 bp fragments. *Step 3*: DNA–protein complexes are selectively immunoprecipitated using an antibody specifically targeting the protein of interest. *Step 4*: Cross-linking reversal is performed to dissociate the complexes. *Step 5*: Extraction of the immunoprecipitated DNA fragments. After the ChIP, The quantity and quality of the resulting DNA is then assessed (*Step 6*). The Input sample indicates that the size of the DNA fragments is suitable for libraries construction. The IP sample (*blue line*) indicates that the quantity of DNA is very low (under the quantification threshold) which guide us to use a protocol for libraries construction dedicated for low DNA concentrations. After libraries construction (*Step7*), the quality of the resulting DNA is controlled again (*Step 8*). The *red line* indicates that the obtained DNA has a size ranging from 200 to 600 bp, which correspond to the size of the immunoprecipitated DNA thus confirming the success of the construction of libraries. Finally, the libraries are multiplexed and sequenced (*Step 9*)

(5) purification of the remaining DNA from the immunoprecipitated complexes.

To perform a genome-wide mapping of the DNA binding sites of a particular protein of interest, ChIP was first combined with DNA microarray technology (ChIP-on-chip). With the significant progress of high-throughput sequencing, the identification of immunoprecipitated DNA by microarray technique tends to be replaced by massive sequencing (ChIPseq). ChIPseq represents various advantages compared to ChIP-on-chip. Mainly, it provides a higher resolution view of protein–DNA interaction sites and reduces considerably the false positive rate of the identified targets [6, 7]. After DNA quality control (Fig. 1, step 6), the procedure for high-throughput sequencing starts with the construction of libraries of the immunoprecipitated DNA (one library per biological sample to be analyzed) (Fig. 1, step 7). The quality checking and the labelling (barcoding) of the different libraries for multiplexing are following. It ends with the final sequencing procedure (Fig. 1, steps 8 and 9).

Sequencing outputs are short sequences (or reads) with associated quality scores [8]. Illumina's technology allows obtaining in a single run, more than 100 millions of reads. The analysis of sequencing output files required computational resources and bioinformatics programs (summarized Fig. 2, steps 1–5)) in order to: (1) control the quality of the sequences, (2) map the reads on the reference genome, (3) perform necessary file format conversions, (4) visualize the data with a genome browser, and (5) detect the peaks, i.e., locate the DNA binding sites of the protein.

We detail here all the steps for ChIPseq experiments and data analyses in yeasts. A practical goal is to optimize the discovery of DNA binding sites by choosing accurate protocols. As yeasts are eukaryotic species with small genomes (<20 Mb), our protocol slightly differs from others used for organisms like mouse or human. Main differences relate to the sequencing depth needed (less than five millions of mapped reads are enough in yeasts) and the required computational resources (all calculations can be performed on a desktop workstation). Working with a reasonable number of reads in genomes with an average complexity represents an important computational advantage, and it is thus critical to use relevant bioinformatics programs. Distinguishing real binding events from intrinsic variability in the sequencing procedure is the main challenge faced by people who perform ChIPseq analyses. It seems important to apply strategies in which lists of peaks identified based on statistical parameters (p-values for instance) are systematically assessed regarding other biologically meaningful information (e.g., detection of overrepresented DNA motifs, proportion of peaks in specific DNA elements, or reference list of positions described in the literature). Illustrations discussed here are in case of the DNA binding sites of specific transcription factors.

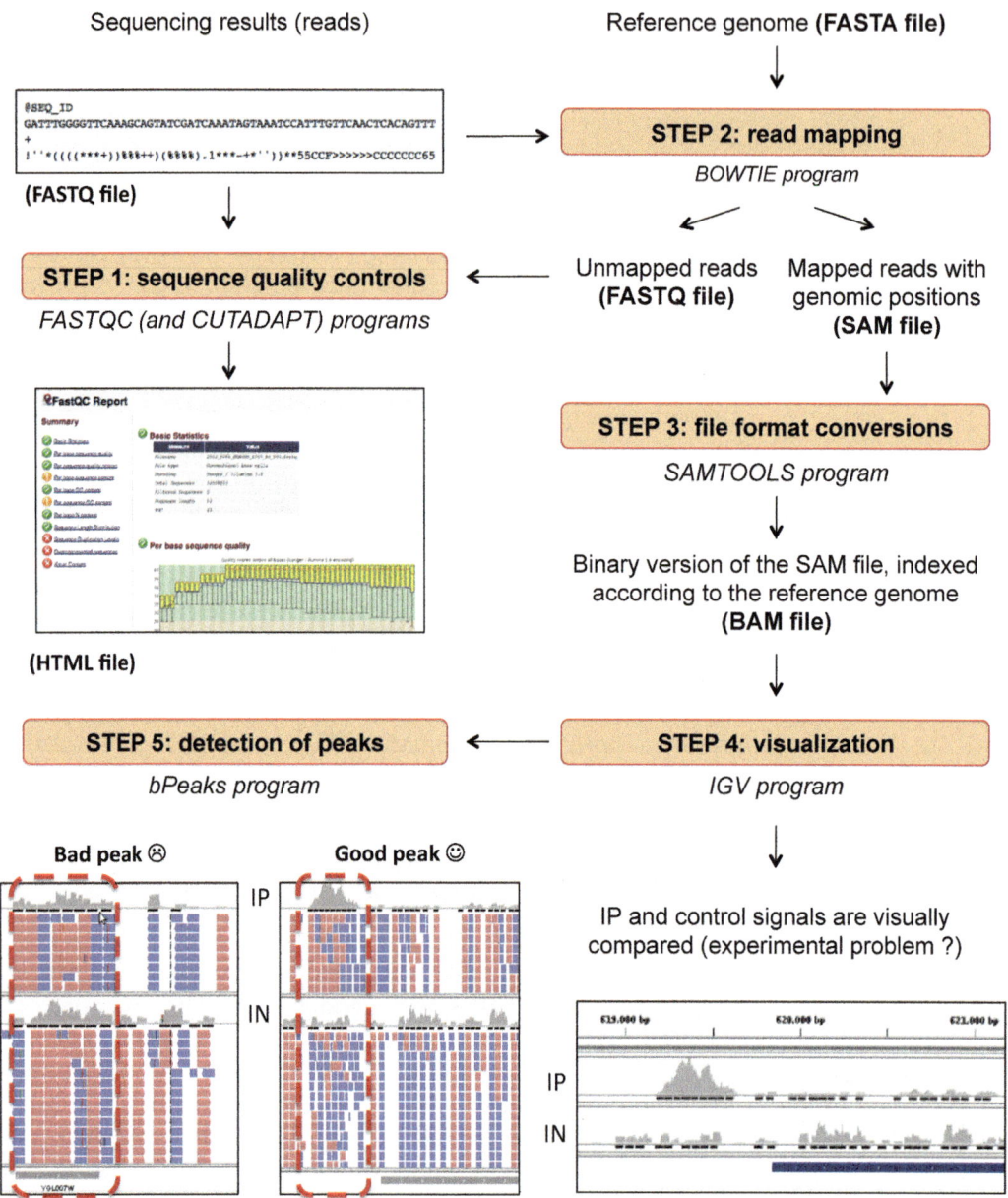

Fig. 2 Procedure for the bioinformatics data analysis. FASTQ files comprise the sequencing results (reads). *Step 1* consists in controlling the quality of the sequencing results and eventually filtering reads with low quality. *Step 2* consists in identifying (or mapping) the position of sequenced reads on the reference genome. *Step 3* consists in converting the mapping outputs in different file formats, which can be used by other programs for visualization (*Step 3*) or peak calling (*Step 4*). Peak calling is clearly the most challenging part of ChIPseq data analysis

2   Materials

2.1   Chromatin Immunoprecipitation (ChIP)

2.1.1   Cross-Linking and Harvesting the Cells

1. Formaldehyde solution: 37 % formaldehyde (to be used under a fume hood).
2. 2.5 M glycine solution.
3. Ice-cold TBS: 20 mM Tris–HCl pH 7.4, 150 mM NaCl.

2.1.2   Preparation of Magnetic Beads

1. "Magnetic beads": Dynabeads Pan Mouse IgG (Dynal).
2. PBS–BSA solution, freshly prepared: 1:10 Phosphate buffered saline 10×, 0.5 % of Bovine serum albumin.
3. Antibody targeting the protein to immunoprecipitate (typically, we usually use the Anti-myc antibody to immunoprecipitate proteins carrying the myc tag).

2.1.3   Cell Lysis and Sonication

1. "Fastprep tubes": 2 ml screw cap tubes for Fastprep bead beater.
2. Acid washed "glass beads" (400–600 μm).
3. Lysis buffer: 50 mM Hepes KOH pH 7.5, 140 mM NaCl, 1 mM EDTA, 1 % Triton X-100, 0.1 % Na-deoxycholate, 1 mM PMSF, protease inhibitor cocktail (*see* **Note 1**).
4. Bead beater Fastprep-24®.
5. Needles.
6. 15 ml polyethylene Falcon tubes (for sonication).
7. Bioruptor® Sonicator (Diagenode).
8. Protease inhibitor cocktail.

2.1.4   Immunoprecipitation, Washes, and Cross-Link Reversal

1. Magnet system (MPC®-S) (Dynal).
2. Lysis buffer supplemented with 360 mM of NaCl.
3. Wash buffer: 10 mM Tris–HCl pH 8, 250 mM LiCl, 0.5 % Na-deoxycholate, 1 mM EDTA (*see* **Note 2**).
4. TE pH 8: 10 mM Tris–HCl pH 8, 1 mM EDTA.
5. TE-SDS solution), freshly prepared: 10 mM Tris–HCl pH 8, 1 mM EDTA, 1 % SDS.
6. Thermomixer®.

2.1.5   Purification of DNA

1. 2× Laemmli buffer: 120 mM Tris-base, 3.4 % SDS, 10 mM EDTA, 15 % glycerol, 0.01 % bromophenol blue, 2.5 % β-mercaptoethanol, protease inhibitors cocktail.
2. Proteinase K mix: 5 % Proteinase K (14–22 mg/ml), 3.5 % glycogen (20 mg/ml) Roche, in TE pH 8.
3. Phenol–chloroform–isoamylalcohol (25:24:1).

4. NaCl solution concentrated to 5 M.

5. Absolute ethanol (*see* **Note 3**).

6. Ethanol 70 % (*see* **Note 3**).

7. RNAse A mix: 3 % RNAseA (10 μg/μl), in TE pH 7.4.

8. Sodium acetate 3 M pH 5.2.

9. QIAquick PCR purification kit (containing buffers PB and PE and the PCR purification columns).

2.2 High-Throughput Sequencing (in Case of Illumina Technology)

2.2.1 Library Construction and Quality Controls

1. 2100 Bioanalyzer instrument (Agilent), Kit RNA 6000 Pico chip and High Sensitivity DNA chip.

2. Qubit 2.0 Fluorometer and Qubit dsDNA HS Assay kit (Life Technologies).

3. NEXTflex™ ChIPseq Kit, NEXTflex is a trademark of Bioo Scientific Corporation.

4. NEXTflex™ ChIPseq Barcodes (*see* **Note 4**).

5. Plate semi-skirted (4titude) or similar, compatible with the thermocycler Adhesive.

6. PCR Fiol Seal (4titude) or similar.

7. Agencourt AMPure XP 5 ml (Beckman Coulter Genomics).

8. Magnetic Stand-96 (Ambion) or similar device.

9. Microcentrifuge for 96-well PCR.

2.2.2 Multiplexing and Illumina Sequencing

1. Solution Tris–HCl Tween 20: 10 mM Tris–HCl, pH 8.5 with 0.1 % Tween 20.

2. NaOH 0.1 N, freshly prepared.

3. Prechilled HT1 (Illumina hybridization buffer).

4. Illumina HiSeq 1500 sequencing instrument (cbot for cluster generation and HiSeq for sequencing), or other high-throughput sequencing system (*see* **Note 5**).

5. TruSeq SR Cluster Kit v3-cbot-HS (Illumina), including the flow cell and reagents for cluster generation on the cbot.

6. TruSeq SBS Kit v3 (Illumina) including sequencing reagents for the HiSeq.

7. Illumina Casava software (for base calling and FASTQ file creation).

2.3 Bioinformatics Data Analyses

2.3.1 Data Storage

Network Attached Storage (NAS) devices (Synology DS713+ or similar), with an internal capacity of around 8 TB (4 TB HDD×2, capacity may vary by RAID types) (*see* **Note 6**).

2.3.2 Workstation

HP Z820 Workstation (Intel Xeon E5-2609 2.4 Ghz CPU and 16 GB DDR3-1600 (8×2 GB) RAM) or similar (*see* **Note 7**).

2.3.3 Software	A list of programs freely distributed to academic users (*see* **Note 8**):

1. FASTQC (http://www.bioinformatics.babraham.ac.uk/projects/fastqc/).
2. CUTADAPT (https://code.google.com/p/cutadapt/).
3. BOWTIE (http://bowtie-bio.sourceforge.net/index.shtml) (*see* **Note 9**).
4. SAMTOOLS (http://samtools.sourceforge.net/).
5. IGV (http://www.broadinstitute.org/igv/home).
6. bPeaks (http://bpeaks.gene-networks.net/) (*see* **Note 10**).

---

3 Methods

3.1 Chromatin Immunoprecipitation

3.1.1 Cross-Linking and Harvesting the Cells

1. Grow an overnight culture (50 ml) of yeast cells in appropriate liquid medium until the A_{600} reaches 0.6–1 (*see* **Note 11**).
2. Add 1.4 ml of formaldehyde solution to the culture and shake occasionally. Incubate at room temperature for 15 min (*see* **Note 12**).
3. Stop the cross-linking by adding 7 ml of glycine solution. Incubate at room temperature for 5 min with occasional agitation. Transfer the culture to a 50 ml Falcon tube (*see* **Note 13**).
4. Centrifuge at $5000 \times g$ for 5 min at 4 °C. Discard the supernatant under the hood.
5. Resuspend the cells in 40 ml ice-cold TBS by inverting vigorously the tube. Recentrifuge as in previous step.
6. Repeat once the **step 5**.
7. Using the remaining liquid, resuspend the cell pellet and transfer to a 1.5 ml tube.
8. Centrifuge 2 min at 4 °C, remove the maximum amount of supernatant by pipetting and freeze immediately the cell pellet at −80 °C (*see* **Note 14**).

3.1.2 Preparation of Magnetic Beads (See Note 15)

1. Vortex the main stock of "magnetic beads".
2. Transfer the volume of magnetic beads needed (50 µl of beads per sample) to a 15 ml Falcon tube and spin at $3000 \times g$ for 1 min (*see* **Note 16**). Discard supernatant.
3. Resuspend the magnetic beads in 10 ml of PBS–BSA solution. Centrifuge as in previous step and discard supernatant.
4. Repeat the **step 3** for a total of two times. Remove completely the supernatant by pipetting.
5. Resuspend the beads in a total volume of PBS–BSA solution corresponding to 250 µl per sample.

6. Add to the beads the antibody targeting the protein to immunoprecipitate (*see* **Note 17**).

7. Incubate the mix containing the magnetic beads and the antibody overnight at 4 °C on a rotating wheel (*see* **Note 18**).

8. After overnight incubation, centrifuge at $2000 \times g$ for 1 min at 4 °C. Discard supernatant.

9. Wash the antibody-coupled beads as in **step 3** for a total of two washes. Remove completely the supernatant.

10. Resuspend the antibody-coupled magnetic beads in a total volume of PBS–BSA solution corresponding to 30 µl per sample.

3.1.3 Cell Lysis and Sonication

1. Prepare "Fastprep tubes" (*see* Subheading 2) containing each 600 µl of acid-washed "glass beads" (*see* Subheading 3). Let them cool down on ice.

2. Thaw the cell pellet on ice.

3. Resuspend the cell pellet in 600 µl of lysis buffer. Transfer the volume to the Fastprep tubes. Place the tube in the FastPrep bead beater at 4 °C (*see* **Note 19**).

4. Beat for 30 s at maximum speed. Place the tubes on ice for 2 min.

5. Repeat the **step 4** for a total of three times (*see* **Note 20**).

6. Punch a hole at the bottom of the Fastprep tubes using a needle (*see* **Note 21**).

7. Fix the Fastprep tubes at the top of 15 ml polyethylene Falcon tubes. Centrifuge at $900 \times g$ for 1 min at 4 °C to collect the lysat in the 15 ml Falcon tubes (*see* **Note 22**).

8. Fill the bath of the sonicator with cold water and some ice. Set at high intensity. Sonicate for four rounds of 30 s on/30 s off.

9. Put the sample on ice for 2 min. Add some ice to the bath of the sonicator and repeat the sonication as in previous step (*see* **Note 23**).

10. Centrifuge for 5 min at $5000 \times g$ at 4 °C. Transfer the supernatant to a new 1.5 ml tube.

11. Centrifuge for 2 min at $30,000 \times g$ at 4 °C. Transfer the maximum amount of supernatant to a new 1.5 ml tube. Add 2 µl of protease inhibitors cocktail. This is the whole cell extract (*see* **Note 24**).

3.1.4 Immuno-precipitation, Washes, and Cross-Link Reversal

1. Transfer 500 µl of the whole cell extract to a new 1.5 ml tube. Add 30 µl of antibody-coupled beads. Incubate overnight at 4 °C on a rotating wheel.

2. Use the magnet system to wash the beads. For each washing, incubate for 1 min at 4 °C on the rotating wheel (*see* **Note 25**).

3. Wash the beads twice with 1 ml of lysis buffer.

4. Wash the beads twice with 1 ml of lysis buffer supplemented with 360 mM of NaCl.

5. Wash the beads twice with 1 ml of wash buffer.

6. Wash the beads once with 1 ml of TE pH 8.

7. Centrifuge at $30,000 \times g$ for 2 min at 4 °C. Remove the maximum amount of the remaining liquid by pipetting. Do not let the beads dry at this point.

8. Add 100 μl of TE-SDS solution to the bead pellet. This is the immunoprecipitated sample (referred to as "IP sample" or IP).

9. In parallel, thaw the whole cell extract from **step 11** (previous section). Transfer 5 μl to a new tube and add 95 μl of TE-SDS. This is the input DNA (referred to as "INPUT sample" or IN).

10. Vortex the tubes (IP and IN) for 1 min.

11. Incubate at 65 °C in a thermomixer® with shaking at $1200 \times g$. After 30 min, decrease the shaking speed to $600 \times g$ and keep the incubation overnight (*see* **Note 26**).

3.1.5   Purification of DNA

1. The next day, vortex the tubes for 1 min. Incubate for another 20 min at 65 °C before proceeding.

2. Centrifuge for 2 min at $30,000 \times g$. Transfer the eluted chromatin to a new 1.5 ml tube.

3. Before proceeding with proteinase treatment, mix 5 μl of the liquid with 5 μl of 2× Laemmeli buffer for subsequent western blot analysis (*see* **Note 27**).

4. Add 150 μl of proteinase K mix to the eluted chromatin. Mix by pipetting and incubate at 37 °C for 2 h.

5. Under a fume hood, add 300 μl of Phenol-chloroform isoamylalcohol (25:24:1).

6. Mix well by vortexing for about 30 s. Centrifuge for 5 min at $13,000 \times g$.

7. Transfer the aqueous phase (upper) to a new 1.5 ml tube.

8. Repeat **steps 5–7** to a total of two times.

9. Add 12 μl of NaCl 5 M (about 1/25th of the volume).

10. Add 750 μl of freeze-cold absolute ethanol (about 2.5 of the volume) and mix by vortexing the tube.

11. Store for 30 min at −80 or at −20 °C overnight (*see* **Note 28**).

12. Centrifuge at $13,000 \times g$ for 40 min at 4 °C.

13. Pour off the liquid and wash with 1 ml of freeze-cold ethanol 70 °C.

14. Centrifuge at $13,000 \times g$ for 15 min at 4 °C (*see* **Note 29**).

15. Pour off the supernatant, spin briefly and remove the remaining liquid with a pipette (*see* **Note 30**).

16. Dry the pellet for 5 min in a vacuum concentrator at 30 °C.

17. Resuspend the pellet in 30 μl of RNAse A mix by vortexing for 5 s.

18. Incubate for 1 h at 37 °C.

19. Add 3 μl of sodium acetate 3 M pH 5.2 and 150 μl of PB buffer (QIAquick purification kit).

20. Mix well by pipetting and transfer the liquid to the QIAquick column.

21. Centrifuge for 1 min at $13,000 \times g$ at room temperature. Discard the liquid from the collect tube.

22. Add 600 μl of buffer PE (QIAquick purification kit) and centrifuge as in previous step. Discard the flow-through from the collection tube. Centrifuge for an additional 1 min at $13,000 \times g$.

23. Place the QIAquick column in a 1.5 ml tube. To elute DNA, add 50 μl of H_2O at 37 °C to the center of the column. Incubate for 1 min and centrifuge for 1 min at $13,000 \times g$.

24. The immunoprecipitated DNA is stored at –20 °C.

3.2 High-Throughput Sequencing

3.2.1 Library Construction and Quality Controls

1. Control of the quality and the length of the immunoprecipitated DNA fragments on a bioanalyzer. Evaluate the quantity using a Qubit 2.0 Fluorometer with a dsDNA HS Assay kit (*see* **Note 31**).

2. Perform the construction of libraries using the NEXTflex™ ChIPseq Kit (*see* **Note 32**). In case of low quantity of DNA, use the particular procedure recommended by the manufacturer (*see* **Note 33**).

3. During the ligation step chose a NEXTflex™ ChIPseq Barcode for each sample (*see* **Note 4**).

4. Estimate the average size of the library using Agilent High Sensitivity DNA chip and the quantity on a Qubit 2.0 Fluorometer (*see* **Note 34**).

The obtained library's structure is compatible with Illumina sequencing. For each sample, 2 nM of library concentration is enough to proceed to sequencing using a HiSeq 1500 Illumina sequencer.

3.2.2 Multiplexing and Illumina Sequencing

1. Normalize the concentration of each library to 10 nM with solution Tris–HCl Tween 20 (*see* **Note 35**).

2. Pool libraries in one tube using 3.4 μl of each sample library and complete the volume to 80 μl with solution Tris–HCl Tween 20. A concentration of 2 nM multipexed libraries is obtained (*see* **Note 36**).

3. Add 10 μl of fresh 0.1 N NaOH to 10 μl of the 2 nM multiplexed libraries to denature DNA template.

4. Incubate 5 min at room temperature.

5. Dilute DNA to 20 pM by adding 980 μl of prechilled HT1 and load 120 μl of this solution in an Illumina cbot instrument (*see* **Note 37**). The cbot enables the cluster generation on a flow cell (*see* **Note 38**).

6. Run a multiplexed Single read 50 on an Illumina HiSeq sequencing instrument.

7. Use Illumina Casava software to perform base calling and generate FASTQ files (one file per multiplexed library).

3.3 Bioinformatics Data Analyses

The bioinformatics analysis starts from the FASTQ file, i.e., the file in which all the raw sequence data (or reads) obtained from the sequencing step are stored [8]. File format and successive steps for the bioinformatics data analysis are summarized Fig. 2.

3.3.1 Quality Controls and Data Filtering

1. Launch the FASTQC program and upload the FASTQ file. A general report regarding read quality is automatically created (HTML file).

2. Open the HTML file with a web browser and check for potential contaminations or experimental bias (*see* **Note 40**).

3. Launch the CUTADAPT program on the FASTQ file to verify that all Illumina adapter sequences (used for the construction of libraries) were removed from you FASTQ file (*see* **Note 41**). Filter all the adapter sequences with CUTADAPT and repeat the **step 1** (*see* **Note 42**).

3.3.2 Mapping on the Reference Genome and Result Visualization

1. Create an indexed version of the reference genome (FASTA file), using the "bowtie-build" function available in BOWTIE program (*see* **Note 43**).

2. Run the BOWTIE program, specify as input the FASTQ file to analyse (*see* **Note 44**) and choose SAM format as output, for the aligned reads with positions on the reference genome.

3. Perform file format conversions using the SAMTOOLS program. First convert the SAM file into a BAM file (a binary compressed version of SAM, *see* **Note 45**) and second, index the BAM file using the reference genome (the reads are ordered following their position on the genome).

4. Launch the IGV program and upload simultaneously BAM files associated to IP and IN samples. Control for possible experimental problems (*see* **Note 46**).

3.3.3 Peak Calling

1. Use the bPeaks program to perform peak calling, i.e., the detection of genomic regions with a significant enrichment of reads in IP sample compared to background noise (*see* **Notes 10** and **47**).

2. Launch the IGV program and upload the BED file with the genomic locations of the detected peaks. Verify that the detected peaks exhibit "good peak" properties as illustrated Fig. 2 (step 5).

3. Evaluate carefully the biological relevance of identified peaks, using other information like detection of overrepresented DNA motifs, proportion of peaks in specific DNA elements (promoters for instance, or reference list of positions described in the literature (*see* **Note 48**).

4   Notes

1. We found that the lysis buffer could be prepared and stored at 4 °C. Only PMSF and protease inhibitors have to be freshly added.

2. Wash buffer could be prepared and stored at 4 °C.

3. Absolute ethanol and ethanol 70 % are stored in advance at −20 °C.

4. In our experiments performed in yeasts *Candidaglabrata*, we multiplex our samples by 6 (six different barcodes are therefore required). This appeared to be a good compromise to obtain enough sequences for successful peak calling analyses and reducing the total cost associated to the project. Note that barcode choice and association have to be performed carefully, following the supplier recommendations (Fig. 2).

5. Illumina HiSeq 1500 is the technology available at the genomics platform at the Ecole Normale Supérieure (http://transcriptome.ens.fr/sgdb/), used in our ChIPseq analysis. We know that it produces between 100 and 150 millions of reads in a single run. To obtain at least five millions of mapped reads in our ChIPseq analyses, the theoretical number of samples that could be sequenced simultaneously is 20 ($100/5 = 20$). Considering the risk of sequencing bias between barcodes (*see* **Note 4**), the necessity to filter low quality reads before peak calling analyses (*see* **Note 42**) and the risk of unmapped reads on the genome (*see* **Note 44**), we decided to multiplex the samples by 6 (*see* **Note 4**). We finally obtained largely more than five millions of reads for all the sequenced libraries, indicating that multiplexing could be certainly increased to 8 or 10. This calculation depends on the sequencing instrument used and can be easily done to evaluate the number of libraries necessary for multiplexing.

6. Applying the protocol presented here to ChIPseq analyses in yeast *Candidaglabrata*, we obtained FASTQ files between 500 MB and 2 GB, with approximately 5–20 millions of reads. Data storage capacity depends on the total number of samples

to be sequenced. It is important to not under estimate this capacity because the different steps of the bioinformatics analysis (Fig. 2) required file duplications and conversions (FASTQ, SAM, BAM, indexed BAM files, etc.). RAID system is recommended, to be able to automatically restore all data files in case of disk crashes.

7. Memory requirements to perform bioinformatics data analyses are proportional to the size of the genome and the sequencing coverage (number of reads per biological sample). Application of the protocol presented here to ChIPseq analyses in yeast *Candidaglabrata* lasted a couple of hours (starting from an initial FASTQ file with five millions of reads). This computational time includes data quality controls with FASTQC (Fig. 2, step 1), read mapping with BOWTIE (Fig. 2, step 2), file format conversion with SAMTOOLS (Fig. 2, step 3) and peak calling with bPeaks (Fig. 2, step 5).

8. We advise to install these programs on a computer with LINUX distribution (UBUNTU for instance is very user friendly, http://www.ubuntu.com/). Some of them can be installed on WINDOWS and MAC OS systems (FASTQC, IGV, and bPeaks).

9. Other programs to align the reads on the reference genome exist (SOAP [9], BWA [10], CGAP-align [11], etc.). An overview and performance comparisons of the available programs can be found here [12].

10. Numerous tools are available for peak calling analyses. Method reviews and comparisons can be found here [13–15]. We developed the bPeaks program for a dedicated use to analyze ChIPseq data in yeasts [16]. We compared bPeaks performances to those of MACS [17], SPP [18], and BayesPeaks [19] and observed that bPeaks is at least as efficient as existing tools in proposing lists of peaks that are enriched in potential targets, but is more precise in defining the peak location.

11. In parallel to the tagged strain, we found that it is necessary to process a ChIPseq of the untagged parental strain to be used for identification of the genomic regions that are unspecifically immunoprecipitated. This control could be done once for each yeast species (it is not necessary to perform it systematically).

12. The time of cross-linking should be adapted depending on the immunoprecipitated protein. Typically, we found that 15 min of cross-linking at room temperature is a suitable time for several transcription factors for yeasts *Saccharomyces cerevisiae* and *Candidaglabrata*.

13. If a time course is performed, at this point samples can be stored in ice up to 30 min before centrifugation.

14. Cell disruption can be processed directly after the centrifugation. Also at this point, cell pellet can be stored at –80 °C for at least 1 month.

15. The necessary amount of magnetic beads is always incubated overnight with antibody, the day before the cell lysis and immunoprecipitation. Hence, cell extract and freshly antibody-conjugated beads are ready the same day.

16. Always consider to prepare for one extra sample to avoid running short of beads.

17. The concentration of antibody must be determined empirically. Typically, for Anti c-myc monoclonal antibody (roche), we found that 4 μg of antibody per 50 μl of magnetic beads (quantity for one sample) is a suitable concentration for efficient conjugation of the beads.

18. We recommend keeping the mix in the 15 ml Falcon tube to insure a good rotation of the liquid.

19. Be sure to screw the cap of the Fastprep tube tightly, to avoid losing the samples during cell breakage.

20. The efficiency of cell disruption is a variant of the equipment, the yeast species and the sample/glass beads ratios. Before proceeding with the subsequent steps of the ChIP, it is highly recommended to optimize the time and intensity of beating needed for an efficient cell disruption. This could be simply checked by looking at the cell under the microscope.

21. Stop when you see a bubble or just until a drop starts to form. Do not go all the way through. It is possible to use the same needle for all the samples but make sure to wipe the needle with a towel (typically a Kimwipes) between samples.

22. Make sure that the whole cell volume was recovered and no more liquid still in the Fastprep tube. The glass beads should be almost dry. If some liquid is still observed, increase the centrifugation speed to $4000 \times g$ and restart until all the liquid goes through the hole.

23. Ideally, the sonicator should be set in a cold room to avoid increasing the samples temperature during sonication. We recommend first performing tests to determine the settings that will generate a majority of DNA fragments in the requested size range ($\approx$500 bp). The size of the DNA fragments generated after sonication should be systematically verified on agarose gel before proceeding in the subsequent steps (Fig. 1, step 2).

24. The whole cell extract could be conserved at –80 °C for at least 1 month.

25. Wait until the beads stick to the wall of the tube then pour of the liquid by inverting the tubes. We found that performing the washing in the cold room helps to avoid protein degradation.

26. The shaking is only necessary during the first 30 min of incubation. In addition to SDS, the shaking participates to the elution of the immunoprecipitated complexes. For the overnight incubation, shaking is not absolutely necessary and it could be done in a hybridization oven without shaking.

27. A western blot must be systematically performed in order to validate the correct immunoprecipitation of the target protein.

28. If you have the choice, we recommend storing the precipitated DNA overnight at −20 °C.

29. In our experience, the pellet should be very small but always visible at this point.

30. At this point, the pellet does not stick well to the tube so be careful to discard supernatant without disturbing it. To avoid losing the pellet we recommend removing the supernatant with a pipette instead of inverting the tube.

31. Immunoprecipitated DNA can have very low concentration and is often under the detection threshold. It is possible to perform the library construction by using the total volume of immunoprecipitated DNA (IP) and 1 ng for the INPUT sample (IN).

32. The immunoprecipitated samples have already been fragmented before immunoprecipitation (Fig. 1, step 2). They do not require further manipulation before construction of libraries (Fig. 1, step 7), following the manufacturer's instructions.

33. There is an Ultra-low input protocol recommended to use in case of very low amount of DNA, less than 1 ng (NEXTflex ChIP-Seq--> Kit Manual_5143-01; v12.10).

34. The concentration of the library is measured on a Qubit 2.0 Fluorometer with a dsDNA High sensitivity Assay kit. It enables to determine the concentration of double strand library.

35. Qiagen buffer EB can be used by adding 0.1 % of Tween 20. The addition of 0.1 % Tween 20 helps to prevent adsorption of the template to plastic tubes upon repeated freeze–thaw cycles.

36. In our experience, libraries can be successfully multiplexed by 6 on a lane of a HiSeq v3 flow cell to obtain between 10 and 30 million of reads which are largely sufficient for subsequent data analysis (*see* **Notes 4** and **5**). Single read of size 50 is also good enough.

37. The quantity used on the sequencer must be adjusted on each HiSeq instrument.

38. More detailed information concerning the Illumina cluster station or cbot and the sequencing procedure can be found online: http://www.illumina.com/.

39. In addition to the identified bases (A, T, C, G), PHRED quality scores are also available. Score values are directly related to the probability of a sequencing error; they are useful to rapidly identify sequencing problems. We typically consider that a score value higher than Q20 is satisfying enough.

40. The HTML report obtained with FASTQC program is easy to understand thanks to summary graphs and tables to quickly assess the data (*see* Fig. 2, step 1). Important information is reported as, for instance, the total number of reads in the initial FASTQ file, sequence length, per base sequence quality and per sequence quality scores (*see* **Note 39**, for quality score information). Per sequence GC content is also compared to a theoretical distribution expected considering the reference genome GC properties. This allows detecting potential contamination problems (read sequences in the FASTQ file that do not match the studied yeast species). Sequence duplication levels are another interesting information because it allows evaluating the library complexity, i.e., the number of unique sequences. A very low complexity in a library can reveal a problem with initial DNA (quantity is too low) or a problem during library construction (problem during the PCR amplification for instance). Interesting ChIPseq guidelines and practices can be found in [20].

41. This step is often automatically performed at the end of the sequencing procedure (during the FASTQ file creation), but we observed in our datasets that adaptor sequences can still remain.

42. If an important number of sequences with poor quality scores is observed (<Q20), we advise to perform a sequence trimming (5′ and 3′ end) or to delete the reads with the lowest average quality. Several user-friendly tools are available on GALAXY website for instance (https://usegalaxy.org/) [21].

43. Pre-built genome indexes can be downloaded here (ftp://ftp.ccb.jhu.edu/pub/data/bowtie_indexes/). If different versions of your reference genome exist, be careful to choose the correct one, i.e., the one you will use for data visualization on a genome browser and peak calling analyses.

44. We use all default parameters for the BOWTIE program, except for the parameter named "-m" settled to 1 in our analyses. This option allows suppressing all alignments for a particular read if more than one significant match exist ("-m 1"). We generally control the unmapped reads, applying the FASTQC program (*see* **Note 40** and Fig. 2, step 2). Note that the BOWTIE program can retrieve results for the read alignment in different file format. We prefer using the file format SAM [22].

45. A BAM file is a binary version of a SAM file. They occupied a lower disk space because of a better compression of the information. We prefer to backup BAM files (instead of SAM files) for saving our resources for data storage (*see* **Note 6**).

46. It is our experience, the visual inspection of the mapping results is very important to detect potential experimental problems. Also it represents a good starting point to define relevant parameters for peak calling programs (Fig. 2, step 5). For that, we superimpose IP and IN mapping results using two different tracks in IGV (as shown Fig. 2, step 4). Protein binding positions are expected to exhibit significant read aggregations in IP signal, whereas we expect to observe uniformly distributed reads in the control signal (IN). Often, distributions of reads in IN are far from being uniform. It is therefore important to consider that regions with high read counts do not necessarily represent DNA binding sites for proteins (as illustrated Fig. 2, step 5). It has been shown for instance that some regions of open chromatin are over represented in the DNA input [20, 23]. Note that with IGV program, it is also possible to colour the reads according to the DNA strand on which they were aligned (as shown Fig. 2, step 5). This can be useful in case of directional protocols to prepare library. The complexity of the sequenced library can also be evaluated regarding the reads aligned at the exact same position (*see* **Note 40**). Finally with IGV, GFF files with information of gene positions can be uploaded to easily locate detected peaks (peaks located in promoters are for instance of particular interests in case of ChIPseq data for a specific transcription factor).

47. Peak calling is clearly the most challenging part in a ChIPseq data analysis. The objective is to correctly estimate enrichment in IP signal compared to IN signal (Fig. 2, step 5). Recently, we developed a simple and robust tool called bPeaks [16], for the detection of transcription factor binding sites from ChIPseq data in small eucaryotic genomes. Interesting genomic regions are identified based on four criteria: (1) a high number of reads in the IP signal, (2) a low number of read in the IN signal, (3) a high value of log fold change (or logFC) between IP and IN signals, and (4) a good sequencing coverage in both IP and control sample. The program bPeaks was optimized for yeast ChIPseq data. A detailed procedure to use bPeaks is available on the website http://bpeaks.gene-networks.net/.

48. In a previous study [16], we presented a general protocol for applying bPeaks to ChIPseq data related to transcription factors Pdr1p (*S. cerevisiae*), Sfl1p (*C. albicans*), and CgAp1p (*C. glabrata*). This protocol consists in first evaluating the influence of bPeaks parameter values on detected genomic regions (or peaks) and second assessing the biological significance of the retrieved lists of peaks. In the specific case of yeast transcription factors, the number of peaks in promoter regions appeared to be very biologically meaningful information, as well as the detection of *cis*-regulatory motifs with "peak-motif" program [24].

References

1. Harbison CT et al (2004) Transcriptional regulatory code of a eukaryotic genome. Nature 431(7004):99–104

2. Lando D et al (2012) Quantitative single-molecule microscopy reveals that CENP-A(Cnp1) deposition occurs during G2 in fission yeast. Open Biol 2(7):120078

3. Barski A et al (2007) High-resolution profiling of histone methylations in the human genome. Cell 129(4):823–837

4. Thurtle DM, Rine J (2014) The molecular topography of silenced chromatin in Saccharomyces cerevisiae. Genes Dev 28(3):245–258

5. Johnson DS, Mortazavi A, Myers RM, Wold B (2007) Genome-wide mapping of in vivo protein-DNA interactions. Science 316(5830):1497–1502

6. Park PJ (2009) ChIP-seq: advantages and challenges of a maturing technology. Nat Rev Genet 10(10):669–680

7. Ho JW et al (2011) ChIP-chip versus ChIP-seq: lessons for experimental design and data analysis. BMC Genomics 12:134

8. Cock PJ, Fields CJ, Goto N, Heuer ML, Rice PM (2010) The Sanger FASTQ file format for sequences with quality scores, and the Solexa/Illumina FASTQ variants. Nucleic Acids Res 38(6):1767–1771

9. Li R et al (2009) SOAP2: an improved ultrafast tool for short read alignment. Bioinformatics 25(15):1966–1967

10. Li H, Durbin R (2010) Fast and accurate long-read alignment with Burrows-Wheeler transform. Bioinformatics 26(5):589–595

11. Chen Y et al (2013) CGAP-align: a high performance DNA short read alignment tool. PLoS One 8(4), e61033

12. Schbath S et al (2012) Mapping reads on a genomic sequence: an algorithmic overview and a practical comparative analysis. J Comput Biol 19(6):796–813

13. Wilbanks EG, Facciotti MT (2010) Evaluation of algorithm performance in ChIP-seq peak detection. PLoS One 5(7), e11471

14. Malone BM, Tan F, Bridges SM, Peng Z (2011) Comparison of four ChIP-Seq analytical algorithms using rice endosperm H3K27 trimethylation profiling data. PLoS One 6(9), e25260

15. Bailey T et al (2013) Practical guidelines for the comprehensive analysis of ChIP-seq data. PLoS Comput Biol 9(11), e1003326

16. Merhej J et al (2014) bPeaks: a bioinformatics tool to detect transcription factor binding sites from ChIPseq data in yeasts and other organisms with small genomes. Yeast 31(10):375–391

17. Zhang Y et al (2008) Model-based analysis of ChIP-Seq (MACS). Genome Biol 9(9):R137

18. Kharchenko PV, Tolstorukov MY, Park PJ (2008) Design and analysis of ChIP-seq experiments for DNA-binding proteins. Nat Biotechnol 26(12):1351–1359

19. Spyrou C, Stark R, Lynch AG, Tavare S (2009) BayesPeak: Bayesian analysis of ChIP-seq data. BMC Bioinformatics 10:299

20. Landt SG et al (2012) ChIP-seq guidelines and practices of the ENCODE and modENCODE consortia. Genome Res 22(9):1813–1831

21. Blankenberg D et al (2010) Manipulation of FASTQ data with Galaxy. Bioinformatics 26(14):1783–1785

22. Li H et al (2009) The Sequence Alignment/Map format and SAMtools. Bioinformatics 25(16):2078–2079

23. Rozowsky J et al (2009) PeakSeq enables systematic scoring of ChIP-seq experiments relative to controls. Nat Biotechnol 27(1):66–75

24. Thomas-Chollier M et al (2012) RSAT peak-motifs: motif analysis in full-size ChIP-seq datasets. Nucleic Acids Res 40(4), e31

Chapter 12

Systematic Determination of Transcription Factor DNA-Binding Specificities in Yeast

Lourdes Peña-Castillo and Gwenael Badis

Abstract

Understanding how genes are regulated, decoding their "regulome", is one of the main challenges of the post-genomic era. Here, we describe the *in vitro* method we used to associate *cis*-regulatory sites with cognate *trans*-regulators by characterizing the DNA-binding specificity of the vast majority of yeast transcription factors using Protein Binding Microarrays. This approach can be implemented to any given organism.

Key words Transcription regulation, Transcription factors, DNA binding domain, *cis*-regulatory element, Enhancers, Binding sites

1 Introduction

Decoding transcription factor (TF)–DNA interaction is one of the crucial steps to understand how genes are regulated. Most known transcription factor binding sites (TFBSs) are short (6–10 bp) and degenerated. In addition, a particular TF may bind multiple binding sites with different affinity. Several factors such as combinatorial action of TFs and chromatin structure regulate gene expression but the first step to understand transcriptional regulation is to characterize individual binding sites.

To address this question, a variety of techniques have arisen in the last two decades; however, few of them are suitable for large-scale studies.

In vivo Chip-derived methods (Chip-Chip [1, 2], Chip-seq [3], Chip Pet [4]) require immunoprecipitation of the TF of interest, and have all been used to characterize numerous TFBSs in several organisms. Drawbacks of these methods are the requirement of specific antibodies, the restriction to TFs expressed and active in experimental conditions, and the likely detection of indirect interactions, which can scramble the motif definition.

Frédéric Devaux (ed.), *Yeast Functional Genomics: Methods and Protocols*, Methods in Molecular Biology, vol. 1361, DOI 10.1007/978-1-4939-3079-1_12, © Springer Science+Business Media New York 2016

In vitro Selex [5] is the oldest low scale method that identifies a set of bound sequences from a random collection of sequences; however, this method is biased by multiple steps of PCR. More modern and powerful versions of Selex have recently been described [6, 7].

Universal Protein Binding Microarray (PBM [8]) is an alternative in vitro method by which most yeast TFBSs have been characterized [9–11]. PBMs have also been used to characterize TFBSs in other organisms [12–14]. In standard PBM experiments, a GST-fused TF is allowed to bind a double stranded microarray containing a representation of all possible 10mer cut in 35mer pieces (see below and in ref. [8] for details). A second step consisting of an antibody labeling highlights spots where the TF is bound. This technique requires no PCR amplification and is highly sensitive and robust. This method is limited by the number of sequences that can be represented on a microarray, which determines the highest complexity of the motifs represented on the array. Consequently, TFs with long binding sites (>10 bp) may be difficult to characterize using this approach.

In this chapter, we provide details of the procedure we used to determine transcription factor DNA-binding specificities for numerous yeast TFs [9] using PBM experiments.

We explain how we rendered this large-scale study feasible, and describe how we computationally processed and analyzed the data.

2 Materials

2.1 Production and Purification of GST-Tagged Proteins

1. C41 DE3 cells.

2. LB^{amp}: 10 g/l Bacto-tryptone, 5 g/l Bacto-yeast extract, 10 g/l NaCl, pH 7.0, [Ampicillin]$_{final}$ = 100 μg/mL.

3. LB^{amp} + glucose: LB^{amp} + 2 g/l glucose.

4. IPTG: stock solution at 100 mM.

5. PBS pH 7.3: 137 mM NaCl, 2.7 mM KCl, 10 mM Na2HPO4, 2 mM KH2PO4.

6. Lysozyme: stock solution at 80 mg/ml.

7. Lysis buffer: 50 mM Tris (pH 8.0), 150 mM NaCl, 2 mM DTT (add fresh).

8. Glutathione sepharose 4B.

9. Wash buffer 1× PBS + 2 mM DTT (add fresh).

10. Elution buffer: 50 mM Tris pH7.5, reduced glutathione 10 mM, cOmplete tablet (Roche), 2 mM DTT (add fresh).

11. Zinc acetate 1 M.

12. ActivePro Kit (Ambion).

Table 1
Oligonucleotide sequences

Name	Sequence
Stilt sequence	5′-CTCACAATCTTGACGGCAGGCATGT-3′
RC Stilt	5′-ACATGCCTGCCGTCAAGATTG-3′

2.2 Protein Binding Microarray

1. Stilt RC primer (*see* Table 1) HPLC-purified (Integrated DNA Technologies).
2. dNTP.
3. Cy3 dUTP.
4. Thermo Sequenase™ DNA Polymerase.
5. Microarray, stainless steel hybridization chamber (Agilent).
6. Four-chamber gasket coverslip (Agilent).
7. LifterSlip coverslips (Erie Scientific).
8. ProScanArray HT Microarray Scanner (Perkin Elmer).
9. GenePix Pro version 6.0 software (Molecular Devices).
10. 10× sequenase reaction buffer (260 mM Tris–HCl, pH 9.5, 65 mM MgCl$_2$) in a total volume of 900 μl.
11. PBS: (phosphate buffered saline) NaCl 137 mM, KCl 2.7 mM, Na$_2$HPO$_4$ 10 mM, KH$_2$PO$_4$ 1.8 mM, pH 7.4.
12. Wash buffer A: PBS + 0.01 % (vol/vol) Triton X-100.
13. Wash buffer B: PBS + 0.1 % (vol/vol) Tween 20.
14. Wash buffer C: PBS + 0.5 % (vol/vol) Tween 20.
15. Wash buffer D: PBS + 0.05 % (vol/vol) Tween 20.
16. Blocking: 2 % (wt/vol) nonfat dried milk dissolved in PBS for 2 h (or overnight) and filtered using a 0.45 μm filter.
17. Alexa 488-conjugated rabbit polyclonal antibody to GST.
18. Salmon testes DNA.
19. Bovine serum albumin.
20. ZnAc 500×: 25 mM Zn acetate, ZnAc 100× = 5 mM Zn acetate.
21. Stripping solution: 10 mM EDTA, 10 % SDS, +210 units protease (Invitrogen, 5.8 units/mg) per 50 ml.

3 Methods

3.1 Experimental Design

The first step of a large-scale characterization of TF–DNA binding affinities is to determine the list of genes to assay and to generate a collection of GST-tagged TFs or DNA-binding domains (DBDs).

In our study, we determined that a region containing the DBD plus 15 flanking amino acids (aa) is sufficient and appropriated for most TFs, as shorter domains are easier to clone and give proteins that are simpler to express and produce. We observe no difference between PBMs obtained from full length or truncated TFs when we compared both; however, the majority of our trials with full length TFs failed to give a sufficient yield to properly run a PBM experiment. Note that the dimerization domain has to be added in the design for TFs expected to dimerize (such as those containing a Helix–Loop–Helix domain).

In order to define the domains to be tested, we selected a list of 36 distinct DBDs containing all the known examples of yeast specific DNA binding transcription factors [9]. In order to catalog all possible yeast transcription factors, we employed the software HMMER (version 2.3.2, available at http://hmmer.janelia.org/) [15] to generate profile hidden Markov models for all DBDs and scanned the yeast genome to detect those DBDs. We also scanned the SMART Database (http://smart.embl-heidelberg.de/) [16] to extend the search and selected a total of 212 independent ORFs containing one or more of the 36 selected domains [9]. Recent reviews [17, 18] estimate the number of known and putative yeast TF to 209.

For flexibility and cost, we created a Donor clone library compatible with the MAGIC system [19] using a ligation independent cloning strategy [20]. Donor clones can be easily transferred by bacterial conjugation into a glutathione-S-transferase (GST) N-terminal tag Recipient vector such as pTH1137, a T7-GST-tagged variant of pML280 [19]. Alternatively, a GATEWAY system [21] or any way to generate GST-fusion protein can be used.

3.2 Microarray Design

Random universal PBM array is a 4X44K customized microarray (Agilent) containing all possible 10-mer within 35-nucleotide probes generated by a De Bruijn sequence of order 10.

The design of this array is described in ref. [8]. The microarray designs we used in our study are variations of the original microarray. Details of the modifications can be found at http://hugheslab.ccbr.utoronto.ca/supplementary-data/yeastDBD/

For each TF, two versions of these arrays (A and B, corresponding to the same complexity) are used to perform replicate PBM experiments with two independently produced GST-tagged proteins. This allows testing the robustness of PBM reproducibility.

3.3 Expression and Purification of GST Tagged DBDs from E. coli

3.3.1 E. coli Cultures and Induction

1. C41 DE3 cells are transformed with a plasmid expressing the GST-fusion gene of interest under the control of a PTAC promoter using standard procedure. 200 ml of LB[amp] + glucose (+ Zn acetate if necessary, *see* **Note 1**) are inoculated with 2 ml of an overnight LB[amp] grown preculture and grown at 25 °C until OD600 is 0.5–0.8. Two milliliters of this "uninduced" culture is set aside in a "negative control" tube.

2. IPTG is added to the main culture to a final concentration of 1 mM.

3. Both cultures are grown at 14 °C overnight shaking. 2 ml of both cultures is saved for further control (*see* **Note 2**).

4. Cultures are centrifuged at 4 °C 15 min at 3200×*g*. Pellets are resuspended in 30 ml ice-cold wash buffer, transferred to a 50 ml Falcon tube and centrifuged at 4 °C 15 min at 3200×*g*.

5. Pellets are decanted and flash-frozen at −80 °C if needed, or can be directly continued to the lysis step described below.

3.3.2 Lysis

1. Pellets are resuspended in 25 ml lysis buffer.

2. From a stock concentration at 80 mg/ml, 160 μl of lysozyme is added so that 12.8 mg of lysozyme is used for a pellet obtained from a 200 ml culture, and incubated in ice 20 min.

3. Cells are lysed by sonication (*see* **Note 3**). Lysates are centrifuged at 4 °C 15 min at 3200×*g*. Cleared lysate are transferred to 50 ml Falcon tubes in ice and NaCl is added to obtain 250 mM final (*see* **Note 4**).

3.3.3 Purification

1. Two hundred microliters of glutathione sepharose beads are equilibrated in 5 ml PBS, rotating at 4 °C for 5 min and centrifuged at 4 °C 5 min at 100×*g*. Supernatants are removed.

2. About 25 ml of lysate are incubated with equilibrated glutathione beads, 1 h rotating at 4 °C, centrifuged at 100×*g* and supernatants are carefully removed.

3. Beads are washed twice with 10 ml PBS wash buffer, 10 min on the rotating wheel at 4 °C, spun down at 100×*g* and cleared from supernatant.

4. Beads are transferred into an Eppendorf tube, spun down at 100×*g* at 4 °C and cleared from supernatant. GST-tagged proteins are eluted with 200 μl elution buffer, 30 min to 1 h at 4 °C rotating.

5. Eluates are collected in a new tube after centrifugation at 4 °C 1 min at 100×*g*. Glycerol is added to each sample to 30 % final. GST-proteins are stoked at a concentration of at least 500 nM when possible. An aliquot is saved for control (by Western blot or SDS-PAGE) and samples are flash-frozen at −80 °C.

3.4 In Vitro Transcription/ Translation

In vitro transcription/translation are performed or proteins unsuitable for *in vivo* purification (such as those forming aggregates). This approach is done using ActivePro Kit and following the Manufacturer's instructions. Glycerol is added to a final concentration of 30 % to IVT samples prior to −80 °C storage. Note that *in vitro* transcribed/translated proteins can be used non-purified (from the kit mixture) in the PBM hybridization.

Molar concentrations of all in vitro translated proteins are determined by Western blot utilizing a dilution series of recombinant GST. Equal volumes of sample and known concentrations of GST are run on a standard Western blot procedure using anti-GST (dilution 1/5000) as a primary antibody, and anti-rabbit IgG-peroxydase (dilution 1/20,000). Concentrations are determined using Quantity One software version 4.5.0 according to the GST standard curve.

3.5   PBM Experiment

3.5.1   Making Agilent Arrays Double Stranded

1. Single-stranded oligonucleotide microarrays are double-stranded by primer extension using 1.17 μM RC stilt primer, 40 μM dATP, dCTP, dGTP, and dTTP, 1.6 μM Cy3 dUTP, 32 Units Thermo Sequenase™ DNA polymerase, and 90 μl 10× reaction buffer. The common primer RC stilt may be labeled (Cy5) to check for uniformity of primer annealing.

2. The reaction mixture, microarrays, stainless steel hybridization chamber, and four-chamber gasket coverslip are pre-warmed to 85 °C in a stationary hybridization oven and assembled according to the manufacturer's protocols.

3. After incubation at 85 °C for 10 min, 75 °C for 10 min, 65 °C for 10 min, and 60 °C for 90 min (*see* **Note 5**), the hybridization chamber is disassembled in 500 ml freshly made wash buffer A at 37 °C. Microarrays are transferred to a fresh dish, washed for 10 min in wash buffer A at 37 °C, washed once more for 3 min in PBS at 20 °C, and spun dry by centrifugation at 40 ×*g* for 1 min (*see* **Note 6**).

4. Double stranded microarrays are scanned for Cy3 (using a resolution of at least 5 μm, excitation 542 nm, emission 570 nm), to check Cy3-dUTP incorporation homogeneity in the reverse strand. Double-stranded microarrays can be stored in dark and dry conditions for months before using for PBM experiments.

3.5.2   Protein Binding Microarray Hybridization

1. Double-stranded microarrays are moistened in fresh wash buffer A for 5 min. Microarrays are blocked with 150 μl Blocking solution under LifterSlip coverslips for 1 h. During blocking, remove materials from freezer to thaw (zinc, BSA, DNA, protein, thaw on ice) and prepare the protein binding mixture.

2. The protein binding mixture is made of the purified TFs diluted to 100 nM (*see* **Note 7**) in a 175 μl final volume containing blocking solution, 51.3 ng/μl salmon testes DNA, and 0.2 μg/μl bovine serum albumin. Resulting mixtures are pre-incubated for 1 h at room temperature

3. Blocking microarrays are washed once with Wash buffer B for 5 min and once with Wash buffer A for 2 min.

4. Pre-incubated protein binding mixtures are applied to individual chambers of a four-chamber gasket coverslip in a steel

hybridization chamber, and the assembled microarrays are incubated for 1 h at room temperature.

5. The hybridization chambers are individually disassembled in 500 ml freshly made Wash buffer A. Microarrays are washed again once with wash buffer C for 5 min and once with Wash buffer A for 2 min.

6. Alexa488-conjugated rabbit polyclonal antibody to GST are diluted to 50 µg/ml in 1 ml blocking buffer and applied to a single-chamber gasket coverslip.

7. The assembled microarrays are again incubated for 1 h at room temperature, then individually disassembled in 500 ml freshly made wash buffer D.

8. Microarrays are then washed twice with wash buffer D for 3 min each, and once in PBS for 2 min. Slides are spun dry by centrifugation at $40 \times g$ for 5 min.

3.5.3 Microarray Stripping

1. After scanning (described below), in order to reuse double stranded microarrays (*see* **Note 8**), bound proteins and antibodies are digested from double-stranded microarrays with 50 ml stripping solution, rotating overnight at 10 rpm in a 50 ml Falcon tube at 37 °C.

2. Microarrays are washed three times for 5 min each in Wash buffer C, once for 5 min in PBS, and rinsed in PBS in a 500 ml staining dish (slowly removed to ensure removal of detergent and uniform drying).

3. Before reuse, slides are scanned once at the highest laser power for Alexa 488 (488 nm excitation (ex), 522 nm emission (em)) to confirm that no protein or antibody signal has remained.

3.5.4 Image Quantification and Data Normalization

1. Protein-bound microarrays are scanned on a ProScanArray HT Microarray Scanner to detect Alexa488-conjugated antibody (488 nm ex, 522 nm em) using three different laser power settings to best capture a broad range of signal intensities and ensure signal intensities below saturation for all spots.

2. Microarray TIFF images are analyzed using GenePix Pro version 6.0 software. Bad spots are manually flagged and removed. The three Alexa488 scans obtained at different laser power settings are combined using masliner software [22] available at http://arep.med.harvard.edu/masliner/supplement.htm.

There are several approaches for normalizing microarray data. Different approaches may be appropriated to PBMs and yield comparable results. In ref. [9], PBM data were normalized using the function justvsn() available in the Bioconductor package vsn [23]. Another normalization procedure applied to PBM data is described in ref. [24].

3.6 Obtaining Probe Sequences

To analyze PBM raw data, one needs to obtain the sequences corresponding to the probes on the microarray. The original universal 10-mer de Bruijn sequence microarrays described in ref. [8] are available via a End-User License Agreement (EULA) at http://the_brain.bwh.harvard.edu/UPBMseqn/UPBMseqn_agreement.html. The microarray designs we used are variations of the original design. All steps henceforth refer to the modified microarray design used in our study [9].

1. Go to http://the_brain.bwh.harvard.edu/UPBMseqn/UPBMseqn_agreement.html and download the excel file if you agree with the EULA.

2. Save the probe identifiers and the probe sequences for array de Bruijn #1.

3. Go to http://hugheslab.ccbr.utoronto.ca/supplementary-data/yeastDBD/ and download the two files with the probe ID mapping.

4. Remove the 25 nucleotides at the end of each sequence (3′ end) in de Bruijn #1 arrays corresponding to the common primer GTCTGTGTTCCGTTGTCCGTGCTGT.

5. Follow the instructions available at http://hugheslab.ccbr.utoronto.ca/supplementary-data/yeastDBD/README to obtain the probe sequences on the two arrays used in ref. [9].

6. Extract the overlapping 8-mers represented on each probe sequence. Note that an 8-mer and its reverse complement are considered to represent the same feature. For example, probe sequences containing either "AAAAAACC" or "GGTTTTTT" are group together as containing the same 8-mer.

7. Write a tab-delimited text file containing the probe identifier in the first column and the 8-mers contained on each probe in the second column (one 8-mer per line). For example, the first six lines of such a file might look as follows:

    ProbeID   Kmer

    TRHyeSpot40330   AAAAAAAA

    TRHyeSpot40330   AAAAAAAA

    TRHyeSpot40330   AAAAAAAA

    TRHyeSpot40330   CAAAAAAA

    TRHyeSpot40330   TCAAAAAA

    TRHyeSpot40330   TTCAAAAA

A Perl script to perform **steps 4–7** is available in the supplementary material provided with this article.

3.7 Obtaining 8-mer Affinity Measurements

Preference of a transcription factor for each 8-mer is represented using three different values: median intensity, robust Z-score [25], and Enrichment-score (E-score [8]). To do all computational steps to obtain these 8-mer based values, we adopted R. Advantages of using R are an integrated interactive environment for analysis and visualization, and the availability of many functions and tools. Furthermore, R has often been adopted for bioinformatics protocols (e.g., [26]). In what follows, all R commands and their output appear in Courier New font. Commands are preceded by a > sign. Note that in this protocol we use the <- notation for variable assignment in R. Computation time is based on a 2-core MacBook Air machine with 8 GB in RAM. If no time is given, the step takes less than 5 min to complete.

1. Read in the Probe to 8-mer mapping file such as the one produced in the previous section (here named ArrayA_probesIDs_2_8mers.txt and assumed to be in a directory called YeastData) by typing in the R console:

   > probe_kmer_mapping <- read.table("YeastData/ArrayA_probesIDs_2_8mers.txt", sep = "\t", stringsAsFactors = FALSE, header = TRUE)
   > head(probe_kmer_mapping)
   R output:
   ProbeID   Kmer
   1 TRHyeSpot40330   AAAAAAAA
   2 TRHyeSpot40330   AAAAAAAA
   3 TRHyeSpot40330   AAAAAAAA
   4 TRHyeSpot40330   CAAAAAAA
   5 TRHyeSpot40330   TCAAAAAA
   6 TRHyeSpot40330   TTCAAAAA

2. Read in the probe intensities file. This file contains a table with the probe IDs as rows and the intensity measurements for each probe per microarray as columns. The file Array_A_35mer_raw_data.txt containing data for 118 arrays available at http://hugheslab.ccbr.utoronto.ca/supplementary-data/yeast-DBD/ is used to demonstrate the following steps.

   > Data <- read.table("YeastData/Array_A_35mer_raw_data.txt", sep = "\t", stringsAsFactors = FALSE, header = TRUE, row.names = 1)
   > dim(Data)
   [1] 43803 118
   > head(rawData[,1:2])
   R output:
   ABF1_4505.2_ArrayA   ABF2_2116.1_ArrayA.1
   TRHyeControl100_DT_100   1433.298   5184.860
     TRHyeControl101_DT_101   2503.233   3372.940
   TRHyeControl102_DT_102   2158.167   6091.378
   TRHyeControl103_DT_103   1255.000   4197.835
   TRHyeControl104_DT_104   1879.434   9360.506
   TRHyeControl105_DT_105   1901.071   4519.405

3. Assemble a table with the probe IDs, corresponding 8-mers and probe intensities.

```
> fullTable <- merge(probe_kmer_mapping, Data, by.x =
"ProbeID", by.y = "row.names")
> head(fullTable[,1:3])
```
R output:

```
ProbeID   Kmer   ABF1_4505.2_ArrayA
1 TRHyeControl1_DT_1   CATCGACC   1843.926
2 TRHyeControl1_DT_1   CCATCGAC   1843.926
3 TRHyeControl1_DT_1   CCCATCGA   1843.926
4 TRHyeControl1_DT_1   CCCCATCG   1843.926
5 TRHyeControl1_DT_1   CCCCCATC   1843.926
6 TRHyeControl1_DT_1   ACCCCCAT   1843.926
```

4. Compute the median intensity and log median intensity for each 8-mer per experiment.

```
> median_intensity <- sapply(3:ncol(fullTable), function(i) {
tapply(fullTable[,i], fullTable[,"Kmer"], median)
})
> colnames(median_intensity) <- colnames(fullTable)
[3:ncol(fullTable)]
> dim(median_intensity)
```
R output:
```
[1] 32896 118
> head(median_intensity[,1:2])
```
R output:

```
ABF1_4505.2_ArrayA   ABF2_2116.1_ArrayA.1
AAAAAAAA   2044.320   7044.084
AAAAAAAC   1844.253   7246.297
AAAAAAAG   2107.263   6254.257
AAAAAAAT   1950.312   7073.151
AAAAAACA   1847.276   7350.742
AAAAAACC   1971.743   7378.000
```

```
> log_median_intensity <- log(median_intensity)
```

5. Calculate the robust Z-score per 8-mer per experiment. The robust Z-score is the number of median absolute deviations (MAD) away from the overall median intensity.

```
> getZscore <- function(mi){(mi - median(mi)) / mad(mi)}
> zscore <- apply(log_median_intensity, 2, getZscore)
> dim(zscore)
```
R output:
```
[1] 32896 118
> head(zscore[,1:2])
```

R output:
ABF1_4505.2_ArrayA   ABF2_2116.1_ArrayA.1

AAAAAAAA   2.942674   -0.079967633
AAAAAAAC   0.468769   0.482131694
AAAAAAAG   3.671098   -2.441892575
AAAAAAAT   1.811884   0.001816074
AAAAAACA   0.508110   0.766348393
AAAAAACC   2.074392   0.839859899

6. Obtain a table with the ranks of the probes in descending order by their intensity (i.e., the rank of probe with the highest intensity is 1) per experiment.
   > assignRanks <- function(intensities){
   length(intensities) - rank(intensities, ties.method = "first") + 1
   }
   > ranksTable <- apply(Data, 2, assignRanks)
   > dim(ranksTable)
   R output:
   [1] 43803 118
   > head(ranksTable[,1:2])
   R output:
   ABF1_4505.2_ArrayA   ABF2_2116.1_ArrayA.1

   TRHyeControl100_DT_100   41017   39360
   TRHyeControl101_DT_101   3093   43704
   TRHyeControl102_DT_102   8135   32012
   TRHyeControl103_DT_103   43547   43029
   TRHyeControl104_DT_104   18268   5116
   TRHyeControl105_DT_105   17199   42280

7. Assemble a table with the probe IDs, corresponding 8-mers and probe ranks.
   > ranksTableFull <- merge(probe_kmer_mapping, ranksTable, by.x = "ProbeID", by.y = "row.names")
   > dim(ranksTableFull)
   R output:
   [1] 1226484 120
   > head(ranksTableFull[,1:3])
   R output:
   ProbeID   Kmer   ABF1_4505.2_ArrayA

   1 TRHyeControl1_DT_1   CATCGACC   19982
   2 TRHyeControl1_DT_1   CCATCGAC   19982
   3 TRHyeControl1_DT_1   CCCATCGA   19982
   4 TRHyeControl1_DT_1   CCCCATCG   19982
   5 TRHyeControl1_DT_1   CCCCCATC   19982
   6 TRHyeControl1_DT_1   ACCCCCAT   19982

8. Calculate the E-score per 8-mer per experiment. The E-score of an 8-mer is the subtraction of the average rank of the top

half of the probes in which the 8-mer is absent minus the average rank of the top half of the probes in which the 8-mer occurs divided by the total number of probes in both top halves [8]. For example, suppose we have 200 probes from which ten contain a given 8-mer. The E-score of this 8-mer is obtained by subtracting the average rank of the 95 brightest probes in which the 8-mer is absent minus the average rank of the five brightest probes in which the 8-mer occurs, and dividing the result of this subtraction by 100 (*see* **Note 9**).

```
#Exact calculation - slow
> get_Escores_exact <- function(ranks, numProbes){
keepFraction <- 0.5
sortRanks <- sort(ranks)
#ranks of background probes; i.e., those without the 8mer
ranksb <- setdiff(1:numProbes, sortRanks)
n <- trunc(length(sortRanks) * keepFraction)
m <- trunc((numProbes - length(sortRanks)) * keepFraction)
pf <- sortRanks[1:n]
pb <- ranksb[1:m]
(sum(pb) / m - sum(pf)/n) / (m+n)
}
```

9. Alternatively the E-score can be approximated by the area under the receiver operating characteristic curve (AUC) minus 0.5 as it is done in the "seed_and_wobble.pl" program accompanying [24]. The following R code is based on the E-score approximation done in the "seed_and_wobble.pl" program. This function is much faster than the exact calculation done in the previous step (Timing ~ 15 min).

```
> get_Escores_approx <- function(ranks, numProbes){
keepFraction <- 0.5
sortRanks <- sort(ranks)
n <- trunc(length(sortRanks) * keepFraction)
m <- trunc((numProbes - length(sortRanks)) * keepFraction)
ranksum <- sum(sapply(1:n, function(i) {
if (sortRanks[i] - i > m) {
m+i-1
} else {
sortRanks[i]
}
}))
((n^2+n)/2 + n*m/2 - ranksum) / (n*m)
}
> E_score <- sapply(3:ncol(ranksTableFull), function(i) {
tapply(ranksTableFull[,i], ranksTableFull[,"Kmer"], get_Escores_approx, nrow(ranksTable))
})
```

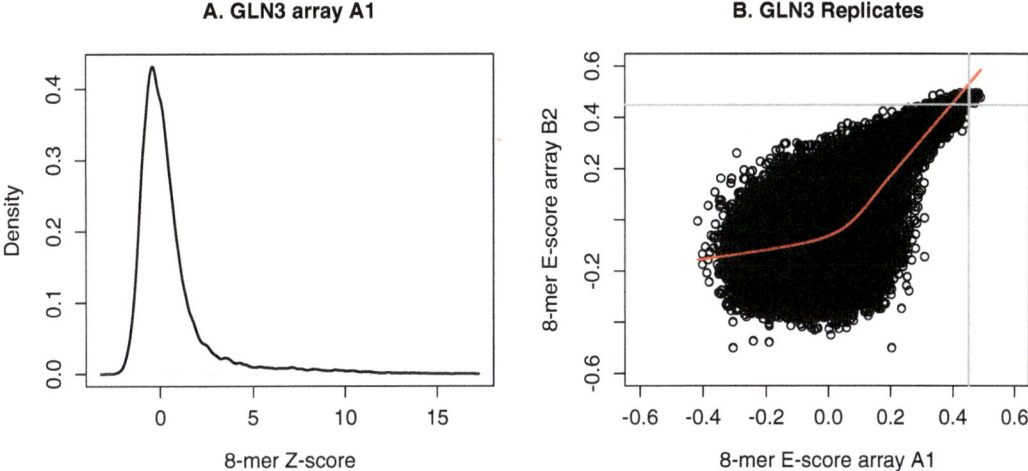

Fig. 1 (**a**) Distribution of Z-scores of a successful array for the TF GLN3. Note the long right tail of the distribution. (**b**) Correlation of 8-mer E-scores for the TF GLN3 obtained from two PBM experiments performed on microarrays of different designs. The *red line* is the loess-smoothed line. The *vertical* and *horizontal gray lines* indicate the 0.45 E-score

> colnames(E_score)   <-   colnames(ranksTableFull)
[3:ncol(ranksTableFull)]
> dim(E_score)
R output:
[1] 32896 118
> head(E_score[,1:2])
R output:
ABF1_4505.2_ArrayA ABF2_2116.1_ArrayA.1
AAAAAAAA   0.2518629   -0.06358507
AAAAAAAC   0.2134302   -0.02259507
AAAAAAAG   0.3254102   -0.14840729
AAAAAAAT   0.2198637   0.09712511
AAAAAACA   0.0913399   0.04167966
AAAAAACC   0.2695739   -0.08644905

10. As a sanity check, check the E-score and Z-score distribution. A PBM experiment is considered successful if it has at least one 8-mer with an E-score above 0.45 and the Z-score distribution shows a long right tail (Fig. 1a). Additionally, Z-scores of independent PBM experiments, done with the same TF, exhibit positive correlation (Fig. 1b).

All 8-mer based values for TFs studied in ref. [9] are available in NCBI Gene Expression Omnibus (GEO) under accession GSE12349.

3.8 Comparing 8-mer Profiles Between TFs of the Same Family

Using the 8-mer profiles of various TFs, we can compare DNA binding specificities of TFs of the same family. For example, Fig. 2 shows a comparison of the 8-mer E-scores for two yeast TFs of the GATA family with distinct motifs, GAT3 and GZF3; while Fig. 3 shows a comparison of the 8-mer E-scores for two yeast TFs of the same GATA family that share the same primary motif, GLN3 and GZF3.

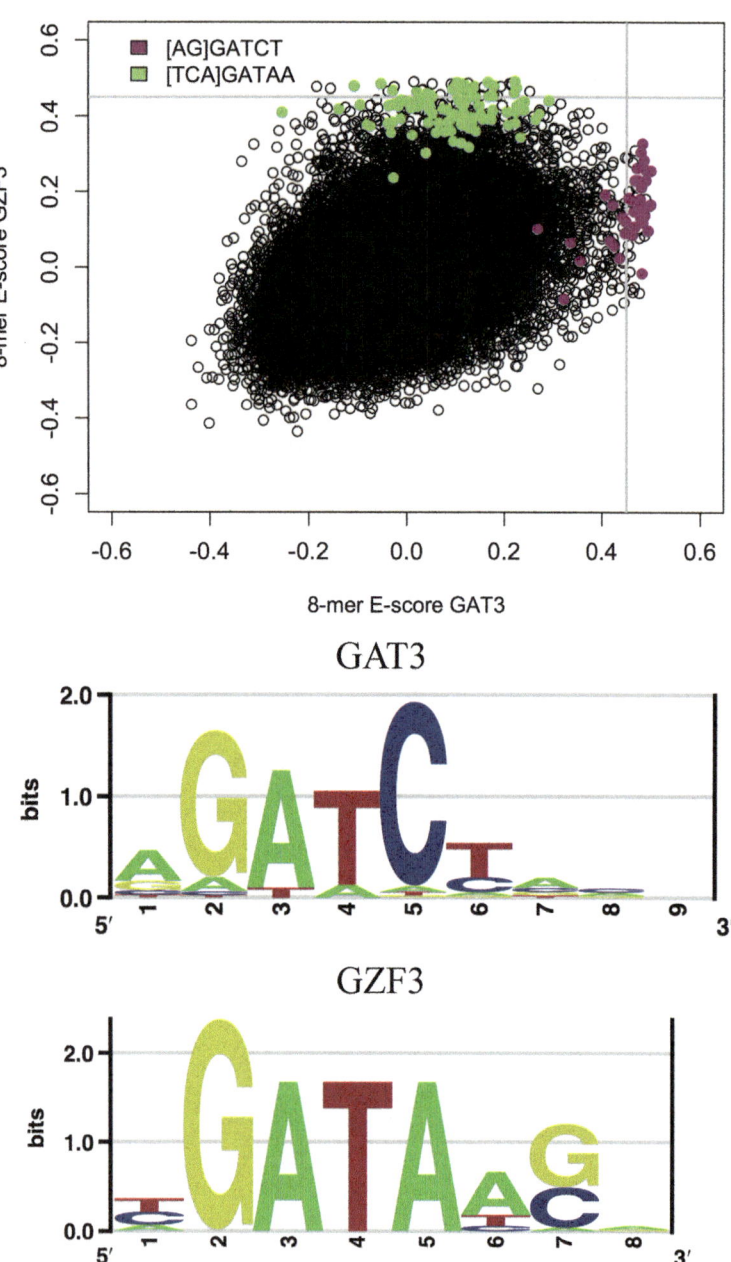

Fig. 2 *Top*: Scatter plot comparing 8-mer *E*-scores for two yeast TFs of the GATA zinc finger family, GAT3 and GZF3. The *highlighted dots* representing 8-mers containing the 6-mers indicated on the *top left* corner of the plot show a clear difference in the sequence preference of these TFs. *Bottom*: Sequence logos of the TFBS of both TFs. Sequence logos were created using enoLOGOS [34]

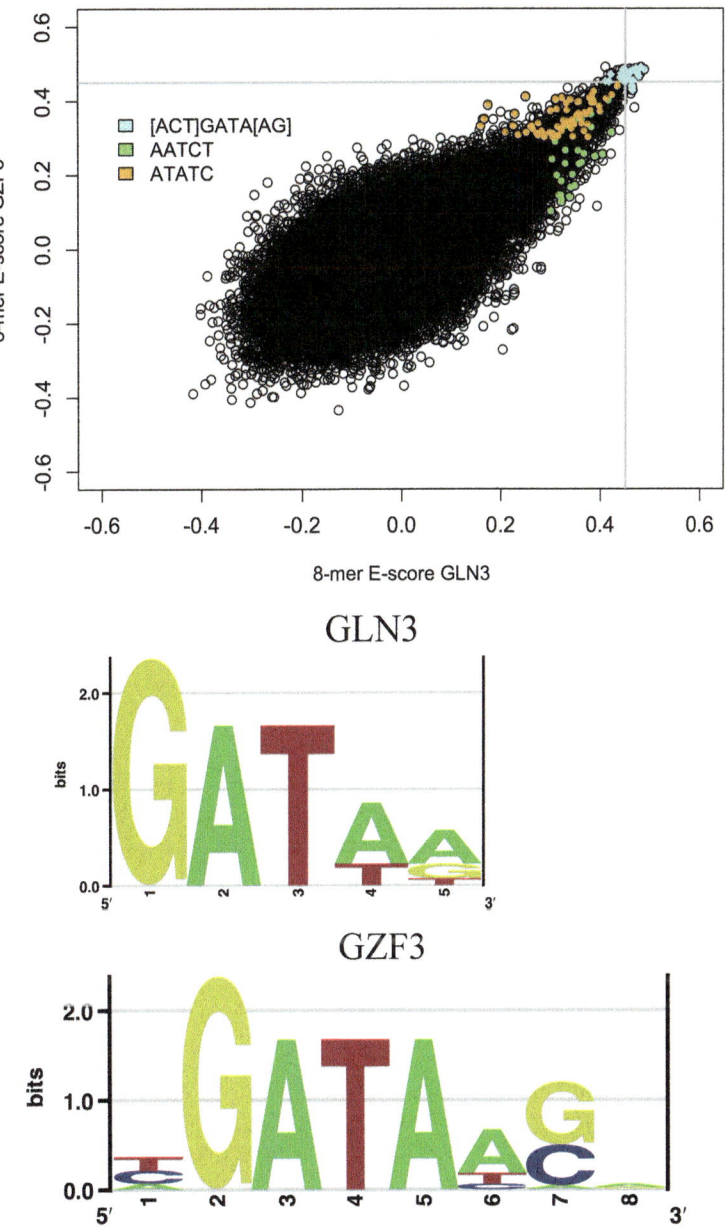

Fig. 3 *Top*: Scatter plot comparing 8-mer E-scores for two yeast TFs of the GATA zinc finger family, GLN3 and GZF3. *Blue dots* represent 8-mers containing "AGATAA", "AGATAG", "CGATAA", "CGATAG", "TGATAA", or "TGATAG" and with an E-score > 0.45 for either of the two TFs. These TFs show identical preferences for the same highest-scoring 8-mers. *Green* and *yellow dots* represent 8-mers containing respectively "AATCT" and "ATATC" with an E-score > 0.3 for either of the two TFs. The distribution of these *dots* in the scatter plot indicates a difference in lower affinity sequence preferences between GLN3 and GZF3. *Bottom*: Sequence logos of the TFBS of both TFs. Sequence logos were created using enoLOGOS [34]

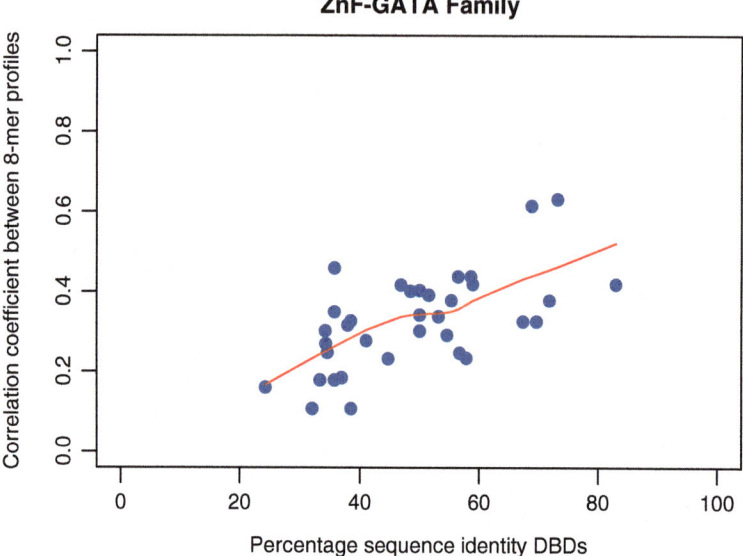

Fig. 4 Similarity between 8-mer profiles across TFs of the GATA zinc finger family as a function of the percentage of sequence identity across the DNA-binding domains of these TFs. The more similar the sequences of the DBDs are, the more similar the 8-mer profiles. The *red line* is the loess-smoothed line

There is a relation between 8-mer profiles and sequence similarity for TFs of the same family. Figure 4 shows this relation for TFS of the yeast zinc finger GATA family. Observation of this fact and the availability of 8-mer profiles produced by PBMs allows to apply machine learning techniques that infer binding preference of a TFs using the k-mer affinity information available for other family members (e.g., refs. [27, 28]). R offers several packages to apply techniques such as random forests (RFs), k-nearest neighbor (KNN), and support vector machines.

3.9 Obtaining DNA Sequences Motifs from Top-Scoring 8-mers

There are several models to represent the DNA sequence specificity of a TF and several methods to obtain such a model from a set of sequences. Position weight matrices (PWMs) are the predominant paradigm to represent DNA motifs bound by a TF. A PWM models the DNA sequence preference of a TF as matrix with a row for each symbol in the alphabet (i.e., A, C, G, and T) and a column for each position of the TFBS (i.e., number of columns is equal to the length of the TFBS). Each column provides a score per nucleotide representing the relative preference for the given base at that position in the binding site. State of the art algorithms and paradigms to represent TFBS have recently been evaluated [29]. Based on this evaluation, the best performing PWM-based method is BEEML-PBM [30] and the best 8-mer based method is FeatureREDUCE (http://bussemakerlab.org/people/ToddRiley/featurereduce.html).

BEEML-PBM is available at http://stormo.wustl.edu/beeml/. This method is written in R and requires as input a two-column table with the normalized intensities and probe sequences, and a PWM as a seed. This seed PWM can be either one obtained by another method, one available in the literature or one from a TF of the same family.

3.10 Seeking Transcription-Factor Binding Sites (TFBS) onto Promoter Region

In addition to determine the sequence specificities of a TF and represent this specificities as a PWM, one usually wants to identify genes being regulated by this TF. Putative targets of a TF can be determined by finding genes whose promoter region contains the motif bound by that TF. It is possible to do all computational steps to identify TFBSs within R. In the following steps, we continue using the same notation as in Subheading 3.4. These steps were adapted from the Bioconductor [31] workflow available at http://www.bioconductor.org/help/workflows/generegulation/

1. Read into R the PWMs of the TFs of interest. An excel file with PWMs for the yeast TFBS determined in ref [9] is available at http://hugheslab.ccbr.utoronto.ca/supplementary-data/yeastDBD/. Assume we have extracted PWMs from this excel file into tab-delimited text files ending with "_PWM.txt" in the directory YeastData. We can then read all these files and converted the PWMs into count matrices by typing into the R console:

   ```
   > files <- list.files("YeastData", pattern = "*_PWM.txt",
   include.dirs = TRUE, recursive = TRUE, full.names = TRUE)
   > PWMs <- sapply(files, read.table, sep = "\t", stringsAsFac-
   tors = FALSE, header = TRUE, row.names = 1)
   > names(PWMs) < gsub(".*/", "", gsub("_PWM.txt", "",
   files), perl = TRUE)
   > PCMs <- lapply(PWMs, function(pwm) {round(100 *
   pwm)})
   > names(PCMs)
   [1] "GAT3" "GLN3" "GZF3"
   > PCMs[["GAT3"]]
   R output:
   ```

	X1	X2	X3	X4	X5	X6	X7	X8	X9
A	65	8	92	12	3	6	51	28	32
C	9	2	0	0	94	27	19	33	25
G	20	88	0	1	1	4	18	21	21
T	6	1	8	87	2	63	12	17	21

2. Obtain the promoter sequence of the genes in whose promoter region one wants to look for a TFBS. Note that the genes must be listed using their systematic name. In this example, we are using 15 genes listed as targets of GZF3 in Saccharomyces Genome Database [32].

```
> ORFs <- read.table("YeastData/GZF3_ORF_targets.txt",
header = FALSE, stringsAsFactors = FALSE)
> ORFs<- ORFs[,1,drop = TRUE]
> ORFs[1:5]
R output:
[1] "YCL025C" "YPR171W" "YBL042C" "YKR039W"
"YFL021W"
> library(GenomicFeatures)
> library(BSgenome.Scerevisiae.UCSC.sacCer3)
> library(TxDb.Scerevisiae.UCSC.sacCer3.sgdGene)
> transcripts_coordinates <- transcriptsBy(TxDb.Scerevisiae.
UCSC.sacCer3.sgdGene, by = "gene")[ORFs]
> promoter.seqs <- getPromoterSeq(transcripts_coordinates,
Scerevisiae, upstream = 1000, downstream = 0)
> head(promoter.seqs, n=3)
R output:
DNAStringSetList of length 6
[["YCL025C"]]     CTGAAAGAGCGCCTTTACCTCAA
CCTACCATGGCAAACATAACAGAAAACATA
AAAAAATTATCCTAGAGCCCAATGTTCCATGAAA
AGAGCTGTGGCAAGGACAGAAACAAAAAAA
AATCAAGAACTCAACATTA...
[["YPR171W"]]     CTGATGTTCAGTAAAGCCGCCT
AGCTTTACGTGCCGAAATATTGATAATA
TGTCTCAGCCACTTCCTGGCTT
AACTATTTAAATGATATTTCTGCATCCATCG
GTATGGCGCACAATAAACGGTAT
CTGAGAATATC...
[["YBL042C"]] GCAATAGTGGCCATATTTTGTTTAAC
TTTATAGTTCAATAGTCTTGGCTACTCTCTTTC
CAACTCAGTTCACCTTGTATTATACCGCTTGT
TTTTGCCACCCTTTGAGTTTCCTCGATCCT
TTAAGTTGGAAAAGAT...
> promoter.seqs <- unlist(promoter.seqs)
> head(promoter.seqs, n = 3)
R output:
A DNAStringSet instance of length 3
width seq names
[1] 1000 CTGAAAGAGCGCCTTTACCTCAACCTACCA
TGGCAAACATAACAGAAAACATAAAAAAAT...
GTTTATTATGTAATCTTTATAGAAGAAGCACGCTAATA
TAGACAAAGATAGCTTCGCACA YCL025C
[2] 1000 CTGATGTTCAGTAAAGCCGCCTAGCTTTAC
GTGCCGAAATATTGATAATATGTCTCAGCC...
ATTCTAATCAATAAAGTCACAGTAACCAGCTTTTCC
TAGCTTTTCGAAGTTTCGGAAGT YPR171W
[3] 1000 GCAATAGTGGCCATATTTTGTTTAACTTTAT
AGTTCAATAGTCTTGGCTACTCTCTTTCC...CATTGCG
GAAATAAAAGGCGGTAACTAGTCCTCTCATTCA
TTAATTCTATATAAGAGAAA YBL042C
```

3. Find matches of the motifs in the promoter sequences obtained in the previous step. After executing the first two commands, pwm.hits contains a list per TF containing the locations of putative TFBSs per gene.

> library(Biostrings)

> pwm.hits <- lapply(PCMs, function(pwm) {

sapply(promoter.seqs, function(pseq, pwm) {matchPWM(pwm, pseq, min.score = "90 %")}, as.matrix(pwm))

})

> names(pwm.hits)

R output:

[1] "GAT3" "GLN3" "GZF3"

> head(pwm.hits[["GAT3"]], n = 2)

R output:

$YCL025C

Views on a 1000-letter DNAString subject

subject: CTGAAAGAGCGCCTTTACCTCAACCTACCAT
G G C A A A C A T A A C A G A A A A C A T A A A A A A T
T A T C C T A G A G C ... A T G T A G A A C A A G T T T
A T T A T G T A A T C T T T A T A G A A G A A G C A C G
CTAATATAGACAAAGATAGCTTCGCACA

views: NONE

$YPR171W

Views on a 1000-letter DNAString subject

subject: CTGATGTTCAGTAAAGCCGCCTAGCTTTACG
T G C C G A A A T A T T G A T A A T A T G T C T C A
G C C A C T T C C T G G C T ... T T T A T A T A T
G A A T T C T A A T C A A T A A A A G T C A C A G T A A C C
AGCTTTTCCTAGCTTTTCGAAGTTTCGGAAGT

views: NONE

> head(pwm.hits[["GZF3"]], n = 2)

R output:

$YCL025C

Views on a 1000-letter DNAString subject

subject: CTGAAAGAGCGCCTTTACCTCAACCTACCA
T G G C A A A C A T A A C A G A A A A C A T A A A A A A
A T T A T C C T A G A G C ... A T G T A G A A C A A
G T T T A T T A T G T A A T C T T T A T A G A A G A A G
CACGCTAATATAGACAAAGATAGCTTCGCACA

views:

start end width

```
[1] 571 578 8 [AGATAAGC]
[2] 747 754 8 [TGATAAGA]
$YPR171W
```

Views on a 1000-letter DNAString subject

```
subject:              CTGATGTTCAGTAAAGCCGCCTAGC
T T T A C G T G C C G A A A T A T T G A T A A T
A T G T C T C A G C C A C T T C C T G G C T ...
TTTATATATGAATTCTAATCAATAAAAGTCACAGTAAC
CAGCTTTTCCTAGCTTTTCGAAGTTTCGGAAGT
```

views:

start end width

```
[1] 43 50 8 [TGATAATA]
> sessionInfo( )
```

R output:

R version 3.0.2 (2013-09-25)

Platform: x86_64-apple-darwin10.8.0 (64-bit)

locale:

```
[1]    en_CA.UTF-8/en_CA.UTF-8/en_CA.UTF-8/C/en_CA.
UTF-8/en_CA.UTF-8
```

attached base packages:

```
[1] parallel stats graphics grDevices utils datasets methods base
```

other attached packages:

```
[1]    BSgenome.Scerevisiae.UCSC.sacCer3_1.3.19   BSgenome_
1.28.0 Biostrings_2.30.1
```

```
[4]              TxDb.Scerevisiae.UCSC.sacCer3.sgdGene_2.9.0
GenomicFeatures_1.12.4 AnnotationDbi_1.24.0
```

```
[7] Biobase_2.22.0 GenomicRanges_1.14.3 XVector_0.2.0
```

```
[10] IRanges_1.20.6 BiocGenerics_0.8.0
```

loaded via a namespace (and not attached):

```
[1]    biomaRt_2.18.0    bitops_1.0-6  DBI_0.2-7  RCurl_1.95-4.1
Rsamtools_1.12.4 RSQLite_0.11.4 rtracklayer_1.20.4
```

```
[8] stats4_3.0.2 tools_3.0.2 XML_3.95-0.2 zlibbioc_1.6.0
```

PWM is a classical model to represent TFBS. It allows summarizing sequence binding information of a TF obtained by various methods into a single motif (*see* JASPAR database as an example [33], and it is easily represented as a sequence logo to visualize the motif with the highest affinity.

The main advantage of PBM experiments is the possibility to generate comprehensive k-mer profiles, which provide more detailed and extensive information on binding affinities. Secondary motifs may be thus revealed by the k-mer profile, exhibiting different sequences and affinities than the main motif (*see* Fig. 3 for an example). These secondary motifs might be excluded from the PWM representation. Such secondary motifs, possibly enhanced by cofactors under physiological conditions, might be relevant in vivo.

4 Notes

1. For zinc finger proteins only, add zinc acetate to all buffers (including LB media, PBS, Wash buffer, *etc.*) to a final concentration of 50 μM.

2. It is important to check on a SDS-PAGE gel the the-GST fusion protein inductions in the IPGT induced and uninduced sample running the crude extract obtain from 2 ml of both cultures on a SDS-PAGE gel.

3. Sonication settings:
   Automatic setting: Pulse 1 s; Rest 3 s. Total pulse time: 2 min, Amplitude 60 min.

   Note that sonication settings depend on the model of sonicator being used. The probe size is usually ½ or ¾ in. Sonication process should be modified for the type of probe, cell, *etc.*

4. The NaCl is used to decrease unspecific ionic binding of proteins to the GSTbeads.

5. Temperature is gradually decreased to ensure proper annealing of the RC stilt primer to template DNA probed on the array. Hybridization can be performed on a Tecan Hybridization station is available, in this case, all the buffers must be filter-sterilized.

6. All washes are performed in a 50 ml Falcon tube at room temperature on a wheel rotating at 10 rpm.

7. The optimal molarity depends on the Kd of each protein. 100 nM is an optimized concentration that we determined experimentally and apply to all our TFs but a range from 5 to 200 nM (empirically determined) is possible, depending on proteins.

8. Microarray can be reused two—without any loss—to up to four times to keep a good quality of signal.

9. The function in Subheading 3.4, **step 8** does the exact E-score calculation; note however that this exact calculation is quite slow (timing ~8 min per TF). We recommend to use instead the function defined in Subheading 3.4, **step 9**.

Acknowledgment

We thank Shaheynoor Talukder for standard operating procedure and Timothy R. Hughes for data availability. We also thank Esther T. Chan for useful comments. G.B. work was supported by the CIHR, the Institut Pasteur and the Centre National pour la Recherche Scientifique. LPC's work was supported by a NSERC Discovery Grant and Memorial University of Newfoundland.

References

1. Ran B, Robert F, Wyrick JJ et al (2000) Genome-wide location and function of DNA binding proteins. Science 290:2306–2309

2. Iyer VR, Horak CE, Scafe CS et al (2001) Genomic binding sites of the yeast cell-cycle transcription factors SBF and MBF. Nature 409:533–538

3. Johnson DS, Mortazavi A, Myers RM et al (2007) Genome-wide mapping of *in vivo* protein-DNA interactions. Science 316:1497–1502

4. Wei C-L, Wu Q, Vega VB et al (2006) A global map of p53 transcription-factor binding sites in the human genome. Cell 124:207–219

5. Oliphant AR, Brandl CJ, Struhl K (1989) Defining the sequence specificity of DNA-binding proteins by selecting binding sites from random-sequence oligonucleotides: analysis of yeast GCN4 protein. Mol Cell Biol 9:2944–2949

6. Zykovich A, Korf I, Segal DJ (2009) Bind-n-Seq: high-throughput analysis of *in vitro* protein-DNA interactions using massively parallel sequencing. Nucleic Acids Res 37, e151

7. Jolma A, Kivioja T, Toivonen J et al (2010) Multiplexed massively parallel SELEX for characterization of human transcription factor binding specificities. Genome Res 20:861–873

8. Berger MF, Philippakis AA, Qureshi AM et al (2006) Compact, universal DNA microarrays to comprehensively determine transcription-factor binding site specificities. Nat Biotechnol 24:1429–1435

9. Badis G, Chan ET, van Bakel H et al (2008) A library of yeast transcription factor motifs reveals a widespread function for Rsc3 in targeting nucleosome exclusion at promoters. Mol Cell 32:878–887

10. Gordân R, Murphy KF, McCord RP et al (2011) Curated collection of yeast transcription factor DNA binding specificity data reveals

novel structural and gene regulatory insights. Genome Biol 12:R125

11. Zhu C, Byers KJRP, McCord RP et al (2009) High-resolution DNA-binding specificity analysis of yeast transcription factors. Genome Res 19:556–566

12. Berger MF, Badis G, Gehrke AR et al (2008) Variation in homeodomain DNA binding revealed by high-resolution analysis of sequence preferences. Cell 133:1266–1276

13. Badis G, Berger MF, Philippakis AA et al (2009) Diversity and complexity in DNA recognition by transcription factors. Science 324:1720–1723

14. Busser BW, Huang D, Rogacki KR et al (2012) Integrative analysis of the zinc finger transcription factor Lame duck in the Drosophila myogenic gene regulatory network. Proc Natl Acad Sci U S A 109:20768–20773

15. Finn RD, Clements J, Eddy SR (2011) HMMER web server: interactive sequence similarity searching. Nucleic Acids Res 39:W29–W37

16. Schultz J, Copley RR, Doerks T et al (2000) SMART: a web-based tool for the study of genetically mobile domains. Nucleic Acids Res 28:231–234

17. Hughes TR, de Boer CG (2013) Mapping yeast transcriptional networks. Genetics 195:9–36

18. de Boer CG, Hughes TR (2011) YeTFaSCo: a database of evaluated yeast transcription factor sequence specificities. Nucleic Acids Res 40:D169–D179

19. Li MZ, Elledge SJ (2005) MAGIC, an *in vivo* genetic method for the rapid construction of recombinant DNA molecules. Nat Genet 37:311–319

20. Aslanidis C, de Jong PJ (1990) Ligation-independent cloning of PCR products (LIC-PCR). Nucleic Acids Res 18:6069–6074

21. Walhout AJ, Temple GF, Brasch MA et al (2000) GATEWAY recombinational cloning: application to the cloning of large numbers of open reading frames or ORFeomes. Methods Enzymol 328:575–592

22. Dudley AM, Aach J, Steffen MA et al (2002) Measuring absolute expression with microarrays with a calibrated reference sample and an extended signal intensity range. Proc Natl Acad Sci 99:7554–7559

23. Huber W, von Heydebreck A, Sueltmann H et al (2002) Variance stabilization applied to microarray data calibration and to the quantification of differential expression. Bioinformatics 18(Suppl 1):S96–S104

24. Berger MF, Bulyk ML (2009) Universal protein-binding microarrays for the comprehensive characterization of the DNA-binding specificities of transcription factors. Nat Protoc 4:393–411

25. Birmingham A, Selfors LM, Forster T et al (2009) Statistical methods for analysis of high-throughput RNA interference screens. Nat Methods 6:569–575

26. Anders S, McCarthy DJ, Chen Y et al (2013) Count-based differential expression analysis of RNA sequencing data using R and bioconductor. Nat Protoc 8:1765–1786

27. Alleyne TM, Pena-Castillo L, Badis G et al (2009) Predicting the binding preference of transcription factors to individual DNA k-mers. Bioinformatics 25:1012–1018

28. Christensen RG, Enuameh MS, Noyes MB et al (2012) Recognition models to predict DNA-binding specificities of homeodomain proteins. Bioinformatics 28:i84–i89

29. Weirauch MT, Cote A, Norel R et al (2013) Evaluation of methods for modeling transcription factor sequence specificity. Nat Biotechnol 31:126–134

30. Zhao Y, Stormo GD (2011) Quantitative analysis demonstrates most transcription factors require only simple models of specificity. Nat Biotechnol 29:480–483

31. Gentleman RC, Carey VJ, Bates DM et al (2004) Bioconductor: open software development for computational biology and bioinformatics. Genome Biol 5:R80

32. Cherry JM, Hong EL, Amundsen C et al (2012) Saccharomyces Genome Database: the genomics resource of budding yeast. Nucleic Acids Res 40:D700–D705

33. Sandelin A, Alkema W, Engström P et al (2004) JASPAR: an open-access database for eukaryotic transcription factor binding profiles. Nucleic Acids Res 32:D91–D94

34. Workman CT, Yin Y, Corcoran DL et al (2005) enoLOGOS: a versatile web tool for energy normalized sequence logos. Nucleic Acids Res 33:W389–W392

Chapter 13

Generation and Analysis of Chromosomal Contact Maps of Yeast Species

Axel Cournac, Martial Marbouty, Julien Mozziconacci, and Romain Koszul

Abstract

Genome-wide derivatives of the chromosome conformation capture (3C) technique are now well-established approaches to study the multiscale average organization of chromosomes from bacteria to mammals. However, the experimental parameters of the protocol have to be optimized for different species, and the downstream experimental products (i.e., pair-end sequences) are influenced by these parameters. Here, we describe a complete pipeline to generate 3C-seq libraries and compute chromosomal contact maps of yeast species.

Key words Yeast, Chromosome conformation capture, 3C, Genome organization, Genome assembly, 3C analysis

1 Introduction

We present a method to characterize the tridimensional (3D) organization of budding yeast genomes (*see* [1], for the first genome-wide analysis performed in *S. cerevisiae*). Using 3C-seq, a derivative of chromosome conformation capture (3C; [2–4], this protocol generates genome-wide contact maps of various yeast species. An interest of the 3C-seq approach, compare to other 3C derivatives such as Hi-C [5, 6], is that it does not use any enrichment for ligation products and can be directly applied to the sequencing and assembly of unknown species, as described in [7]. Briefly, 3C gives access to the contact frequencies between restriction fragments (RFs) along a chromosome, reflecting the average chromosome organization within the nuclei within a population [2] and, eventually, unveiling functional reorganization upon changes in DNA-related metabolic processes such as DNA repair [8], homolog pairing during meiosis [2], or transcription [9]. Several methods have been published regarding the generation and analysis of 3C

Frédéric Devaux (ed.), *Yeast Functional Genomics: Methods and Protocols*, Methods in Molecular Biology, vol. 1361,
DOI 10.1007/978-1-4939-3079-1_13, © Springer Science+Business Media New York 2016

libraries, including a recent comprehensive discussion that recapitulates the overall experimental approach and analysis [10]. In its classical version, a cellular culture of a species of interest is treated with a cross-linking agent (typically formaldehyde) that generates covalent bounds between proteins and between DNA and proteins [2]. In each cell, cellular components, including the chromosomal set, will "freeze" in a disposition that is assumed to reflect the physiological configuration. To quantify the contacts between different DNA regions of the genome, two steps are necessary. First, the cells (and nuclei) are gently lysed and the cross-linked chromatin is digested with a carefully chosen restriction enzyme. The insoluble part of the raw chromatin extract is then isolated through centrifugation, diluted, and incubated in presence of DNA ligase. Using the insoluble fraction diminishes the background by removing small DNA molecules that were not cross-linked in large complexes [11, 12]. Ligating under diluted conditions aims in turn at alleviating the relegation of molecules which are trapped in the different cross-linked complexes. Following the ligation step, the cross-link is then reversed and the DNA purified. The resulting 3C library consists of a mix of different ligation products whose relative abundance reflects their average spatial proximity within the cell population at the time of the fixation step. The different religation events within a 3C library can be quantified using pair-end (PE) sequencing and genomic contacts maps generated through a variety of protocols [1–3, 5, 6, 13].

This section describes the experimental protocol for generating and sequencing a 3C library of a yeast species. The experimental part is then followed by a brief overview of the computational analysis necessary to extract meaningful contact information from the raw data sequencing. Generationand analysis of 3C libraries do not require special equipment (except obviously access to a sequencing apparatus able to process a large number of PE sequences). However, the preparation of the assay requires careful planning. The choice of the restriction enzyme and of the cross-linking conditions is critical for the success of the experiment, and must be thoughtfully envisioned before starting (*see* **Notes 1** and **2**).

2   Materials

2.1   3C Library Components

1. 50 mL disposable conical tubes.

2. Filtration unit 0.22 μm.

3. 1.5 and 2 mL lo-binding microcentrifuge tubes (Eppendorf, Hamburg, Germany).

4. VK05 Precellys tube (Bertin Corp, Rockville, Maryland, USA).

5. Yeast species of interest (genome sizes, ~10–15 Mb).

6. Restriction enzyme and corresponding restriction enzyme buffer (*see* **Note 2**).

7. 5 U/μL T4 DNA ligase (Weiss Units).

8. 20 mg/mL proteinase K in water.

9. 10 mg/mL DNAse-free RNAse A in water.

10. 37 % formaldehyde solution (v/v) (Sigma-Aldrich, Saint Louis, Missouri, USA).

11. 2.5 M glycine: weigh 75.07 g of glycine and transfer to a 1 L cylinder. Add water to a volume of 400 mL and dissolve glycine using a magnetic stirrer and a stir bar (*see* **Note 3**). Filtrate on a 0.22 μm filtering unit and store at room temperature (RT).

12. 10 % sodium dodecyl sulfate (w/v) (SDS) in water. Add 20 mL of 20 % SDS (*see* **Note 4**) in a 50 mL disposable conical tube. Add 20 mL of water. Mix gently by returning tube several times. Store at RT.

13. 20 % Triton X-100 (v/v) in water. Add 10 mL of Triton X-100 in a 50 mL falcon. Add 40 mL of water and incubate in a 37 °C water bath until complete dissolution (it can take several hours). Store at RT.

14. 10× ligation buffer (without ATP): 500 mM Tris HCl pH 7.4, 100 mM $MgCl_2$, 100 mM DTT. Add 100 mL of Tris–HCl pH 7.5, 20 mL of $MgCl_2$ 1 M, and 10 mL of DTT 2 M to a 500 mL cylinder. Add water to reach 200 mL, mix and filtrate on 0.22 μm filtering unit. Split as 10 mL aliquot and store at –20 °C.

15. 10 mg/mL bovine serum albumin (BSA) in water. Store as 1 mL aliquots at –20 °C.

16. 100 mM adenosine triphosphate (ATP) pH 7.0 in water. Weigh 1 g of ATP and transfer to a 50 mL falcon. Add 14 mL of water. Add 1.6 mL of NaOH 1 M. Complete to 16.7 mL with water. Check that the pH is around 7.0. Filtrate on 0.22 μm filtering unit. Store as 1 mL aliquots at –20 °C (*see* **Note 5**).

17. 500 mM EDTA in water, pH 8.0.

18. 3 M sodium acetate in water, pH 5.2. Weigh 204.12 g of sodium acetate and transfer to a 1 L cylinder. Complete with water to 400 mL, and adjust pH to 5.2 with acid acetic 100 %. Complete to 500 mL with water. Filtrate on a 0.22 μm filtering unit and store at RT.

19. Isopropanol.

20. 10:9:1 phenol–chloroform–isoamylalcohol pH 8.2.

21. 100 % Ethanol.

22. TE buffer, pH 8.0. Add 5 mL of TE 10× to a 50 mL falcon. Add 45 mL of water and filtrate on a 0.22 μm filtering unit. Store at RT.

23. Precellys (Precellys®24) (Bertin Corp, USA) (*see* **Note 6**).

24. 16 °C water bath.

25. 65 °C oven.

26. Magnetic stirrer and stir bar.

27. Variable temperature incubator (25, 30 and 37 °C).

28. Dry bath at 65 °C.

29. Refrigerated tabletop centrifuge (for 50 mL falcon tubes).

2.2 NGS Library Processing Components

1. Covaris S220 instrument (Covaris Ltd., Woburn, Massachusetts, USA).

2. Snap Cap microTUBE for Covaris (Covaris Ltd.).

3. Column PCR purification Kit (QIAgen, Venlo, Netherlands) (*see* **Note 7**).

4. Column MinElute PCR purification Kit (QIAgen) (*see* **Note 7**).

5. 1.5 mL lo-binding microcentrifuge tubes (Eppendorf).

6. Illumina paired-end adapters and amplification primers (*see* **Note 8**; Illumina, San Diego, California, USA).

7. Tabletop centrifuge.

8. NanoDrop (Thermo Fisher Scientific).

9. 10× ligation Buffer (New England Biolabs, Ipswich, Massachusetts, USA—NEB): 500 mM Tris–HCl (pH 8.0), 100 mM $MgCl_2$, 100 mM DTT, 10 mM ATP.

10. 10× NEBuffer 2 (NEB): 500 mM NaCl, 100 mM Tris–HCl (pH 7.9), 100 mM $MgCl_2$, 10 mM DTT.

11. 10 mM deoxynucleotide triphosphates in water. Add 100 μL of each dNTP (dNTP set 100 mM) in a microcentrifuge tube 1.5 mL. Complete to 1 mL with water. Make 50 μL aliquots and store them at –20 °C (*see* **Note 9**).

12. 1 mM deoxyadenosine triphosphate in water. Add 10 μL of dATP 100 mM (from the dNTP set) in a microcentrifuge tube 1.5 mL. Complete to 1 mL with water. Make 50 μL aliquots and store them –20 °C (*see* **Note 9**).

13. 10 U/μL T4 polynucleotide kinase.

14. 1 U/μL T4 DNA polymerase.

15. 5 U/μL Klenow DNA polymerase.

16. 5 U/μL Klenow (exo-) DNA polymerase.

17. 400 U/μL T4 DNA ligase (Cohesive End Unit).

18. Phusion polymerase (Thermo Fisher Scientific).

2.3 Data Processing

1. Computer with a UNIX system (Linux, MacOSX, Ubuntu). Large memory space and multiprocessor core are needed for efficient reads alignments.

2. Alignment program (for instance, Bowtie2).
   http://bowtie-bio.sourceforge.net/bowtie2/index.shtml

3. A script language like Bash or python to manipulate files.

4. A tool to visualize big matrices like Matlab (license needed) or Octave (free).

3 Methods

3.1 Generation of a 3C Library of Mixed Species

The generation of the 3C library takes 3 days, and the generation of the sequencing library an additional 2–3. The 3C library can be stored at –80 °C and therefore the two processes can be separated. Whereas it remains difficult to prepare more than four libraries at a time, processing the samples for sequencing can be performed at a larger scale (up to eight libraries), the limiting step being then, to some extent, the purification of molecules of a size appropriate for sequencing (*see* **Note 10**). Therefore, timing is important criteria when planning to do the experiment! The overall schedule will require for an experienced experimentalist an afternoon (partly), a morning–afternoon (partly), a morning (full), followed by 2 full days (with several incubations steps).

3.1.1 Culture Fixation

1. Start culture of yeast species in your favorite medium. For instance, strains can be grown at 30 °C in 100 mL BMW medium [14] up to 1×10^7 cells/mL (this quantity will allow to realize two libraries) (*see* **Note 11**).

2. Add 8.5 mL of the fresh formaldehyde solution (i.e. 37 %) to the culture (final concentration of 3 %) (*see* **Note 2**).

3. The cells are incubated for 30 min at room temperature (RT) under gentle agitation with a magnetic stirrer.

4. Move the cell culture at 4 °C for another 30 min under gentle agitation.

5. Transfer the culture at RT and add 25 mL of Glycine 2.5 M (final concentration: 470 mM) to quench the remaining formaldehyde; incubate under agitation for 5 min at RT.

6. Relocate the culture at 4 °C and keep them under gentle agitation for an extra 15 min.

7. Pellet the fixed cells at 4 °C ($3500 \times g$—10 min).

8. Wash the cells with 10 mL of the initial medium.

9. Pellet the fixed cells at 4 °C ($3500 \times g$—10 min).

10. Suspend the cells into 2 mL of medium and transfer them into 2×1.5 mL microcentrifuge tubes.

11. Pellet the cells at 4 °C ($3500 \times g$—10 min).

12. Remove the supernatant and flash freeze the pellet (i.e. in liquid nitrogen or dry-ice + ethanol).

13. Store pellets at −80 °C until use.

NB: Do not store pellet for more than 6 months (*see* **Note 12**).

3.1.2 3C Library Generation

Day one

1. Thaw the pellet on ice for 1 h.

2. Resuspend the cells in 4.5 mL of 1× restriction buffer (*see* **Notes 1** and **2**).

3. Transfer the cell suspension into 3× VK05 tubes (Precellys) (*see* **Note 6**).

4. Lyse the cell using the following program: 9 cycles × (6500 × g— 30 s ON/60 s OFF).

5. Transfer lysate into 8 × 1.5 mL microcentrifuge tube (500 μL per tube).

6. Add 15 μL of 10 % SDS per tube (final concentration: 0.3 %).

7. Incubate tubes in a dry bath at 65 °C for 20 min.

8. Promptly transfer tubes on ice and incubate for 1 min.

9. Incubate tubes for 30 min at 37 °C and under agitation.

10. Add 50 μL of Triton X-100 20 % and 6 μL of 10× restriction buffer per tube.

11. Incubate tubes for 30 min at 37 °C and under agitation.

12. Put one tube aside as a non-digested control.

13. Add 150 units of restriction enzyme in each of the 7 remaining tubes.

14. Incubate overnight at the appropriate temperature for the chosen restriction enzyme.

Day two

15. The next morning, take the non-digested control and one of the digested samples (non-digested and digested controls, respectively). Add 100 μL of SDS 10 % and 30 μL of proteinase K to each tube and incubate them at 65 °C overnight (these controls will then be furthered processed at **step 30**).

16. Centrifuge the 6 remaining tubes at 16,000 × g for 20 min at temperature in order to isolate the insoluble fraction of the cross-linked chromatin [11].

17. Remove the supernatant and suspend each pellet in 500 μL of H_2O.

18. Pool the pellets three by three and dilute the two samples in 22.5 mL of a precooled (4 °C—on ice) ligation reaction mix (10× ligation buffer 2.4 mL, BSA 10 mg/mL 240 μL, ATP 100 mM 240 μL, water) in 50 mL conical tubes.

19. Add 125 units of T4 DNA ligase.

20. Homogenize the reaction by inverting the tubes 2–3 times.

21. Incubate for 4 h in a 16 °C water bath.

22. Transfer to a 25 °C water bath for an extra 45 min.

23. Add 200 μL of EDTA 500 mM per tube to stop the reaction.

24. Add 200 μL of proteinase K (20 mg/mL) and incubate the tube overnight at 65 °C.

Day three

25. The next morning, cool down the tubes at room temperature and transfer the solution to new 50 mL conical tubes.

26. Add 2.4 mL of 3 M Na Acetate pH 5.0 and 24 mL isopropanol and incubate at −80 °C for 1 h in order to precipitate DNA (*see* **Note 13**).

27. Centrifuge the tube in an appropriate centrifuge at $10,000 \times g$ for 20 min.

28. Remove the supernatant and dry the pellet on the bench (*see* **Note 14**).

29. Suspend each pellet in 900 μL of TE buffer 1× and transfer them in 2×2.0 mL microtube.

30. Perform a DNA extraction for each tube using 900 μL of phenol–chloroform. Also extract the DNA from control samples from **step 15** using 500 μL of phenol–chloroform–isoamylalcohol.

31. Recover 2×400 μL of the aqueous phase (upper phase) for each tube (800 μL per tube in total) (and 1×400 μL for control tubes and transfer them into 1.5 mL microcentrifuge tube.

32. Add 40 μL of 3 M Na Acetate pH 5.0 and 1 mL of cold ethanol to each tube.

33. Vortex the tubes and incubate at −80 °C for 30 min.

34. Centrifuge the tubes at $16,000 \times g$ for 20 min; discard the supernatants.

35. Wash each DNA pellet with 500 μL of cold 70 % ethanol.

36. Centrifuge tubes at $16,000 \times g$ for 20 min and remove supernatant.

37. Dry pellet by incubating them on a 37 °C dry bath.

38. Suspend each pellet in 30 μL TE buffer 1× supplemented with RNAse A (0.1 μg/mL final concentration).

39. Incubate at 37 °C for 45 min.

40. Pool the tubes containing the 3C libraries.

41. Estimation of the quality and quantity on a 1 % agarose gel (Fig. 1; *see* **Note 15**).

42. *Optional: store at −80 °C as ~6 μg DNA aliquots* .

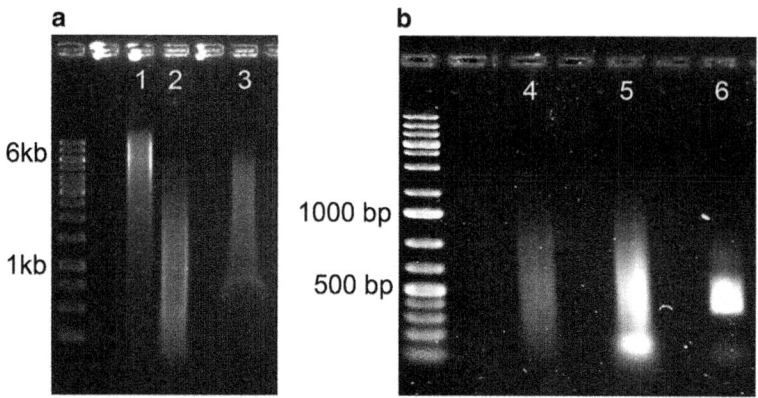

Fig. 1 Photography of gel electrophoresis migration of DNA at various steps of a 3C library construction. (**a**) Non digested control (*1*); Digested control (*2*); 3C library (*3*). (**b**) Processing of a 3C library for Illumina sequencing. Profile after shearing (*4*); profile following reparation, addition of 3′ A-tail and ligation of PE adapters (*5*); profile after size selection (400–800 bp) and PCR amplification (*6*)

3.1.3 Processing the 3C Library for Deep-Sequencing

The current protocol applies if the sequencing apparatus is an Illumina sequencer. For other brands/technologies, refer to the manual to design an appropriate protocol.

1. *Optional:* thaw a dry 3C sample on ice for 30 min.

2. Adjust the volume of a melted ~6 μg aliquot of a 3C library to 130 μL with water. With less DNA the protocol can nevertheless be pursued (down to 500 ng in our experience) but more amplification cycles will be necessary at the end.

3. Shear the library using your favorite instrument. For instance, we use a Covaris with the following settings: Peak Power: 105, Duty Factor 5 %, Cycles per Burst 200, Treatment time (s) 80 s, to obtain DNA fragments between 300 and 1500 bp (*see* **Note 16**).

4. Purify the DNA on a QIAquick column and elute with 5 μL of elution buffer (EB).

5. Quantify the DNA on a NanoDrop apparatus and prepare a tube with 5 μg of DNA and adjust the volume to 80 μL with water.

6. Add 12 μL of 10× ligase buffer (NEB), 4 μL of dNTP 10 mM, 15 μL of T4 DNA polymerase, 5 μL of T4 polynucleotide kinase, 1 μL of Klenow DNA polymerase) and complete to 120 μL with H_2O. Incubate at RT for 30 min.

7. Purify on QIAgen MinElute column and recover 30 μL of DNA in EB (to do so, add 31 μL of EB on the column in order to recover 30 μL).

8. Add 5 μL of 10× buffer NEB2, 10 μL of dATP 1 mM, 3 μL of Klenow DNA polymerase (exo minus) and complete to 50 μL with water. Incubate at 37 °C for 30 min followed by 20 min at 65 °C.

9. Purify on QIAgen MinElute column and recover 20 μL of DNA in EB.

10. Add 3 μL of 10× ligation buffer (NEB), 4 μL of adapters 10 μM, and 3 μL of T4 DNA ligase (NEB). Incubate at room temperature for 2 h (it is also possible to incubate overnight at 4 °C).

11. Purify fragment between 400 and 800 bp with your favorite method (Gel, Pippin Prep, caliper—*see* **Note 10**). Recover DNA in a volume of 40 μL in EB or TE.

12. Determine the optimal number of cycle and quantity of matrix to generate enough library for sequencing. Prepare several PCR reaction (phusion DNA polymerase—Volume of 50 μL per reaction) with different amount of library (Typically 1 and 2 μL). Temperature profile of the reaction is as follow: 30 s at 98 °C followed by 9, 12, or 15 cycles of 10 s at 98 °C, 30 s at 65 °C, 30 s at 72 °C, and a final 7 min extension at 72 °C.

13. Run the PCR reaction on a 1 % agarose gel and determine the optimal conditions.

14. Prepare 8 PCR reaction using the determine conditions and run them.

15. Purify on two QIAquick MinElute columns and recover around 40 μL of DNA.

16. Quantify on NanoDrop and check the profile on gel.

17. Run your library on an Illumina sequencing platform.

3.2 Analysis of Pair-End Sequencing Reads

Each 3C based protocol present peculiarities likely to generate noise or specific biases in the data. These caveats can be attenuated by an appropriate, specific preprocessing of the data and by proper normalization. Several approaches have been described that aim at correcting biases, or alleviating it to improve the quality of subsequent analysis [15–17]. This part presents the main steps to process the 3C-seq data described above. We provide commands and software's that we currently use. This description is only an illustration of what can be done, since many other bioinformatics tools exist and are available to the community.

3.2.1 Mapping Along the Genomes of the Mixed Species Yeast

1. If there is a reference genome for the species you are studying, recover or generate the fasta file (*see* **Note 17**). If the genome is unknown, *see* Marbouty et al. [7]. For the sake of illustration, we will use in the following the file "*genomes_yeasts.fa*" as the name of the reference genome.

2. Indexing the reference genome. Aligner software such as Bowtie 2 or BWA are needed for this task and routinely used [18]. The first step is to index your reference genome.
   > *bowtie2 -build genomes_yeasts.fa genomes_yeasts_index*

3. Align each mate (from the file sequences_mate1.fastq) independently using the most sensitive mode of the alignment software. We recommend being very stringent when mapping the reads against the genome (whether from mixed samples of from unique samples) to minimize alignment mistakes (*see* **Note 18**).
   > *bowtie2 -x genomes_yeasts_index -p6 --sam-no-hd --sam-no-sq --quiet --local --very-sensitive-local -S p1.sam sequences_mate1.fastq*
   > *bowtie2 -x genomes_yeasts_index -p6 --sam-no-hd --sam-no-sq --quiet --local --very-sensitive-local -S p2.sam sequences_mate2.fastq*

4. The alignment software generates a SAM file (Sequence Alignment/Map format). You can select and keep the fields relevant for subsequent analysis to save memory space using awk. For instance,
   > *awk '{print $1,$3,$4,$2,$5;}' p1.sam > p1.sam.select*
   To recover the pair-end information, a convenient and fast way is to use the bash commands "sort" and "paste":
   > *sort -T /path_to_temporary_repository p1.sam.select > p1.sam.select.sorted*
   > *sort -T /path_to_temporary_repository p2.sam.select > p2.sam.select.sorted*
   > *paste p1.sam.select.sorted p2.sam.select.sorted > p1_p2.select.merged*
   where /path_to_temporary_repository points to a temporary repository where storage space is available for the sort command to be executed.

5. Finally, complete the mapping procedure by filtering ambiguous hits on the genome. Only the pairs of mapped reads with a quality above a certain threshold will be retained (*see* **Note 19**).
   > *awk '{$5 >= 40 && $10 >= 40) print $0;}' p1_p2_merged > p1_p2_merged.MQ40*

3.2.2 Building the Contact Network

Assign every pair of mapped read on their restriction fragment by crossing the coordinates of the mapped reads with the restriction map of the genome. This can be done for instance with the "restrict" function from the bioinformatics suite EMBOSS (*see* **Note 20**).

3.2.3 Filtering Out of Non-informative Contacts and Construction of the Contact Map

1. A possibility that always arise at the ligation step is the formation of a loop from a long DNA molecule that contains a successive of restriction fragments that are not cut by the restriction enzyme. These events can lead to the detection of false long

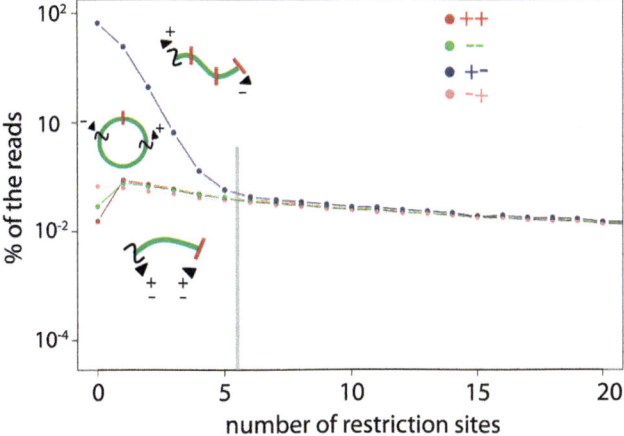

Fig. 2 Distribution of the different types of molecules (*green lines*) sequenced from a 3C library, plotted as a function of the number of restriction sites (*red bars*) between the two pair-reads (*black arrowheads*). Sonicated sites are indicated with *black twisted lines*. The directionality of the reads according to the coordinates of the reference genome are indicated by + and − symbols and are used to characterize different sub-population in the library (*see* Cournac et al. [16] for details). If the religation events were truly random between two RFs, each one of the four extremities of two restriction fragments would have the same probability to be ligated with each of the three others. However, because restriction efficiency is far from being 100 % and because of the occurrence of other types of religation events (circularization etc.), neighboring RFs present variations in the distribution of ligation events, with "uncuts" or "loops" events being overrepresented which do not reflect "contact" information but biochemical or physical biases. The *grey bar* indicates the number of restriction sites needed before all the different categories of events are equally represented within the population, corresponding to molecules that are retained for further analysis

range contacts and to avoid them, it is necessary to filter out some contacts ([16]; Fig. 2). Once this threshold is estimated PE reads that do not present this significant number of RF between them are discarded from the analysis (*see* **Note 21**).

2. To reduce the dimension of the contact map, and alleviate some local variations resulting from the size of RF, neighboring RF can be pooled into "bins" regrouping the sum of contacts of successive RF. These bins can be made either of a fixed number of successive RF, or in a window constant in size (*see* **Note 22**).

3.2.4 Normalization and Representation of the Contact Map

1. The raw matrix is then normalized to attenuate biases inherent to the protocol (*see* **Note 23**). Several methods can be used to achieve this step, including the Sequential Component Normalization (SCN; [16]; *see* **Note 24**).

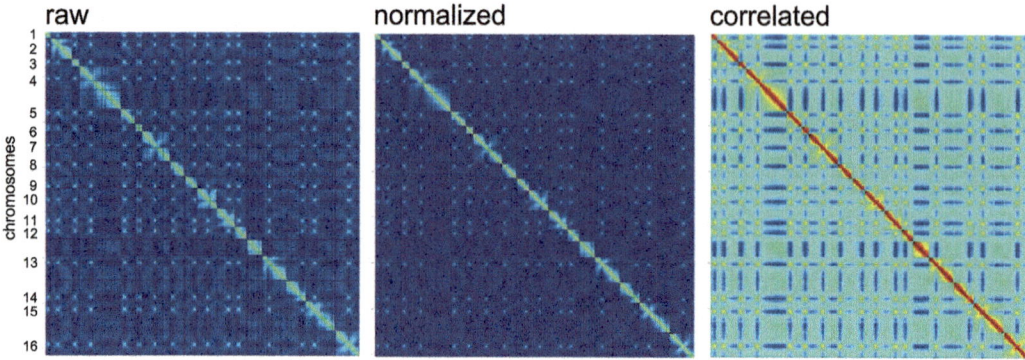

Fig. 3 (**a**) Raw genomic contact map of *Saccharomyces cerevisiae*, with each vector of the matrix representing ten restriction fragments. (**b**) The same matrix after SCN normalization. (**c**) Pearson correlation representation of the normalized matrix

For instance, using MatLab (Mathworks, Natick, MA, USA), the contact map Mat is normalized through several iterations of the following commands:

```
> for i = 1:1:n1
> Mat2(:,i) = Mat(:,i)/norm(Mat(:,i));
> End
> for i = 1:1:n1
> Mat_scn(i,:) = Mat2(i,:)/norm(Mat2(i,:));
> end
```

2. The correlation contact map of the normalized matrix can also be computed (*see* **Note 25**; Fig. 3). Using Matlab, the corrcoef function calculates the Pearson correlation coefficient between each line and column of the normalized map.

```
> Mat_corr = corrcoef(Mat_scn);
```

3. To visualize the contact map, the imagesc function from Matlab is a convenient tool. The contrast can be improved by raising all the elements of the matrix to the power n, with for instance n=0.4 (*see* **Note 26**).

```
> figure, imagesc(mat.^0.4);
```

3.2.5 Statistical Analysis

1. Although genomic contact maps unveil the global genome organization of a population of cells, such as centromere clustering that reveal the position of centromeric sequences in yeast species [1, 4], a large fraction of the information contained in the data is not directly visible on the contact map and need to be statistically exploited. The statistical analysis aims at determining whether the contacts observed between two or more DNA regions of interests is higher than expected by chance. One way to do that is to calculate the mean of normalized interactions between the different members of the group

of interest and compare them to a random set to evaluate significance (*see* **Note 27**). This implies carefully designing a null model taking into account the specificity of the global chromosome organization.

4 Notes

1. The choice of the restriction enzyme and buffer is a key component of this experiment. Several restriction enzymes become inactive under the experimental conditions described in the protocol (i.e. cellular brut extract of yeasts). The cheapest enzymes—usually the best characterized—provide the best candidates to generate a 3C library. Consequently, we highly recommend choosing «classical» restriction enzymes when designing the experiment. However, it is still possible that the enzyme selected is not active enough (for instance SacII works pretty bad in our hands; MM, personal communication). Restriction buffer used to generate 3C libraries have to contain DTT. Consequently, we strongly suggest avoiding new NEB buffer (NEBuffer 1.1, 2.1, 3.1 and CutSmart).

2. Building a 3C library is also entirely dependent on the match between the restriction enzyme chosen and the condition of the fixation step (Fig. 4). The likelihood for a RF to be cross-linked is dependent on the probability for one bp to be cross-linked and thus on the incubation parameters in the presence of a fixative agent level [16]. Notably, a 4-cutter (restriction enzyme recognizing a 4pb site) will require a higher concentration of cross-linking agent (or longer incubation, to

Formaldehyde concentration

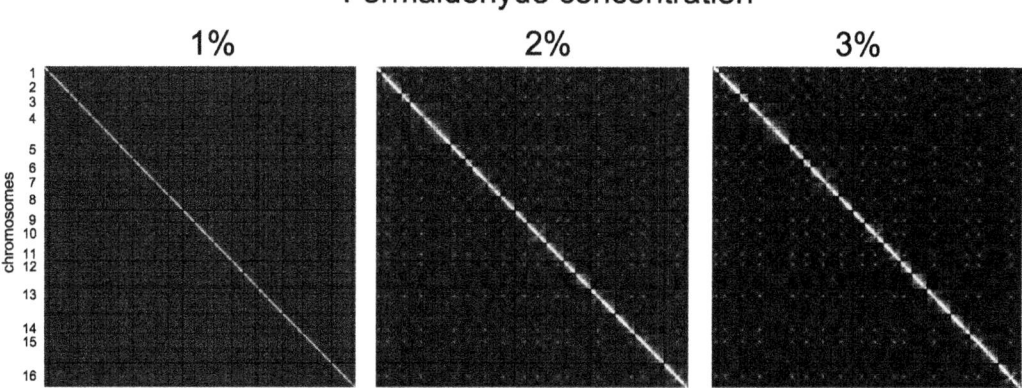

Fig. 4 Illustration of the influence of cross-link concentration on the genome-wide contact profile. *S. cerevisiae* contact maps obtained as described in this methods, but with varying formaldehyde concentrations (1, 2, and 3 %). Each contact map contains approximately the same amount of PE reads

some extent) than a 6-cutter. The protocol described in this article is designed for enzymes that generate RFs with an distribution average lower than 500 bp (± 200 bp). In general, we do not recommend enzymes that generate RFs with a distribution average lower than 300 bp.

3. Dissolving glycine at such concentration can take several hours. The process can be accelerated by gently warming the solution (40–50 °C).

4. SDS treatment is a critical step. We noticed an important drop in the quality of the library when the SDS begins to precipitate (warming prior use does not solve the problem). Therefore the SDS solution has to be changed immediately if signs of precipitation are visible.

5. ATP is a critical cofactor of the ligase reaction. In order to avoid any problem due to ATP degradation, discard the thawed aliquot after use.

6. Some yeast species—or some metabolic states—appear somehow resistant to zymolyaze treatment. Lysis is thus obtained through mechanical treatment using for instance a Precellys (Bertin Corp, Rockville, Maryland, USA).

7. We have noticed a decrease in the efficiency of the QIAgen PE buffer (i.e., wash buffer) over time. To avoid this problem we strongly recommend preparing the required amounts extemporarily to the experiment.

8. Sequences of Pair-End Adapters (with index) and Pair-End Amplification Primers can be found at http://support.illumina.com/downloads/illumina-customer-sequence-letter.html.

9. As for ATP, discard aliquot once thawed and used.

10. For size selection, we routinely use a Pippin Prep apparatus (Sage Science), though gel purification works well.

11. The mixed culture of 100 mL with a concentration of 1×10^7 cells of genome sizes ~10–15 Mb is sufficient to generate two libraries in the conditions described in the protocol. For other conditions, the cross-linking step will have to be adapted, as it will change the DNA–protein–formaldehyde ratio (*see* **Notes 1** and **2**).

12. We have noticed a quality decreased after storage of more than 6 months.

13. After 1 h at –80 °C, solution will froze. Prompt freezing is necessary for a good recovery of libraries.

14. The pellet does not have to be entirely dry, since DNA will be subsequently extracted by a phenol–chloroform step and re-precipitated.

15. Quantify the libraries on a gel using an image quantification software (such as Image J or Quantity One). Indeed, large DNA fragments and impurities prevent the use of NanoDrop or Qbit quantification.

16. Shearing can also be done using Bioruptor or nebulizer.

17. Importantly, the paired-end mode of the software Bowtie2 must not be used to align 3C or Hi-C data. Indeed, this mode sometimes favors wrong positions for ambiguous reads. Notably, it can favor a *cis* position for two ambiguous PE reads against a distant or *trans* alignment. This leads to "speckles" in *trans* positions between the regions involved, which often correspond to repeated sequences (for example ribosomal protein encoding genes) and can generate artefacts when analyzing co-localization of DNA regions.

18. Based on our experience, the Mapping quality threshold must be set as high as possible. Generally use a quality value of 40. Even then, incorrect mapping can still be detected. Another possibility to filter ambiguous reads is to keep the reads that do present a 'XS' field in the SAM file. The 'XS' option contains the score of the second best alignment and therefore is an indicator of a nonunique alignment. This approach is used in the python library hiclib [17]. However, in our hands, this approach is relatively less stringent than putting a threshold on the Mapping Quality.

19. A typical 3C-seq library contains a large amount of molecules that do not result from religation of two non-adjacent RF, and therefore that do not bring information about the 3D genome structure and have to be removed. To do so, plot the distribution of events according to the orientation of the pair-end reads compare to the reference genome coordinates. We distinguish at least three categories [16]. First, «uncut» events that correspond to mate pairs separated by none, or a few consecutive RFs, most likely not digested. "Loops", that correspond mostly to one or several consecutive RFs circularized during the ligation step and subsequently sheared. Finally, "weird" events which are pairs of reads belonging to the same restriction fragment but with same directions with respect to the reference genome, and may eventually be exploited to look at sister chromatids or homologs behavior. Indeed, thistype of events is largely accountable from the presence of several copies of the same DNA molecule within the same cellular compartment, as shown notably by the small number of religation events between DNA molecules belonging to different cellular compartments [7]..

20. We assign each read to its restriction fragment along the genome using a custom-made C routine. The C language is

fast and allows precise allocation of memory. At this step, the distance between each read and the associated restriction site and keeps as well the size of the associated restriction fragment can also be calculated. Several other filters can be applied at this stage to remove incorrect events. Notably, the size of the fragment sent to the sequencer can also be calculated and the distribution can be checked to be in accordance with the experimental one (*see* **step 11** selection with Pippin Prep). A filter can also be applied to reads too close of their associated restriction site.

21. The "visibility" of RFs in the PE reads will depend principally of their size, and whether they contain repeated sequences or not. This variability will impact on the visibility of the bins. Therefore, working with fixed size bins may reflect this variation in visibility. There is no perfect solution, since working with bins made of successive RFs (and therefore representing regions of different sizes) can also generate visual discrepancies between bins: for instance, a bin made of a successive long RFs will be represented in the contact map the same way as a bin made of successive small RFs. Working with fixed size bins can also simplify the analysis and comparison between contact maps. In this case, the genome is divided in equal size bins and every reads is attributed to the bin where its start position belongs.

22. The rationale behind the normalization procedure is that each bin has an equal probability to be detected overall. Plotting the distribution of contact per bin reveal a subset of elements that are undetectable or present only a handful of contacts. These bins can either result from repeated sequences that prevent confident mapping of reads in these regions, but potential structural variations between the genome of the strain being tested and the reference genome can also generate a similar outcome (for instance, deletion can easily be detected with such approaches, since consecutive bins will present no contacts; [19]. The tail of this distribution has to be removed since these bins correspond clearly to "invisible" regions. The remaining graph presents a clear Gaussian distribution, meaning that some bins are less detectable than others. Such differences result notably from variations in the distribution of RS if the bins in the contact map are made of constant sizes, or from differences in the sizes of RF binned together if this binning approach has been retained. To attenuate the differences of detection, divide each matrix element by the sum of elements of the line it belongs then do the same by dividing each matrix element by the sum of the column it belongs. Iterate this process until the matrix converges to a stable one. To our experience, a few iterations (5–10) is sufficient to have a stable

matrix. The normalization procedure ensures that the sum over the column and lines of the matrix equals 1, which reduces the noise and biases inherent to the protocol. You can do that in a matlab script, C code, or python code.

23. The SCN procedure is based on similar approaches and mathematical operations than the ICE procedure published concomitantly [17]. To our knowledge, it gives similar results.

24. Each element of the correlation matrix corresponds to the Pearson coefficient between the line and column vectors. Correlation map are to be handled with care, since they do not reflect necessary important contacts between two elements, but provide indications about their behavior similarity. This representation will increase the contrast between elements presenting similar neighbors, and others positions, but will not provide indications on the strength of the contacts between these elements.

25. "Beautification" of the contact map can be increased by applying a blurring effect on the matrices. Applying a convolution matrix with as kernel the 3×3 matrix [0.05 0.05 0.05; 0.05 0.05 0.05; 0.05 0.05 0.05] to the contact map will result in such effect. The convolution has to be repeated to emphasize the structures. You can as well display the matrices with R which contains several Gaussian filters built in functions or in python with the tool *imshow* which contains an interpolation function as default. It has to be noted that this type of image processing adds information to the initial data, and that statistical analysis cannot be done with such processed matrixes.

26. Choosing a relevant null model is not trivial. The minimalist null model must respect the distribution along the different chromosomes as proposed in [20]. A more stringent null model has to take into account the global organizational features of the genome being studied. For yeasts, whose chromosomes are organized under a Rabl organization with centromeres co-localizing and chromosome arms extending from there in the nuclear space, the positions along the genome of the elements being studied has to take into account their distance from the centromeres and, eventually, from the subtelomeric regions. Otherwise, if a subset of genes appear to be positioned at equal distances from their respective centromeres they will mechanically present enriched contacts due to the constraint imposed by centromere clustering. For instance, studying colocalization of coregulated, paralogous genes in *S. cerevisiae* raises this question accurately, since many of those genes originated from whole genome duplication events and have remained at relatively equal distances from centromeres. Failing to take into account this disposition in the null model

will lead to the conclusion that coregulated genes are colocalizing in space, which may be true, but impossible to assert since this colocalization can alsosimply reflect the distance separating them from their respective centromeres. Similar precautions apply when studying the colocalization of regions positioned along the same chromosome, which will mechanically present enriched contacts compared to the average contacts over the entire genome, if the distance separating them not included in the null model.

Acknowledgements

This research was supported by funding to R.K. from the European Research Council under the 7th Framework Program (FP7/2007-2013)/ERC grant agreement 260822. M.M. is the recipient of an Association pour la Recherche sur le Cancer fellowship 20100600373.

References

1. Duan Z, Andronescu M, Schutz K et al (2010) A three-dimensional model of the yeast genome. Nature 465:363–367

2. Dekker J, Rippe K, Dekker M et al (2002) Capturing chromosome conformation. Science 295:1306–1311

3. Sexton T, Yaffe E, Kenigsberg E et al (2012) Three-dimensional folding and functional organization principles of the Drosophila genome. Cell 148:458–472

4. Marie-Nelly H, Marbouty M, Cournac A et al (2014) Filling annotation gaps in yeast genomes using genome-wide contact maps. Bioinformatics 30(15):13

5. Lieberman-Aiden E, van Berkum NL, Williams L et al (2009) Comprehensive mapping of long-range interactions reveals folding principles of the human genome. Science 326:289–293

6. de Laat W, Dekker J (2012) 3C-Based technologies to study the shape of the genome. Methods 58:189–191

7. Marbouty M, Cournac A, Flot J-F et al (2014) Metagenomic chromosome conformation capture (meta3C) unveils the diversity of chromosome organization in microorganisms. eLife 3, e03318

8. Oza P, Jaspersen SL, Miele A et al (2009) Mechanisms that regulate localization of a DNA double-strand break to the nuclear periphery. Genes Dev 23:912–927

9. O'Sullivan JM, Tan-Wong SM, Morillon A et al (2004) Gene loops juxtapose promoters and terminators in yeast. Nat Genet 36:1014–1018

10. Lajoie BR, Dekker J, Kaplan N (2015) The Hitchhiker's guide to Hi-C analysis: practical guidelines. Methods 72:65–75

11. Gavrilov AA, Gushchanskaya ES, Strelkova O et al (2013) Disclosure of a structural milieu for the proximity ligation reveals the elusive nature of an active chromatin hub. Nucleic Acids Res 41(6):3563–3575

12. Nagano T, Lubling Y, Stevens TJ et al (2013) Single-cell Hi-C reveals cell-to-cell variability in chromosome structure. Nature 502:59–64

13. van de Werken HJG, Landan G, Holwerda SJB et al (2012) Robust 4C-seq data analysis to screen for regulatory DNA interactions. Nat Methods 9:969–972

14. Thompson DA, Roy S, Chan M et al (2013) Evolutionary principles of modular gene regulation in yeasts. eLife 2, e00603

15. Yaffe E, Tanay A (2011) Probabilistic modeling of Hi-C contact maps eliminates systematic biases to characterize global chromosomal architecture. Nat Genet 43:1059–1065

16. Cournac A, Marie-Nelly H, Marbouty M et al (2012) Normalization of a chromosomal contact map. BMC Genomics 13:436

17. Imakaev M, Fudenberg G, McCord RP et al (2012) Iterative correction of Hi-C data reveals hallmarks of chromosome organization. Nat Methods 9:999–1003

18. Langmead B, Salzberg SL (2012) Fast gapped-read alignment with Bowtie 2. Nat Methods 9:357–359

19. Marie-Nelly H, Marbouty M, Cournac A et al (2014) High-quality genome (re)assembly using chromosomal contact data. Nat Commun 5:5695

20. Witten DM, Noble WS (2012) On the assessment of statistical significance of three-dimensional colocalization of sets of genomic elements. Nucleic Acids Res 40:3849–3855

Chapter 14

A Versatile Procedure to Generate Genome-Wide Spatiotemporal Program of Replication in Yeast Species

Nicolas Agier and Gilles Fischer

Abstract

Here, we describe a complete protocol, comprising both the experimental and the analytical procedures, that allows to generate genome-wide spatiotemporal program of replication and to find the location of chromosomally active replication origins in yeast. The first step consists on synchronizing a cell population by physical discrimination of G1 cells according to their sedimentation coefficient. G1 cells are then synchronously released into S-phase and time-point samples are regularly taken until they reach the G2 phase. Progression through the cell cycle is monitored by measuring DNA content variation by flow cytometry. DNA samples, covering the entire S-phase, are then extracted and analyzed using deep sequencing. The gradual change of DNA copy number is measured to determine the mean replication time along the genome. A simple method of peak calling allows to infer from the replication profile the location of replication origins along the chromosomes. Our protocol is versatile enough to be applied to virtually any yeast species of interest and generate its replication profile.

Key words Yeast, Replication, Origins, G1 synchronization, Elutriation, Gradient of Percoll, Flow cytometry, Deep sequencing, DNA copy number

1 Introduction

The first genome-wide spatiotemporal program of replication in eukaryotes was achieved in 2001 in *Saccharomyces cerevisiae* by using a method based on the semi-conservative replication of DNA described by *Meselson and Stahl* in 1958 [1, 2]. Since then, other methods allowing to generate replication profiles and to localize replication origins were developed [3, 4]. Genome-wide replication profiles are now being established in a growing number of yeast species [4–11] and the protocol presented in this chapter has been specially developed to be easily applied to any yeast species of interest.

The first step to generate a replication timing profile is to isolate S-phase replicating cells. This can be achieved either by directly isolating a S-phase cellular fraction using a cell-sorting device [5–7] or by synchronizing cells in the G1 phase of the cycle and synchronously

Frédéric Devaux (ed.), *Yeast Functional Genomics: Methods and Protocols*, Methods in Molecular Biology, vol. 1361,
DOI 10.1007/978-1-4939-3079-1_14, © Springer Science+Business Media New York 2016

releasing them into S-phase [1, 4, 8–10]. A more direct alternative to these approaches, consisting on using exponentially growing cells instead of S-phase replicating cells, was recently described [11]. All these methods give access to the genome-wide localization of chromosomally active replication origins. However, only the synchronous release of G1 cells into S-phase will give access to other important replication features such as timing, fork velocity and origin efficiency (i.e., the proportion of cells in the population which actively fires each origin) [9, 10].

The synchronization of cells into G1 can be achieved either by a chemical process such as an alpha factor treatment, but this has turned out to be inefficient for many yeast species outside the *Saccharomyces sensu stricto* group, or by the physical discrimination of cells according to their sedimentation coefficient. Physical separation is a more versatile method that allows generating homogeneous G1 daughter cell samples and has proven to be very efficient and suitable for many yeast species [9, 12]. The physical separation of G1 cells can be achieved by using a dedicated centrifugal elutriation device or by centrifuging cells into a density gradient, which only requires standard molecular biology equipment.

G1 cells are synchronously released into S-phase and timepoint samples are regularly taken until they reach the G2 phase. DNA samples, covering the entire S-phase, are then extracted and analyzed using deep sequencing. The gradual change of DNA copy number along the genome is measured to determine the mean replication time along the genome (also called Trep). All the experimental procedures and the computational analysis steps required to generate a genome-wide replication profile are presented below.

2   Materials

All solutions are prepared with ultrapure water (18 MΩ cm sensitivity at 25 °C) and legal waste disposal regulations must be followed for propidium iodide and sodium azide containing solutions.

2.1   Yeast Growth Culture

1. YPD growth medium: 2 % peptone, 1 % yeast extract, 2 % glucose, pH 6.5. Add 50 g of Difco YPD broth powder in 1 L of purified water (*see* **Note 1**). Mix thoroughly and autoclave at 110 °C for 30 min.

2. YPD-Agar growth medium: 2 % peptone, 1 % yeast extract, 2 % glucose, 1.5 % agar, pH 6.5. Add 65 g of Difco YPD Agar powder in 1 L of purified water (*see* **Note 1**). Mix thoroughly and autoclave at 110 °C for 30 min (*see* **Note 2**).

3. Microbiological incubator.

4. Multitron standard shaking incubator (INFORS).

5. Sterile plastic loop of 1 µL.

2.2 Cell Synchronization	1. Phosphate-Buffered Saline (PBS; 1×): 137 mM NaCl, 2.7 mM KCl, 8 mM Na_2HPO_4, 2 mM KH_2PO_4, pH 7.4.
2.2.1 Centrifugal Elutriation	2. Beckman elutriation system (Beckman Coulter): Avanti J-26 XP centrifuge modified with the Upgrade J-26 XP Kit , JE-5.0 Elutriator Rotor , 40 mL Elutriation chamber , Pump drive (Masterflex, Cole-Parmer) and standard pump head (Masterflex, Cole-Parmer).
	3. JLA 10.500 rotor and adapted 500 mL centrifugal bottles.

2.2.2 Percoll Gradient

1. Percoll solution: 70 % Percoll (GE Healthcare Life Sciences), 0.15 M NaCl. Add Percoll and NaCl solutions to sterilized water and then shake vigorously (*see* **Note 3**).
2. Tris solution: 50 mM Tris-HCl, pH 7.
3. Avanti J-26 XP centrifuge (Beckman Coulter).
4. JA25.50 fixed-angle rotor (Beckman Coulter).
5. Megafuge 40R (Heraeus).
6. TX-750 Swinging Bucket Rotor (Thermo scientific).
7. 50 mL polycarbonate centrifuge tubes (Beckman Coulter).

2.3 S-phase Time Course Experiment

1. Sodium azide solution: 1 % sodium azide.
2. Water bath.
3. Magnetic stirrer.
4. Nineteen 15 mL falcon tubes containing 2.3 mL of absolute ethanol stored overnight at 4 °C.
5. Seventeen 50 mL falcon tubes containing 3 mL of the sodium azide solution stored overnight at 4 °C.
6. A 1 L Erlenmeyer flask containing 460 mL of YPD growth medium and a magnetic stir bar.
7. A 100 mL Erlenmeyer flask containing 40 mL of YPD growth medium.
8. 1 L of ultrapure water stored overnight at 4 °C.

2.4 Flow Cytometry Analysis

1. Sodium citrate solution: 50 mM sodium citrate, pH 7.
2. Staining solution: 50 mM sodium citrate, 40 μg/mL propidium iodide.
3. RNAse A stock: 100 mg/mL RNAse A.
4. Flow cytometer: MACSQuant Analysers (Miltenyi Biotec).
5. MACSQuantify software (Miltenyi Biotec).

2.5 DNA Extraction

1. Genomic DNA buffer set (QIAGEN).
2. Genomic-tip 20/G (QIAGEN).
3. NanoDrop (Thermo scientific).

3   Methods

Carry out all experimental procedures at room temperature unless otherwise specified. For yeast manipulations, perform the inoculation steps in sterile conditions. Cells are pelleted by centrifugation at $3500 \times g$ for 5 min or at $16,000 \times g$ for 2 min when in Flacon or Eppendorf tubes, respectively. The flowchart of the complete protocol is presented in Fig. 1. Cell synchronization can be achieved by two alternative methods, centrifugal elutriation (*see* Subheading 3.2) or gradient of Percoll (*see* Subheading 3.3).

3.1   Yeast Growth Conditions

1. *Day 1*: Spread yeast cells directly from the −80 °C stock onto YPD-agar petri dishes and incubate for 2 days at 30 °C, in the microbiological incubator, until a confluent lawn of cells has grown on the plate (*see* **Note 4**).

2. *Day 3*: Scrape cells from the YPD-agar plate with a sterile plastic loop of 1 μL and transfer them into an Eppendorf tube containing 1 mL of sterile water.

3. Measure the optical density (OD, 600 nm) of the cell suspension to estimate cell concentration and use 10^9 cells to inoculate 500 mL of YPD-growth medium. Yeasts are then grown in the Multitron incubator for 18 h at 30 °C with shaking ($160 \times g$) until the beginning of stationary phase (*see* **Note 5**).

4. *Day 4*: Measure the OD (600 nm) of the resulting asynchronous culture. OD should be around 25–30.

5. *Day 4*: Proceed either to Subheading 3.2 or 3.3.

3.2   Cell Synchronization in G1 Phase Using Centrifugal Elutriation

1. *Day 4*: Prepare the Beckman elutriation system following manufacturer's instructions.

2. *Day 4*: Transfer 1 mL of the asynchronous cell culture in 2.3 mL of cold ethanol to fix the cells (*see* **Note 6**).

3. *Day 4*: Pellet 8×10^{10} cells from the asynchronous culture by centrifugation of the corresponding volume of culture. Wash the cells once in an equivalent volume of PBS 1×. Pellet the cells and resuspend them in 50 mL of PBS 1× (*see* **Note 7**).

4. *Day 4*: Load the cells in the elutriator flow chamber at a flow rate of 22 mL/min, with a rotor speed of $1175 \times g$. Increase gradually the flow rate (by 2 mL/min steps) until the cells reach the top of the chamber. Then let the elutriator system run in those conditions for 1 h in order to equilibrate the chamber.

5. *Day 4*: Every 5 min, increase the flow rate by a 2 mL/min step until the cells come out from the elutriator (*see* **Note 8**).

6. *Day 4*: The first cell fraction of 800 mL is then recovered into two 500 mL centrifugal bottles. Increase the flow rate by a

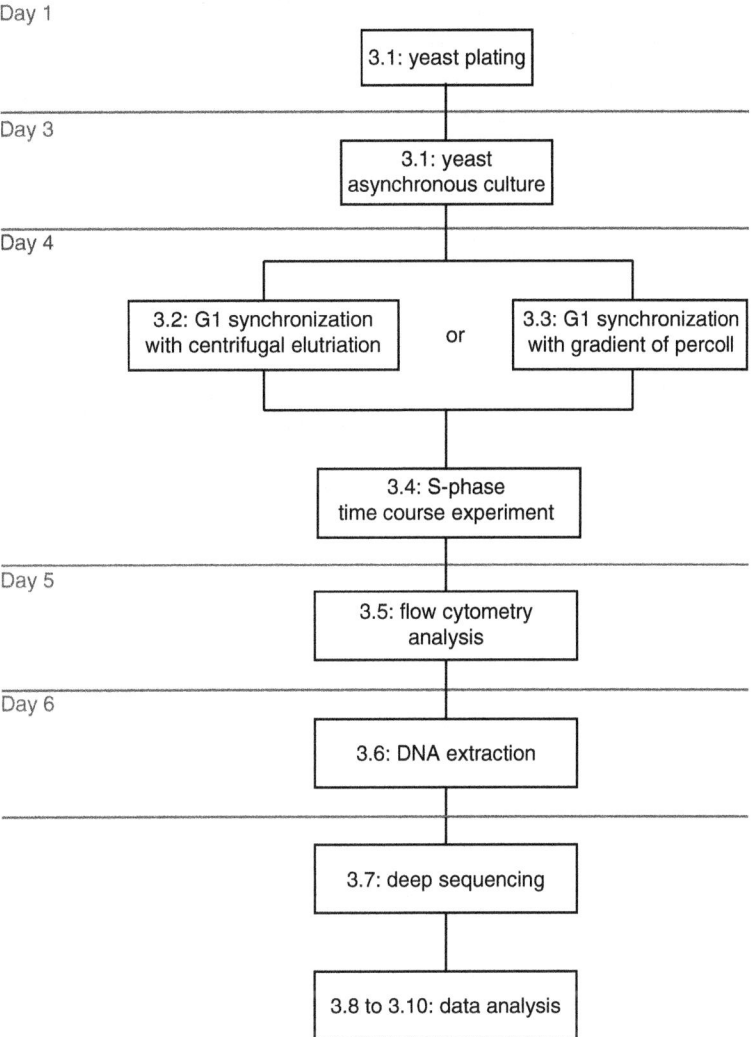

Day 1

3.1: yeast plating

Day 3

3.1: yeast
asynchronous culture

Day 4

3.2: G1 synchronization
with centrifugal elutriation   or   3.3: G1 synchronization
with gradient of percoll

3.4: S-phase
time course experiment

Day 5

3.5: flow cytometry
analysis

Day 6

3.6: DNA extraction

3.7: deep sequencing

3.8 to 3.10: data analysis

Fig. 1 Flowchart of the complete protocol

2 mL/min and start recovering the second fraction. A complete fraction takes between 20 and 40 min to come out from the elutriator. During this time, take an aliquot of 1–2 mL from the first fraction and determine the proportion of G1 cells under a microscope. In parallel, estimate the number of cells by measuring the OD of the first fraction.

7. *Day 4*: Keep on recovering and analyzing the next fractions as above until you recover a single fraction of 800 mL containing at least 90 % of G1 cells and totaling at least 10^9 cells (*see* **Note 9**).

8. *Day 4*: Transfer 1 mL of this G1 synchronized fraction in 2.3 mL of cold ethanol to fix the cells.

9. *Day 4*: Proceed to Subheading 3.4 (*see* **Note 10**).

3.3 Cell Synchronization in G1 Phase Using Gradient of Percoll

1. *Day 4*: Fill eight polycarbonate centrifugal tubes of 50 mL with 30 mL of the Percoll solution prepared extemporaneously (*see* **Note 11**).

2. *Day 4*: Centrifuge the tubes for 15 min at $19,300 \times g$ in a fixed angle rotor (34°) to generate the gradient of Percoll.

3. *Day 4*: Pellet 4×10^{10} cells from the asynchronous culture by centrifugation and resuspend them in 16 mL of Tris solution.

4. *Day 4*: Gently dispense 2 mL of the cell solution at the top of the gradient of Percoll in each of the eight tubes (*see* **Note 12**).

5. *Day 4*: Centrifuge the eight tubes for 30 min at $400 \times g$ in a swinging rotor.

6. *Day 4*: Gently pipet up five fractions of 6 mL from the top of each gradient, using 1 mL tips and a smooth pipetman.

7. *Day 4*: Discard the first fraction of each tube and determine the proportion of G1 cells under a microscope for all other fractions.

8. *Day 4*: Select all the fractions that contain at least 90 % of G1 cells (usually one or two fractions per gradient) and pool them into Falcon tube(s) to produce a single fraction of G1 synchronized cells (*see* **Note 13**).

9. *Day 4*: Transfer 200 µL of this G1 synchronized fraction in 500 µL of cold ethanol to fix the cells.

10. *Day 4*: Proceed at once to Subheading 3.4.

3.4 S-phase Time Course Experiment

1. *Day 3*: Prepare all solutions and growth media (*see* Subheading 2.3). Store sodium azide and absolute ethanol solutions at 4 °C and YPD solutions at 23 °C.

2. *Day 4*: At least 1 h before the beginning of the time course experiment, put the Erlenmeyer flask containing the 460 mL pre-warmed YPD into the hot water bath equilibrated at 23 °C in the cold chamber (Fig. 2) (*see* **Note 14**). Start the agitation using the magnetic stirrer.

3. *Day 4*: Pellet the G1 synchronized cells (*see* **Note 15**). Resuspend the cells into the 40 mL of YPD pre-warmed at 23 °C and transfer them in the Erlenmeyer flask equilibrated in the hot water bath. This step defines the starting point of the time course experiment (time 0, *see* **Note 16**).

4. *Day 4*: Every 10 min, rapidly take out one aliquot of 1 mL (with a 1 mL pipette) and transfer it into a 15 mL Falcon tube containing cold ethanol (for flow cytometry analysis) and one aliquot of 30 mL (with a 50 mL pipette) and transfer it into a 50 mL Falcon tube containing the sodium azide solution (for DNA extraction, *see* **Notes 17** and **18**). During the course of the experiment, take an additional sample of 1 mL to estimate

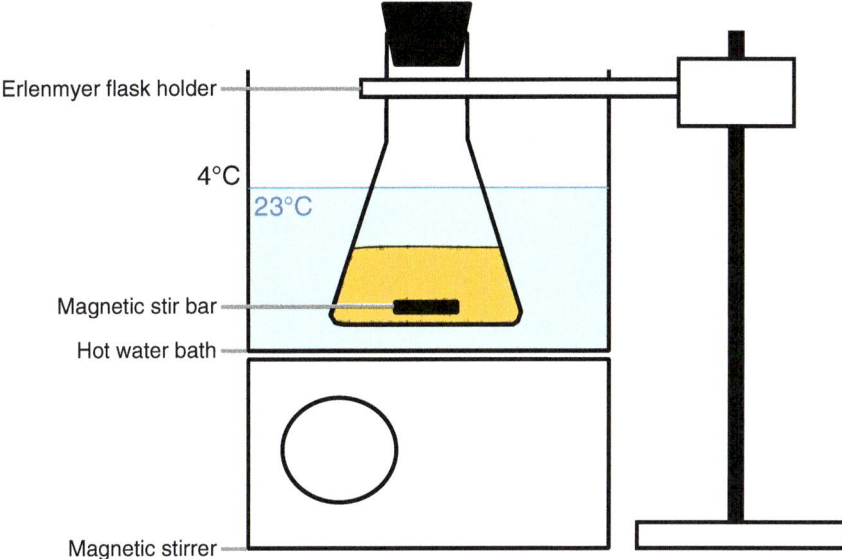

Fig. 2 Installation diagram of the incubation system used to perform the time course experiment

cell concentration by measuring OD at 600 nm. The entire time-course experiment runs for 160 min (*see* **Note 19**).

5. *Day 4*: Store all ethanol aliquots for one night at 4 °C before performing flow cytometry analysis (*see* Subheading 3.5).

6. Pellet the cells from the sodium azide samples by centrifugation and wash them two times with 30 mL of cold water. Cell pellets are then frozen at −80 °C and kept into the freezer until DNA extraction (*see* Subheading 3.6).

3.5 Flow Cytometry Analysis

1. *Day 5*: Pellet the cells fixed in ethanol by centrifugation and resuspend them into 1 mL of sodium citrate solution (*see* **Note 20**). Pellet the cells again and resuspend them into sodium citrate solution at the final concentration of 10^7 cells/mL (based on the cell concentration estimated during the time-course experiment).

2. *Day 5*: Transfer 100 μL of each aliquot into a 1.5 mL Eppendorf tube and add 1 μL of RNAse A stock solution. Incubate for 2 h at 37 °C in a hot water bath (*see* **Note 21**).

3. *Day 5*: Add 400 μL of the staining solution into each 1.5 mL Eppendorf tube and incubate for 40 min at room temperature. Pellet the cells by centrifugation and remove the supernatant containing the staining solution. Resuspend the cells in 100 μL of sodium citrate solution. Samples are ready for flow cytometry analysis.

4. *Day 5*: Quantify the level of propidium iodide fluorescence for at least 30,000 cells from each sample with the flow cytometer.

5. *Day 5*: Determine the proportion of G1 and G2 cells for each sample by plotting the distribution of the fluorescence levels with a data analysis software (e.g., MACSQuantify), In an asynchronous population, two main peaks should be identified, the first one corresponds to G1 cells and the second one to G2 cells. The median fluorescence value of the second peak should be about two times higher than the median fluorescence value of the first one (Fig. 3a). Determine for each time point the proportion of G1 and G2 cells (Fig. 3b). The average DNA content (*N*) can be calculated with the following formula:

$$N = \frac{G1 + 2 \times G2}{G1 + G2}$$

Where G1 and G2 are the proportions of G1 and G2 cells in the population, respectively.

6. *Day 5*: Plot the evolution of DNA content (*N*) as a function of time to visualize the progression of S-phase during the time-course experiment (*see* **Note 22**).

7. *Day 5*: Choose the eight samples that best cover the entire S-phase (red dots in Fig. 3c, *see* **Note 23**).

3.6 DNA Extraction

1. *Day 6*: Extract genomic DNA from all eight samples using the Qiagen Genomic-tip 20/G kit, following manufacturer's instructions (*see* **Note 24**).

2. *Day 6*: Check DNA quantity and quality using a NanoDrop apparatus and on an agarose gel.

3.7 Deep Sequencing

1. Prepare eight multiplexed libraries from at least of 300 ng of DNA from each sample, using the Illumina technology. The eight multiplexed libraries must be pooled together before any PCR amplification step in order to avoid differential PCR biases between samples.

2. Run Single-End Illumina sequencing of the pooled libraries to produce about 10 M 50 bp reads per library (*see* **Notes 25** and **26**).

3.8 Data Processing

1. Libraries must be de-multiplexed and adaptor sequences removed from the reads.

2. For each library, check the read quality. Quality scores across all bases have to be higher than 28 and the sequence content across all bases have to respect the Chargaff's rule (the amount of guanine is equal to the amount of cytosine and the amount

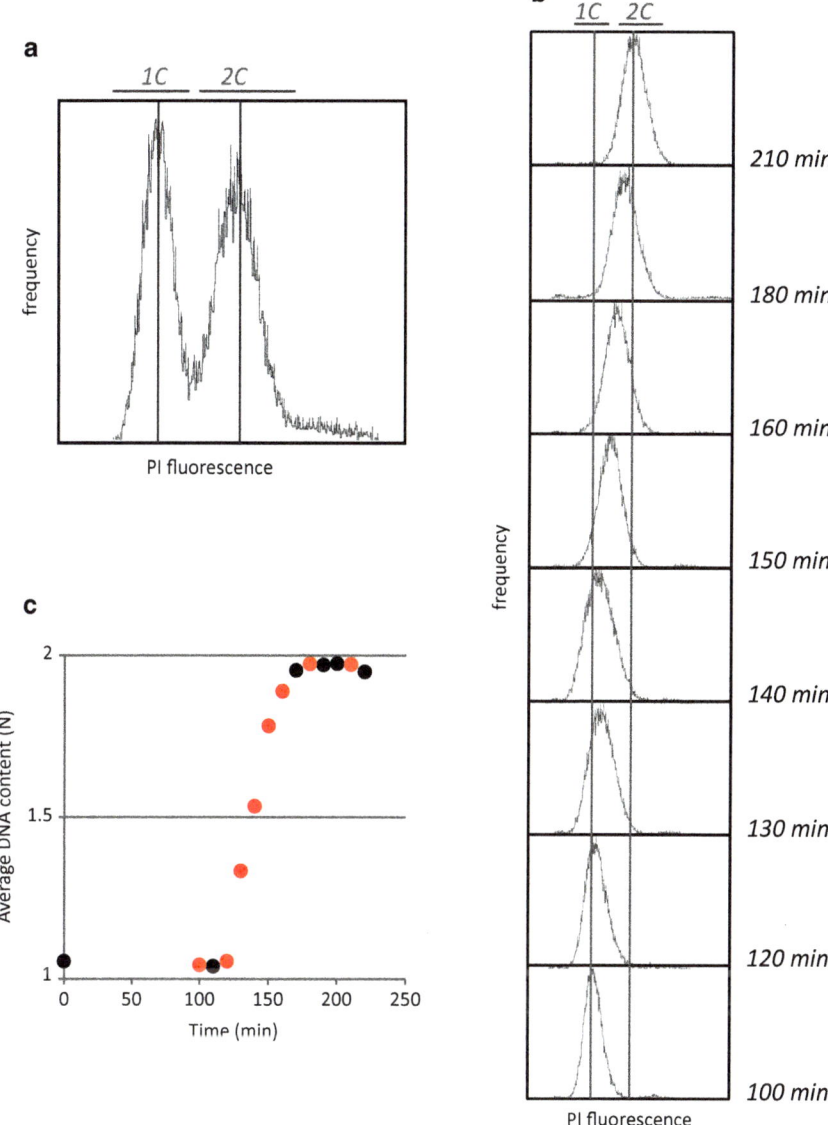

Fig. 3 Flow cytometry analysis. (**a**) Distribution of the Propidium iodide (PI) fluorescence in an asynchronous cell population. PI fluorescence is proportional to DNA content. 1C: one genome DNA content (G1 cells) and 2C: two genome DNA content (G2 cells). (**b**) Distribution of the PI fluorescence for aliquots taken during time course experiment. As cells go throughout S-phase, PI fluorescence increases from 1C to 2C. (**c**) Variation of average DNA content (*N*) during time course experiment. Samples covering the entire S-phase and selected for DNA extraction are indicated as *red dots*

of adenine is equal to the amount of thymine). If needed, sequences can be trimmed using fastx_trimmer (http://hannonlab.cshl.edu/fastx_toolkit/).

3. Align the reads (from the my_reads.fastq files) on the reference genome (genome.fasta file) using BWA [13], allowing zero

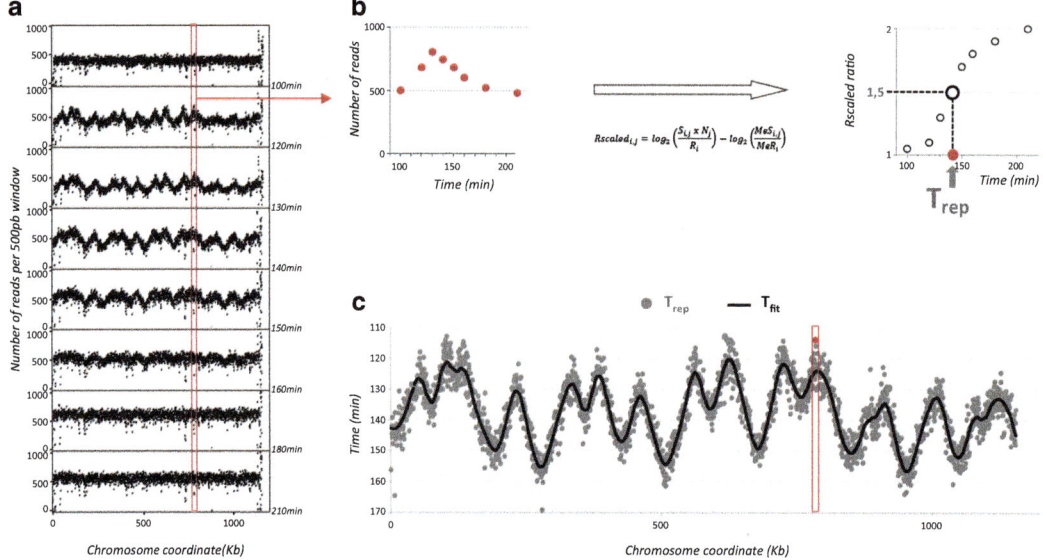

Fig. 4 Analysis pipeline to determine the replication profile from the sequencing data. (**a**) Number of reads for 500 bp window for 8 time points along one chromosome. (**b**) Calculation of *Rscaled* ratio and T_{rep}. (**c**) Spatiotemporal replication profile for one chromosome

mismatch and no gap (given in Appendix). Remove all reads from the results.sam files that map to multiple locations and that have quality mapping scores (MAPQ) lower than 37 (i.e., base call accuracy less than 99.98 %).

3.9 Calculating Replication Profile from Processed Sequencing Data

1. For each time point, determine the total number of reads that start in each of the 500 bp nonoverlapping window that cover the entire the genome (Fig. 4a, *see* **Note 27**).

2. Choose the reference sample where the total number of reads remains constant in all nonoverlapping windows along the genome (usually the G1 or the G2 sample, *see* **Note 28**) indicating that no replication is ongoing.

3. Calculate the *Rscaled* ratio for each window *i* along the genome and for each sample time point *j* using the following formula:

$$Rscaled_{i,j} = \log_2\left(\frac{S_{i,j} \times N_j}{R_i}\right) - \log_2\left(\frac{MeS_{i,j}}{MeR_i}\right)$$

where N is the average DNA content calculated with flow cytometry, S and R are the total number of reads for the sample and the reference, respectively, and MeS and MeR are the

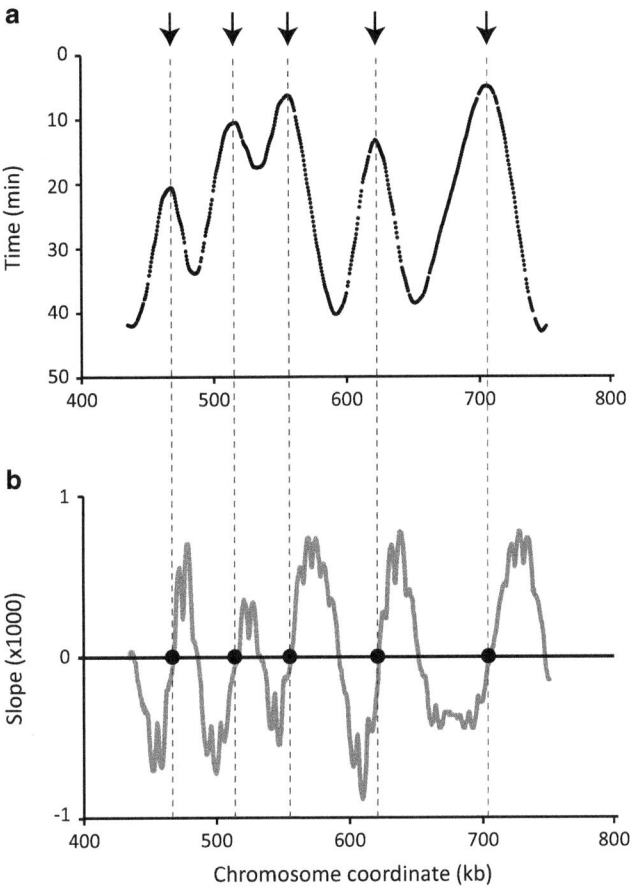

Fig. 5 Peak calling procedure. (**a**) T_{fit} plotted as a function of chromosome coordinates. *Arrows* indicate the position of the replication origins. (**b**) Slope curve calculated from the T_{fit} curve. The *black dots* represent the coordinates where the slope curve intersects 0 from negative to positive value. They correspond to the peaks on the T_{fit} curve

medians of the total number of reads measured for sample and reference, respectively.

4. Estimate the time (T_{rep}) when the *Rscaled* ratio is equal to 1.5 for each window along the genome (Fig. 4b) with the R script provided in Appendix.

5. Determine the spatiotemporal replication profile of the genome by plotting the T_{rep} as a function of chromosome coordinates and by fitting a loess regression curve to the data (*see* **Note 29**) with the R script provided in Appendix (Fig. 4c). The resulting fitted values of T_{rep} are denoted in the output of the R script as T_{fit} values.

1. For each coordinate x of the window i along the genome, calculate the slope of the T_{fit} curve using the following formula:

$$\frac{T_{fit_i} - T_{fit_{i-1}}}{x_i - x_{i-1}}.$$

2. Plot the slope values as a function of the chromosomal coordinates. The resulting slope curve is an estimate of the first derivative of the T_{fit} curve (Fig. 5).

3. Select the chromosomal coordinates where the slope curve intersects 0, from negative to positive values. These coordinates give the localizations of the replication initiation points called replication origins (*see* **Note 30**).

4 Notes

1. Small differences in growth medium composition have a strong impact on yeast growth. To avoid this source of variability, it is recommended to prepare the total volume of required medium at the same time. It is recommended to put water at the bottom of the cylinder before adding the powder to help dissolution.

2. After autoclave, bottles of hot YPD-Agar medium can be transferred in a hot-water bath, at 50 °C for 1 h to cool down and directly poured into 94 mm petri dishes (25 mL of YPD-Agar per plate).

3. The Percoll solution must be prepared just before use.

4. Always use the –80 °C stock rather than old plates stored at 4 °C to inoculate YPD-Agar plates with yeast cells.

5. The correspondence between OD and cell concentration must be calibrated. In our conditions, 1 OD unit corresponds to 7×10^6 cells/mL. A minimum of 8×10^{10} cells is required for the cell cycle synchronization by elutriation [12]. Growth conditions can be adapted to reach this threshold at the beginning of stationary phase.

6. Ethanol cell fixation is achieved by overnight incubation at 4 °C. Fixed cells can then be stored for at least 1 month before performing the flow cytometry analysis.

7. A this step, cells are highly concentrated (about 1.6×10^9 cells/mL) and can aggregate, which would compromise efficient discrimination of G1 cells with elutriation. This can easily be checked by observing an aliquot (5 µL) under microscope. Vortexing at full speed for 1 min can solve this problem.

8. The optimal flow rate parameters depend on cell size and morphology. As a result they can be very different from one yeast species to another.

9. For species where the size difference between G1 and S or G2 cells is limited, it can be difficult to reach a level of synchronization of 90 %. In those cases, we found that a minimum of 75 % of G1 synchronized cells can be sufficient to obtain a replication profile.

10. Synchronized cells in PBS can be kept at 4 °C for one night before performing the S-phase time course without affecting the experiment.

11. In our tests, for each 30 mL gradients of Percoll, about 10^8 G1 cells are recovered. A complete S-phase time course experiment requires 10^9 G1 cells, which corresponds to 10 gradients of Percoll. However, the yield of the gradient of Percoll and the position of the G1 fractions in the gradient can change from one species to another.

12. The best way to add the cell solution on the top of the Percoll gradient without disturbing it is to put the extremity of the tip in contact with the tube's wall, 1 cm above the surface of the Percoll solution.

13. Percoll is nontoxic; there is no need to wash the cells.

14. To ensure a constant temperature of 23 °C during the whole S-phase time course experiment, we prefer a hot water bath set at 23 °C in a cold chamber rather than a microbiological incubator. The hot water bath is placed on a magnetic stirrer and shaking is carried out by a sterile magnetic stir bar directly put in the growth medium (Fig. 2).

15. At this step, cell pelleting can be inefficient due to low cell concentration. To overcome this problem, transfer the G1 cell solution in 50 mL falcon tubes and centrifuge at $3500 \times g$ for 10 min.

16. To perform a full S-phase time course experiment in the presence of dedicated drugs (e.g., hydroxyurea), one should add the drug solution directly in the 460 mL YPD medium few minutes before transferring the cells and starting the time course experiment.

17. Flow cytometry and DNA extraction aliquots must be taken almost immediately one after the other. During the time course experiment, both ethanol and sodium azide falcon tubes must be kept on ice.

18. For budding yeast, the progression of cells through S-phase can be monitored by observing a small aliquot under the microscope: the beginning of S-phase is usually concomitant with the appearance of the bud. However, this is not true for all species and the analysis by flow cytometry remains the best way to monitor the progression of the cell cycle.

19. There is a delay before the G1 synchronized cells enter into S-phase. Note that the duration of this delay changes from one

species to another. As a consequence, the duration of the entire time course experiment and the time points chosen to take aliquots have to be defined for each new species.

20. Resuspending cells that have been fixed in ethanol can be very difficult. To help resuspension, remove the maximum amount of ethanol by placing the tubes upside down on a towel paper and letting them dry for 5 min. Add the 1 mL sodium citrate solution and let the pellet rehydrate for 10 min before resuspending.

21. After incubation, cells in sodium citrate with RNAse A can be stored for one night at 4 °C before performing the next steps in the flow cytometry analysis.

22. Two parameters should be considered when qualifying a time course experiment: the proportion of G1 cells at time 0 and the completion of the S-phase. Use the plot of the average DNA content as a function of time to check these two parameters. The plot should follow a sigmoid curve with two well-defined plateaus. The first plateau, at the beginning of the experiment, corresponds to G1 cells and should be close to 1. The second plateau, at the end of the experiment, reflects the completion of S-phase and should be as close as possible to 2 and almost flat.

23. It is very important to choose the first and the last time points for sequencing in the G1 and G2 phases, respectively, where there is no DNA replication.

24. Adding isopropanol to spool the DNA may be inefficient. To overcome this problem, mixes well the tubes containing eluted fractions plus isopropanol and centrifuge them for 1 h at $16,000 \times g$ and 4 °C.

25. We found that, for a 10–12 Mb genome, a minimum of 10 M sequences per sample is needed to achieve a 500 bp resolution when calculating the replication profile from sequencing data.

26. A more straightforward alternative to the sequencing of all eight libraries is to pool the genomic DNA from the six replicating time points. Then, only two DNA samples need to be sequenced: the "replicating pool" and one of the non-replicating reference time point (i.e., the G1 or G2 sample). Only 10 M of reads are needed for each of these two samples. An even more straightforward alternative that was recently described consists on sequencing DNA from an exponentially growing cell sample and from a stationary phase sample [11]. In this case, it was reported that 120 M of reads for each sample was required. However we found that only 15 M of reads are sufficient to generate a suitable replication profile. For these two alternative protocols, the replication program is obtained by calculating, along the genome, the ratio between

the number of aligned reads in the "replicating pool" (or the exponential sample) and in the reference sample (or the stationary phase sample). This ratio however does not provide the real replication timing of the chromosomes and does not allow inferring origin efficiency [10] or fork velocity [9]. To access such information, all time samples have to be independently sequenced, as described in step 3.6 of the main protocol.

27. The resolution of the replication profile is directly linked to the number of reads obtained for each sample. We found that a minimum of 500 reads in each window is needed to achieve a good resolution. Therefore, for a genome of 10 Mb and a resolution window of 500 bp, a minimum of 10 M reads for each time point is required. With a 1000 bp window this value drops to 5 M reads.

28. Choosing the first time point for which DNA will be sequenced can be difficult because it can happen that replication has already started at this point. This can be easily seen by the presence of some small peaks near the origins when plotting the number of sequences along the genome for this time point. In that case, this time point should not be used as the reference and we recommend using instead the last sample (G2) as the reference time point.

29. We find that the best window for the loess smoothing function is 60 kb. It gives the best compromise between sensitivity and noise correction.

30. Note that the coordinates where the slope curve intersects 0 from positive to negative values indicate the position of the termination regions.

---

5 Appendix

List of commands for BWA:
    'genome indexing
        ./bwa index genome.fasta
    'read mapping
        ./bwa aln -n 0 -o 0 genome.fasta my_reads.fastq > results_temp.sai
    'format as sam file
        ./bwa samse -n 0 -f results.sam genome.fasta results_temps.sai my_reads.fastq
    'filter the results.sam file
        awk '{if($5==37 && $12=="XT:A:U") print}' results.sam > results_filtered.sam
    R script for Trep estimation:
        'Prepare one Input table for each chromosome. The first column named "ID" contains a unique 'ID for each 500 window

and the following columns (one column per time point) contain the 'Rscaled ratios. The following script is given for 8 times points j = [25, 30, 35, 40, 45, 50, 55, 60] 'and the corresponding columns containing the Rscaled ratio are named T1 to T8.

```
t = read.table("Input table", header=TRUE)
lev = levels(t$ID)
lev = lev[2:length(lev)]
tab = matrix(nrow = length(lev), ncol = 3)
for(i in 1:length(lev))
{
    if(lev[i] != "")
    {
        print(paste("Sonde ", as.character(i), " / ", as.character(length(lev)), " : ",
lev[i], sep = ""))
'format the data.
        me = seq(1,8)
me[1] = t$T1_Adj[t$ID == lev[i]]
me[2] = t$T2_Adj[t$ID == lev[i]]
me[3] = t$T3_Adj[t$ID == lev[i]]
me[4] = t$T4_Adj[t$ID == lev[i]]
me[5] = t$T5_Adj[t$ID == lev[i]]
me[6] = t$T6_Adj[t$ID == lev[i]]
me[7] = t$T7_Adj[t$ID == lev[i]]
me[8] = t$T8_Adj[t$ID == lev[i]]
'Here enter the list of values for j manually
        time=c(25,30,35,40,45,50,55,60)
'Loess fitting of scaled data for each 500 bp window
        l = loess(me~time, span = 0.75, family =
"gaussian")
'Estimation of time (Trep) when the Rscaled ratio is equal to 1.5.
        pred = -1
'Enter inf value manually (inf value = min(j))
        inf = 25
'Enter sup value manually (sup value = max(j))
        sup = 60
        lim = 0.03
        while(pred == -1)
        {
            temps = mean(c(inf, sup))
            if(sup - inf < 0.03)
            {
                pred = predict(l, temps)
                break
            }
            if(predict(l, temps) > 1.5)
                sup = temps
```

```
        else if(predict(l, temps) < 1.5)
            inf = temps
        else
            pred = predict(l, temps)
    }
    print(paste("Point ", as.character(pred), " trouve a
", as.character(temps),
    sep = ""))
        tab[i, ] = c(lev[i], temps, pred)
    }
}
colnames(tab) = c("Probe", "Trep", "Rscaled")
write.table(tab, "My_results.txt", row.names = FALSE, sep
= "\t", quote = FALSE)
```

R script for spatiotemporal replication profile:

'Prepare one Input table for each chromosome. Each input table contains 3 columns: 'Chromosome, Position, Trep. Position are expressed in base pair.

```
t = read.table("Input table", header = TRUE)
threshold = 60000
sp = threshold/max(t$Position)
l = loess(t$Trep~t$Position, span = sp)
    tabresult = matrix(nrow=l$n, ncol=3, data=c(l$x, l$y,
l$fitted))
colnames(tabresult) = c("Position", "Trep", "Tfit")
Result = as.data.frame(tabresult)
write.table(Result, file = "profile.txt",row.names = FALSE,
sep=",")
```

References

1. Raghuraman MK, Winzeler EA, Collingwood D et al (2001) Replication dynamics of the yeast genome. Science 294:115–121

2. Meselson M, Stahl F (1958) The replication of DNA in Escherichia coli. Proc Natl Acad Sci U S A 44:671–682

3. Feng W, Collingwood D, Boeck ME et al (2006) Genomic mapping of single-stranded DNA in hydroxyurea-challenged yeasts identifies origins of replication. Nat Cell Biol 8:148–155

4. Yabuki N, Terashima H, Kitada K (2002) Mapping of early firing origins on a replication profile of budding yeast. Genes Cells 7: 781–789

5. Müller CA, Nieduszynski CA (2012) Conservation of replication timing reveals global and local regulation of replication origin activity. Genome Res 22:1953–1962

6. Koren A, Tsai HJ, Tirosh I et al (2010) Epigenetically-inherited centromere and neo-centromere DNA replicates earliest in S-phase. PLoS Genet 6, e10011068

7. Liachko I, Youngblood RA, Tsui K et al (2014) GC-rich DNA elements enable replication origin activity in the methylotrophic yeast Pichia pastoris. PLoS Genet 10, e1004169

8. Di Rienzi SC, Lindstrom KC, Mann T et al (2012) Maintaining replication origins in the face of genomic change. Genome Res 22: 1940–1952

9. Agier N, Romano OM, Touzain F et al (2013) The spatiotemporal program of replication in the genome of Lachancea kluyveri. Genome Biol Evol 5:370–388

10. Hawkins M, Retkute R, Müller CA et al (2013) High-resolution replication profiles

define the stochastic nature of genome replication initiation and termination. Cell Rep 5:1132–1141

11. Müller CA, Hawkins M, Retkute R et al (2014) The dynamics of genome replication using deep sequencing. Nucleic Acids Res 42:1–11

12. Marbouty M, Ermont C, Dujon B et al (2014) Purification of G1 daughter cells from different Saccharomycetes species through an optimized centrifugal elutriation procedure. Yeast 31:159–166

13. Li H, Handsaker B, Wysoker A et al (2009) The sequence alignment/map format and SAMtools. Bioinformatics 25:2078–2079

Chapter 15

Single-Step Affinity Purification (ssAP) and Mass Spectrometry of Macromolecular Complexes in the Yeast *S. cerevisiae*

Christian Trahan, Lisbeth-Carolina Aguilar, and Marlene Oeffinger

Abstract

Cellular functions are mostly defined by the dynamic interactions of proteins within macromolecular networks. Deciphering the composition of macromolecular complexes and their dynamic rearrangements is the key to getting a comprehensive picture of cellular behavior and to understanding biological systems. In the last decade, affinity purification coupled to mass spectrometry has emerged as a powerful tool to comprehensively study interaction networks and their assemblies. However, the study of these interactomes has been hampered by severe methodological limitations. In particular, the affinity purification of intact complexes from cell lysates suffers from protein and RNA degradation, loss of transient interactors, and poor overall yields. In this chapter, we describe a rapid single-step affinity purification method for the efficient isolation of dynamic macromolecular complexes. The technique employs cell lysis by cryo-milling, which ensures nondegraded starting material in the submicron range, and magnetic beads, which allow for dense antibody-conjugation and thus rapid complex isolation, while avoiding loss of transient interactions. The method is epitope tag-independent, and overcomes many of the previous limitations to produce large interactomes with almost no contamination. The protocol described here has been optimized for the yeast *S. cerevisiae*.

Key words Proteomics, Single-step affinity purification, Cell lysis, Cryo-milling, Yeast, Mass spectrometry

---

1 Introduction

In the last decade, many researchers have put considerable effort into elucidating protein–protein interaction networks in order to better understand the interplay between proteins and gain a comprehensive picture of dynamic pathways that govern cellular processes [1–3]. The methods to study macromolecular complexes within these networks have improved greatly as part of the ever-expanding proteomics field, and affinity purification–mass spectrometry (AP-MS) has become a powerful approach to study

Frédéric Devaux (ed.), *Yeast Functional Genomics: Methods and Protocols*, Methods in Molecular Biology, vol. 1361, DOI 10.1007/978-1-4939-3079-1_15, © Springer Science+Business Media New York 2016

interactions within a variety of cellular complexes in many organisms [1, 4–6].

However, determining the interactions between macromolecules in a cell is a formidable undertaking for several reasons. The number of interacting entities is huge. In *Saccharomyces cerevisiae* alone, there are ~6200 open reading frames that code for proteins [7, 8]. Moreover, proteins do not work in isolation but within organized complexes along pathways, forming intricate information networks, and are present in a broad range of abundance (10^1–10^6 copies/cell) [9–11]. Finally, protein interactions have a wide range of affinities [12].

Many of the early affinity purification approaches came with a number of limitations that were not up to handling these many variations, which make up the very nature of dynamic macromolecular complexes and pathways. Heat-generating lysis methods, long incubation times due to slow affinity kinetics, and resins with limited pore dimensions all resulted in partial complex degradation, loss of interactors, association, and identification of a wide range of contaminants, and overall poor yields and low signal to noise [13–15]. Here, we describe a faster method, which employs cell lysis by cryo-milling followed by a rapid single-step affinity purification (ssAP) using solely one epitope tag and magnetic resin, which allow us to preserve the integrity of the isolated complexes.

The biggest challenge of any affinity purification experiment is to preserve the interactions within the targeted complex, or pool of complexes, with a high degree of fidelity. Cell lysis disrupts the cellular environment and causes the intermingling of components that are not normally exposed to one another, and thus the resultant possibility of aberrant molecular interactions—a major source of "nonspecific background." Another undesired result of this unnatural intermingling is the exposure to degradation enzymes, such as proteases, RNAses, and DNAses, that are normally kept at bay in a living cell. Contrary to other cell disruption methods, cryolysis better preserves the integrity of complexes as well as its transient and weak interactors as the cell breakage occurs in liquid nitrogen, below –196 °C, and keeps cellular components in a solid state, thus preventing rearrangements of complexes during the lysis of cells and subsequent disruption of the cellular compartments. The method also protects complexes from the activity of nucleases and proteases released during lysis. Cryolysis is efficiently performed using a planetary ball mill, which yields a homogenous cell grindate with particles of approx. 1–2 μm in diameter [16].

Besides many stable interactions, there are proteins, subcomplexes, and macromolecules that constantly associate and dissociate in the form of dynamic or transient interactions. Fourth, while not directly interacting, there are vicinal macromolecules—i.e., macromolecules that are particular to the environment of the tagged macromolecule, and which help to define the physical and chemical characteristics of the environment. After cell lysis, the goal is to isolate these complexes very rapidly, before many macromolecular assemblies degrade, and transient and dynamic interactions fall apart.

Commonly used tandem affinity epitope tags often require long incubation times and several steps of purification, which result in the loss of dynamic and transient interactions. Single-step affinity tags reduce necessary steps, and thus time and handling, resulting in faster isolations and conservation of these interactions. While either Protein A (PrA), GFP, FLAG, HA, or myc are now commonly used, PrA tag in particular has high recovery rates due to the presence of multiple IgG-binding moieties, which ensures faster kinetics during the binding step, and high affinity for rabbit IgG (~10 nM) [17].

In addition, affinity resins can also pose a limiting factor. Sepharose-based affinity isolations are still the most widely used outside of our laboratory and tend to be slow (4–14 h). Moreover, the resin pore dimension limits the upper size of any isolated subcomplexes, which is a particular issue for affinity purification of macromolecular complexes. On the other hand, magnetic resin is small (1 nm–2.8 μm diameter) and nonporous, which allows the dense conjugation of antibody to its surface. This ensures a large surface-to-volume ratio, allowing for short incubation times (5–30 min) for efficient, above 90 %, complex recovery, regardless of the nature or cellular abundance of the bait protein or size of complexes [16]. Short incubation times also have the advantage of minimizing nonspecific binding and loss of significant transient or weakly associated interactors [14, 16]. After complex isolation, the affinity-purified proteins are then either trypsin digested directly on the magnetic resin ("on-bead") to prepare them for mass spectrometry, or eluted from the resin under denaturing conditions for SDS-PAGE and western blotting analysis.

Sample preparation protocols and mass spectrometry (MS) methods have also advanced significantly in the past decade. In particular, a number of new proteolytic approaches for peptide digestion have emerged in the last few years. While for many years in-gel trypsin digestion was the standard for MS sample preparation, with the emergence of gel-free methods such as in-solution and, most recently, on-bead trypsin digest, sample handling has become much faster and simpler. In the gel electrophoresis-based approach, the proteins are separated in one or two dimensions (1D/2D) on a gel and enzymatic digestion is performed in-gel, which is a time-consuming and tedious process [18]. In the gel-free or in-solution-based approach, the proteins or peptides, or both, are separated chromatographically using on-line LC systems and the proteins are digested in-solution [19]. More recent protocols use on-bead trypsin digest, where proteins are digested directly on the resin in lieu of elution [20]. The in-solution and on-bead-based approaches tend to be the simplest in terms of sample handling and speed.

In this chapter, we describe a step-by-step protocol for affinity-purifying epitope-tagged dynamic macromolecular complexes using a single-step approach in the yeast *S. cerevisiae*. We first describe cell growth harvesting, cell material and buffer selection, and antibody conjugation to the magnetic resin. We then guide through the different steps: cryo-milling, affinity purification, on-bead trypsin digest, and mass spectrometry. Finally, we will briefly describe aspects of data analysis.

2 Materials

2.1 Conjugating Magnetic Dynabeads with Rabbit IgG (See Note 1)

1. Dynal Dynabeads M-270 Epoxy (Life Technologies).
2. IgG from rabbit serum (Sigma-Aldrich).
3. Polypropylene tubes: 4×15 ml, 2×50 ml.
4. Magnetic holder for 15 ml-tubes (e.g. DynaMag™-15).
5. Nutator or any slow rotating platform or wheel.
6. Millex-GP Syringe Filter Unit, 0.22 μm, polyethersulfone.
7. 20 ml-syringe.
8. 1.5 ml polypropylene tubes.
9. Tabletop centrifuge ($\geq 13,000 \times g$).
10. Vortex.
11. Parafilm.
12. 0.1 M $NaPO_4$, pH 7.4.
13. 3 M $(NH_4)_2SO_4$, pH 7.4.
14. $1 \times$ PBS pH 7.4: 10 mM $NaPO_4$ pH 7.4, 150 mM NaCl.
15. 100 mM Glycine pH 2.5.
16. 10 mM Tris pH 8.8, store up to 1 year at RT.
17. 100 mM Triethylamine (make fresh) (*see* **Note 2**).
18. $1 \times$ PBS, 0.5 % Triton X-100.
19. 10 % NaN_3 (*see* **Note 2**).

2.2 Harvesting and Freezing Yeast

1. Epitope-tagged yeast strain (here we use W303) grown to mid-log phase $OD_{600nm} \sim 0.8$ (*see* **Note 3**).
2. Liquid N_2 (~1 l/strain).
3. Cryo-gloves.
4. Spatula, cooled in liquid N_2.
5. 50 ml polypropylene tubes.
6. 20 ml-syringe.

7. 1 M DTT.

8. 100× Solution P: 100 mM PMSF, 0.4 mg/ml Pepstatin A; store at −20 °C for up to 3 weeks.

9. Protease inhibitor cocktail.

10. Resuspension buffer: 20 mM Hepes pH 7.4, 1.2 % PVP-40, store at 4 °C.

11. Centrifuge(s) for 1 l and 50 ml tubes, capable of $3500 \times g$.

12. Clean and dry Styrofoam box big enough to fit a 9×50 ml tube rack.

13. 50 ml tube rack with open sides that can withstand liquid nitrogen.

2.3 Cryolysis (Cryomilling) (See Note 4)

1. Yeast noodles (frozen yeast pellet).

2. 10 l/strain of Liquid N_2.

3. Cryo-gloves.

4. Retsch PM-100 Planetary Ball Mill.

5. For 20–50 ml of "noodles": 125 ml stainless-steel jar and 12×20 mm stainless-steel balls. For less than 20 ml of "noodles": 50 ml stainless-steel jar and 4×20 mm stainless-steel balls.

6. Spatula.

7. Clean and dry Styrofoam box deep enough to immerse a 9×50 ml tube rack, with lid, or covered with aluminum foil.

8. 50 ml tube rack that can be immersed in liquid N_2.

9. 50 ml polypropylene tube pierced with a spatula (Fig. 1).

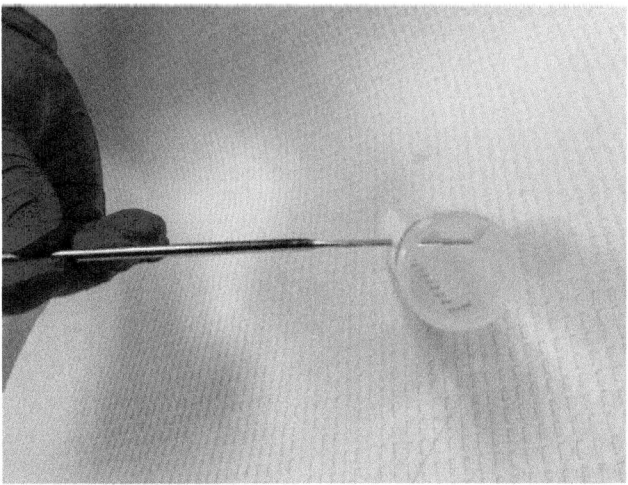

Fig. 1 Homemade liquid N_2 decanter

2.4 Weighing Yeast Cell Grindate

1. 1 l Liquid N_2.

2. Yeast cell grindate (from Subheading 2.3) (*see* **Note 5**).

3. 50 ml polypropylene tubes (*see* **Note 5**).

4. Spatula.

5. Clean and dry Styrofoam box deep enough to immerse a 9×50 ml tube rack.

6. Two 9×50 ml tube racks that can be immersed in liquid N_2.

2.5 Single-Step Affinity Purification

1. Pre-weighted yeast cell grindate.

2. Pre-determined ssAP buffer prepared with MS grade water (*see* **Note 6**).

3. Vortex.

4. Ice and ice bucket.

5. Polytron homogenizer equipped with a 7 mm probe.

6. Centrifuge for 15 and 50 ml tubes capable of $3500 \times g$.

7. Conjugated IgG magnetic beads (*see* **Note 1**).

8. Acetonitrile-washed 1.5 ml low-bind tubes.

9. Timer.

10. Nutator or any slow rotating platform or wheel.

11. Cold room.

12. DynaMag-2, -15 and/or -50 Magnets.

13. Last Wash Buffer (make fresh): 0.1 M NH_4OAc, 0.1 mM $MgCl_2$, 0.02 % Tween-20 in MS grade water.

14. Last Wash Buffer without detergent (make fresh): 0.1 M NH_4OAc, 0.1 mM $MgCl_2$ in MS grade water.

15. 20 mM Tris pH 8.0.

16. Elution buffer (make fresh): 0.5 M NH_4OH, 0.5 mM EDTA.

17. Vacuum concentrator (SpeedVac) (*see* **Note 7**).

18. Solution A: 0.5 M Tris–HCl pH 8.0, 5 % SDS.

19. Solution B: 75 % glycerol, 124.5 mM DTT, 0.05 % Bromophenol Blue; store at 4 °C.

20. Antifoam B Emulsion (Sigma-Aldrich).

2.6 Trypsin Digest

1. Protein LoBind tubes 1.5 ml.

2. Acetonitrile diluted in MS-grade water to a final concentration of 70 % (*see* **Note 6**).

3. Vortex with 1.5 ml tubes holder.

4. Beaker.

5. Aluminum foil.

6. Trypsin.

7. Thermomixer for 1.5 ml tubes.

8. Formic acid diluted 50 % in MS grade water.

2.7 Mass Spectrometry

1. Tryptically digested samples (from ssAP).

2. Mass spectrometer (at least LTQ, Velos, or Q-Exactive Orbitrap).

3. Column Easy Column Proxeon C18 (10 cm, 75 μm i.d., 120 A).

4. Solvent A (0.1 % formic acid in MS grade water).

5. Solvent B (100 % acetonitrile, 0.1 % formic acid).

2.8 Sample Analysis

1. Protein database library (we use NCBInr protein database with the *Saccharomyces cerevisiae* taxonomy).

2. Analysis Software (Mascot, X!Tandem, Sequest, or similar).

3. Scaffold Software (optional).

3 Methods

3.1 Conjugating Magnetic Dynabeads with Rabbit IgG

1. *Resuspend and wash the magnetic beads*: Resuspend 300 mg of Dynabeads in 16 ml of 0.1 M $NaPO_4$, pH 7.4 (vortex, 30 s). Divide the mixture evenly in four 15 ml polypropylene tubes (4 ml/tube). Incubate the solution with mild rotation (RT, 10 min). Recover the beads using a magnetic holder and remove the buffer (*see* **Note 8**). Resuspend the beads once again in 4 ml of 0.1 M $NaPO_4$, pH 7.4 (vortex, 15 s).

2. *Reconstitute the antibodies*: Rabbit IgG is received lyophilized (100 mg). Solubilize the antibodies in ultrapure water (7 ml) to obtain a final concentration of ~14 mg/ml. Aliquot the solution (1 ml) into 1.5 ml tubes and store at –80 °C.

3. *Prepare the antibodies for conjugation*: Take 4 ml of antibodies prepared above, and centrifuge (10 min, V_{max}) using a tabletop centrifuge. In a 50 ml polypropylene tube, mix 3.525 ml of the supernatant with 9.85 ml of 0.1 M $NaPO_4$, pH 7.4 (*see* **Note 9**). Slowly add 6.65 ml of 3 M $(NH_4)_2SO_4$ drop by drop to avoid local precipitation of the antibody and mix the solutions by gentle agitation (tapping the side of the tube). After all $(NH_4)_2SO_4$ has been added, briefly vortexing the tube for 2 s. Filter the solution into another 50 ml tube (0.22 μm filter fitted on a 20 ml syringe) to eliminate precipitate. The antibody solution is now ready for conjugation.

4. Magnetically recover the beads, and wash them once more with 4 ml 0.1 M NaPO$_4$, pH 7.4 buffer (vortex, 15 s).

5. *Conjugation*: Magnetically recover the beads, remove the buffer, and resuspend them in 5 ml of antibody solution. Close the tubes, seal them with parafilm, and incubate overnight (18–24 h, 30 °C, mild rotation).

6. *Washes*: Successively wash the conjugated beads with the following solutions (*see* **Note 10**): 1 × 4 ml of 100 mM Glycine, pH 2.5 (do not leave the beads in this buffer for too long).

7. 1 × 4 ml of 10 mM Tris–HCl, pH 8.8.

8. 1 × 4 ml of 100 mM Triethylamine (do not leave the beads in this buffer for too long!).

9. 1 × 4 ml of 1× PBS, 0.5 % Triton X-100, rocking for 5 min.

10. 1 × 4 ml of 1× PBS, 0.5 % Triton X-100, rocking for 15 min.

11. *Final resuspension and storage*: Pool the beads into 1 ml of 1× PBS, 0.02 % NaN$_2$ in a 5 ml tube. Use another 1 ml to recover any beads left behind in the tubes. The beads are now resuspended in 2 ml of 1× PBS, 0.02 % NaN$_3$. Seal the tube with parafilm to prevent buffer evaporation and store at 4 °C.

The conjugation provides densely antibody-coated Dynabeads (160 µg of rabbit IgG per mg of Dynabeads) with a final concentration of ~0.15 µg of beads per µl of solution. The beads can be stored for several months at 4 °C without loss of activity (*see* **Note 11**).

3.2 Harvesting and Freezing of Cells

1. Grow yeast cells to no more than a density of $0.866–2.7 \times 10^7$ cells/ml (OD$_{600}$ ~ 0.8–1.2; early log phase) (*see* **Note 3**).

2. Harvest cells by centrifugation (4 °C, 5–10 min, 3500 × g).

3. Resuspend and wash the cell pellet(s) with ultrapure water and transfer them into a 50 ml tube.

4. Pellet the cells by centrifugation (4 °C, 5 min, 3500 × g).

5. Wash the cells once more with an isovolume of ultrapure water and centrifuge once more.

6. While the cells are centrifuging, supplement an isovolume of cold resuspension buffer with DTT (1:1000), PIC (1:100), and solution P (1:100).

7. Remove the supernatant from **step 5**, resuspend the cell pellet in the supplemented Resuspension Buffer prepared in **step 6** and vortex.

8. Pellet the cells by centrifugation (4 °C, 5–15 min, 3500 × g), remove the supernatant and centrifuge the cells again (4 °C, 5–15 min, 3500 × g).

9. While the cells are centrifuging, place a 50 ml tube rack with an empty 50 ml polypropylene tube into the Styrofoam box.

Fill both the Styrofoam box and the 50 ml polypropylene tube with liquid N_2. As the liquid nitrogen will rapidly evaporate from the tube during cooling, add more liquid N_2 until the temperature has equilibrated. Unplug the syringe and ready the spatula.

10. After centrifugation, remove any remaining buffer from the top of the cell pellet. Use the spatula to fill a 20 ml-syringe with the paste-like yeast pellet. Cool a fresh spatula in liquid N_2.

11. Put the plunger back into the syringe, and push the yeast pellet through the syringe into the liquid N_2-filled 50 ml tube, filling the tube to the rim. Yeast cells will freeze instantly. Using the cold spatula, push the noodles to the bottom of the tube and break them (see **Note 12**). Repeat this step until there is no more pellet left in the syringe.

12. Using a 18 G 1½ needle, make a few holes into the 50 ml-tube screw cap and close the noodle-filled tube, before rapidly inverting the tube over the Styrofoam box to remove any remaining liquid N_2. Care should be taken at this step, as liquid N_2 will burst from the tube when inverted, and a cryo-glove should be used.

13. Tubes with frozen yeast noodles can then be transferred to a –80 °C freezer, and the punctured screw caps changed after 15–30 min to allow for complete evaporation of liquid N_2, and prevent condensation during indefinite storage.

3.3 Cryogenic Cell Lysis (Cryo-Milling)

1. Prechill the metal jar and metal beads by filling them in liquid nitrogen inside a Styrofoam box Liquid N_2 levels in the Styrofoam box should not exceed 2/3rd of the jar's height (see **Notes 13** and **14**).

2. Remove any liquid N_2 from inside the stainless steel jar, add your yeast noodles and the prechilled stainless steel balls until the jar is full, leaving just enough space to completely close the lid (see **Note 14**).

3. Weigh the jar containing the noodles and metal balls wearing cryo-gloves.

4. Put the closed jar back into the liquid N_2-filled Styrofoam box.

5. Match the jar weight with the appropriate counterbalance weight on the Retsch PM-100 ball mill.

6. Program ball mill to perform the following cycle: $400 \times g$ rotations for a total of 3 min per cycle; intervals of 1 min between alternating rotation directions.

7. Secure the jar tightly on the ball mill with the provided clamp. Close the cover on the machine. Once "Start" is pressed, the machine will ask whether you have tightly secured the jar. Press "Yes" to start cycle. This is cycle 1 (see **Notes 15–17**).

8. Once the machine stops, cool the jar in liquid N_2 for at least 2 min or the rigorously bubbling stops. Pour a small amount of liquid N_2 on the closed jar to cool it further.

9. Repeat **steps** 7 and **8** until you have performed eight full cycles.

10. On the last cycle, put a 50 ml rack into a clean Styrofoam box filled with liquid N_2, immerse a clean 50 ml tube labeled with the sample information, and add a spatula in the rack to prechill.

11. Open the stainless steel jar and place the balls into the liquid N_2 using tweezers. The cryo-lysed yeast cell grindate will be visible and compacted at the sides and bottom of the jar. To obtain a fluffy cell grindate, use the prechilled spatula to scrape the cell grindate off the side and bottom of about ¼ of the jar, place five prechilled balls back into the jar, cool the jar and repeat **step** 7 stopping the milling cycle after ~15 s. Make sure that the tips of the pair of tweezers and spatula have been left in liquid nitrogen to keep cold.

12. Remove the balls with the tweezers. Open the prepared 50 ml tube and make sure that there is no liquid N_2 in it. Wearing a cryo-glove, tilt the jar over the 50 ml tube. With the other hand, using the prechilled spatula, transfer the yeast cell grindate from the jar to the tube. Cool the tip of the spatula in liquid N_2 from time to time to avoid sticking of the cell grindate to the spatula. Close the tube (hand-tight) and store the cryo-lysed cell grindate at –80 °C for an indefinite time.

3.4 Single-Step Affinity Purification (See Note 18)

3.4.1 Weighing Out the Yeast Cell Grindate

1. Prepare the 50 ml tubes (*see* **Note 5**), and immerse them in liquid N_2 Prechill the tip of the spatula and place one rack into liquid N_2, and the other on the balance.

2. Place the tube containing the cell grindate into liquid N_2.

3. Completely remove the liquid N_2 from one of the empty tubes, place it on the balance and zero the balance.

4. With the prechilled spatula, quickly transfer the required amount of grindate for the ssAP into the prechilled tube, and put the closed tubes back in liquid N_2. Cool the sample in between if there are many strains or a large quantity of cell grindate to weigh out. Dip the spatula in liquid nitrogen from time to time to prevent the sticking of the grindate. Use a fresh spatula for each strain. Move the tubes containing the yeast cell grindate stock back to –80 °C for storage.

3.4.2 Affinity Purification

The amount of material can be adjusted according to need. For example, for buffer optimization experiments, 0.2 g of grindate is sufficient to perform a silver-stained SDS-PAGE and western blot for a bait protein present in ~3300 copies/cell (Fig. 3) (*see* **Note 6**).

1. Leave the weighted yeast cell grindate to warm up on ice until it has an ice cream look.

2. For every gram of grindate, add 9 ml of ssAP buffer.

3. Vortex at maximum speed for 30–60 s to resuspend the cell grindate.

4. Put the tube on ice and further homogenize the grindate using the polytron at ~2/3rds of the maximum speed for 30 s being careful not to create bubbles.

5. Centrifuge the tubes in a tabletop centrifuge at 4 °C, $5000 \times g$, 10 min.

6. While the lysate is clearing in the centrifuge, take an appropriate amount of IgG-conjugated magnetic beads per sample and put them into an acetonitrile-washed 1.5 ml tube and add 1 ml of ssAP buffer; for example, use 100 µl of beads for 2 g of grindate, or 10 µl for 0.2 g; adjust the quantities according to Fig. 3.

7. Invert tubes until the beads are homogeneously dispersed.

8. Magnetically recover the beads, and proceed with two more washes using the ssAP buffer and leave the beads in the last ssAP buffer wash. If more than one sample is used, beads can be pooled for washes, but should be divided in different tubes for the last wash.

9. When the sample has finished centrifuging, pour the supernatant into a new tube and put the tube on magnet a holder. Aspirate the last wash from the beads using a magnet holder, then remove the washed bead tube from the magnet, and use the cleared lysate to resuspend and transfer the beads to the clear lysate tube. Rinse the pipette tip with the clear lysate that is still on magnet and wash the 1.5 ml tube used for washing the beads to recover all beads.

10. Incubate 30 min at 4 °C with slow rotation.

11. Prepare the Last Wash and Elution buffers, and prepare the 1.5 ml tubes (two tubes for on-beads trypsin digestion, and two tubes for sample elution).

12. At RT, magnetically recover the beads (see **Note 8**).

13. Use 1 ml to recover the beads, and transfer them in a 1.5 ml tube sitting on a magnet. Wait for beads to clear from the sample, and use the cleared sample to wash and recover the beads left in the tubes used for the target binding.

14. Wash the beads with 1 ml of ssAP buffer. This can be done by slowly resuspending the beads with a P-1000 pipette, then putting the tube on the magnet to wait for the beads to clear the sample, and then rinsing the pipette tip with the cleared sample while the tube is still on the magnet. Alternatively, the tube can be inverted several times to resuspend and wash the

beads before putting it back on the magnet. If inversion is used to resuspend and wash the beads, after putting the beads back on the magnet and sample has cleared, the tubes should be slowly inverted to recover all the beads located beneath the cap of the tube. Proceed to washing the beads another time with ssAP buffer.

15. Wash the beads for 5 min with the Last Wash Buffer.

16. Wash the beads quickly three times with the Last Wash Buffer without detergent.

17. Wash beads with the Last Wash Buffer without detergent for 5 min, rocking on the nutator or similar device.

18. Remove 100 μl of the bead suspension to a new tube for elution and subsequent PAGE gel and Western blot (Subheading 3.6).

19. Wash the beads with 1 ml of 20 mM Tris–HCl, pH 8.0.

20. Resuspend beads in 50 μl 20 mM Tris–HCl, pH 8.0 and continue with trypsin digest (*See* **Note 19**).

3.5 Trypsin Digest

3.5.1 Preparations

1. Add 500–1000 μl of 70 % Acetonitrile in each 1.5 ml LoBind tube. Vortex for 5–10 min at RT. Discard Acetonitrile 70 % and repeat this step two more times.

2. Put the 1.5 ml tubes in a beaker, close it loosely with aluminum foil and let dry overnight.

3. The following day, completely cover the beaker with the aluminum foil. The tubes can be stored this way for an indefinite amount of time. Alternatively, leave the tubes open on a rack with clean Kimwipes on top until they are dry and ready to use. For ssAP of samples destined for mass spectrometry (MS), only use acetonitrile-washed low bind 1.5 ml tubes.

4. Solubilize 20 μg of lyophilized trypsin in 20 μl 20 mM Tris–HCl pH 8.0 and aliquot them in 1–3 μl aliquots, at a concentration of 1 μg/μl. Store them at –80 °C. 1 μg is needed to digest one sample.

3.5.2 On-Bead Digestion (See Notes 20 and 21)

1. Spike the beads with 500 ng of trypsin (0.5 μl). Incubate the spiked sample (4 h, 37 °C, $900 \times g$).

2. Transfer the supernatant into another tube and spike it with another 500 ng. Let incubate overnight (12–16 h, 37 °C, shaking optional).

3. Quench the trypsin reaction with 2 μl of a 50 % formic acid solution (Final concentration ~2 %).

4. While the sample can be stored at –20 °C, it is preferable to analyze it directly by mass spectrometry.

3.6 Protein Elution, Coomassie Gel, and Western Blot

1. After the last wash, elute the purified complexes from the beads with 500 µl of elution buffer (20 min, RT, slow rotation).

2. Transfer the eluate in a new tube and keep on ice.

3. Proceed with a second elution as above (500 µl Elution Buffer, 20 min, RT, slow rotation). A third elution might be necessary.

4. Pool the eluates, and make a hole into the lid of each 1.5 ml tube, using an 18 G 1½ needle.

5. Speed vacuum the samples overnight at room temperature (do not set any temperature on the SpeedVac).

6. The next morning, solubilize the lyophilized proteins by washing the inside wall of each 1.5 ml tube with 8 µl of Solution A.

7. Quickspin 15 s, $V_{max.}$

8. Place the tubes in the magnetic rack, transfer the supernatants into new tubes and add 8 µl of Solution B to the supernatants.

9. Denature your sample at 70 °C, for 10–20 min.

10. Load 8 µl on a SDS-PAGE gel for Silver or Coomassie staining. While the Coomassie stain used has a sensitivity of about 30 ng/µl using brilliant blue R-250, silver staining is the method of choice, since it has a sensitivity of less than 1 ng.

11. Load 1/200th of your total sample on an SDS-PAGE gel for subsequent western blotting and probing.

3.7 Mass Spectrometry

1. To load the sample onto a column, centrifuge the sample at $13,000 \times g$ for 5 min to remove any debris that may otherwise clog the HPLC column.

2. Clean the sample on a ZipTip or an MCX column (*see* **Note 22**).

3. Transfer the supernatant to a fresh tube and load it directly onto a C18 reversed phase column for LC–MS/MS analysis (*see* **Note 23**).

4. Set the flow rate to 300 nl/min and run a 20 min gradient from 95 % Solvent A to 25 % Solvent B.

5. With the same flow rate, continue a to 45% B over 40 min.

6. With the same flow rate, finish with a 10 min elution gradient up to 80 % B.

7. Set up the collision-induced dissociation (CID) activation to fragment peptides with a collision energy of 40 %, an activation Q of 0.25 for 10 ms.

8. Acquire the MS spectrum in full ion scan mode from m/z 400–1800 at a resolution of 30,000.

9. In a data-dependent mode, switch to MS/MS for the 20 most intense precursor ions. After MS/MS analysis of a precursor ion, include a dynamic exclusion of 20 s and an exclusion mass width of 10 ppm.

10. Repeat **steps 2–9** for the untagged strain, blank, PrA-tag alone (*see* **Note 24**). Inject the same volume for the control strains.

3.8 Data Analysis

3.8.1 Data Search

1. Raw data is initially processed using Proteome Discoverer 1.2 or Xcalibur™ 2.1 Software (Thermo Fisher Scientific).

2. Data analysis can be performed using one of several search programs: X!tandem, Mascot, and Sequest; convert your .RAW file into the appropriate file format that can be read by the search program: .mzXML for X!Tandem, .mgf for Mascot, or .dta file for Sequest (*see* **Note 25**).

3. In your search settings, consider the following modifications in mass calculation:

    Oxidation (M), Phosphorylations (STY), Mass tolerance of fragment ions set to 10 ppm for precursor ions, and 0.6 Da for fragments, Limit to two maximum number of missed cleavages for trypsin digestion.

4. Match the peptides against *Saccharomyces cerevisiae* taxonomy in a protein database library (we use NCBInr protein database, but SwissProt can also be used).

5. Each sample should be run in triplicate or at least in duplicate.

3.8.2 Data Analysis

1. Comparing results obtained for several samples simultaneously is possible and best done in Scaffold, if available. Scaffold software allows comparing peptide counts, ion fragmentation, and spectra (Fig. 2).

2. Any given protein identified must the two to threefold more abundant in the tagged sample than in the control samples to be considered above background. For a list of nonexhaustive contaminants, *see* **Note 26**.

3. Proteins identified with low abundance peptide numbers should be considered only if they have been associated with exclusive unique and consistent ion fragmentation patterns (Fig. 3; *see* **Note 27**).

4. For quantitative analysis spectral counting should be performed [21].

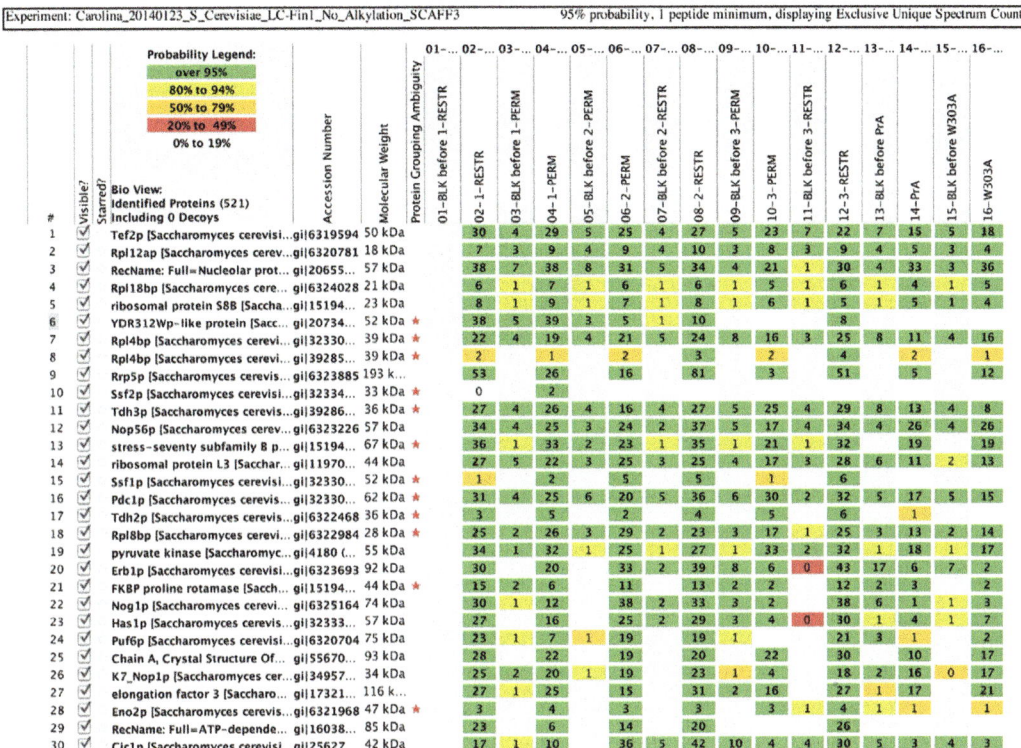

Fig. 2 Scaffold Viewer. Scaffold screen view of the exclusive unique spectral counts for several samples including the two controls and blank runs between samples

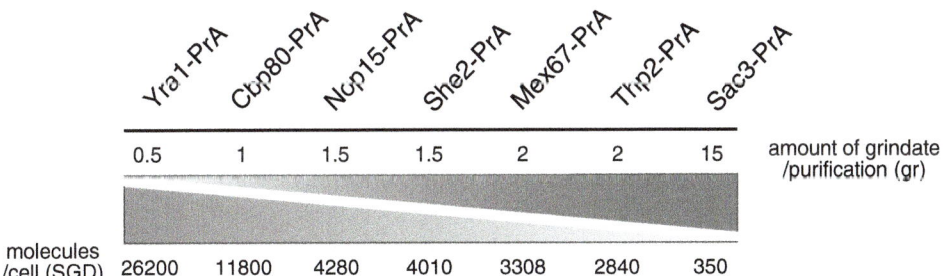

Fig. 3 Sample amounts. The amount of yeast cell grindate required to recover similar amounts of bait protein depends on the cellular abundance of the bait protein

4   Notes

1. Our laboratory uses rabbit IgG from total serum as it has high affinity with ProteinA (PrA). Guinea IgG from serum could be used with similar results [22]. Subclasses of IgG which could be used with similar results include human IgG_1, IgG_2, and IgG_4 as well as mouse IgG_{2a} [22]. IgG from other species

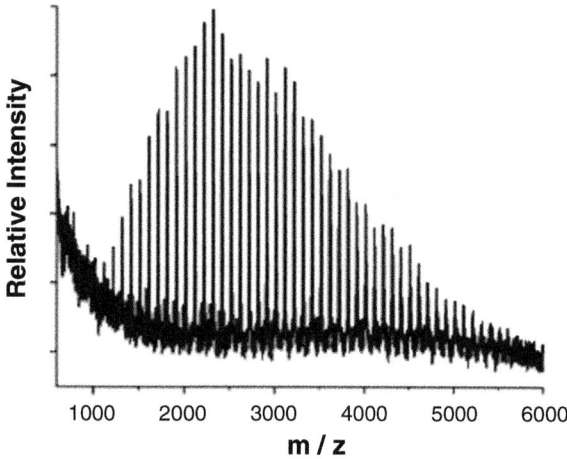

Fig. 4 Polymer contamination. Example of a polymer containing mass spectrum. Figure adapted from [30]

should not be used as they bind PrA with lower affinities. The IgG from goat, rat, and chicken should be absolutely avoided as they have a low affinity for PrA [22]. Once conjugated, IgG-coupled magnetic beads last for up to 6 months at 4 °C. Caution should be taken, however, with GFP-conjugated beads, as these are only active for ~2 weeks stored at 4 °C.

2. Refer to MSDS specifications sheet and handle these reagents with caution.

3. For these experiments, usually 6 l (3 × 2 l) of culture are grown to OD_{600} ~0.8–1.2. This will produce a reasonable amount of material to work with.

4. Alternatively to planetary ball mill grinding, cell cryolysis can also be achieved using a mortar or a grinder under cryo-freeze conditions; however, the efficiency of cell breakage will be reduced and less homogenous.

5. *Sample Amount*: The amount of cell grindate required to recover an amount of bait protein sufficient for mass spectrometric analysis needs to be taken into account. Amounts should be based on molecules of protein/cell, which can be obtained from SGD (yeastgenome.org). And protein with an abundance of ~3300 molecules/cell requires 1.5–2 g of cell grindate per ssAP. Examples are shown in Fig. 4. For 0.5–1 g of cell grindate, use a 15 ml polypropylene tube. When 1.5–2 g of cell grindate is needed, divide the cell grindate into two 15 ml tubes. If 15 g of cell grindate and more are required, divide into three or more 50 ml polypropylene tubes.

6. *Buffer Selection*: The buffers used for ssAPs vary and need to be determined empirically as their choice mainly depends on

the nature of the bait protein and the stability of the complex it forms. For example, our usual buffers are made containing 20 mM HEPES-KOH pH 7.4, a salt, and two detergents. Salt concentrations can vary from 0 mM to 1 M. Different salts can be used according to their stabilization properties as determined by the Hofmeister Series: $KOAc > NH_4OAc > NaCl > Na$ Citrate [23–25]. Our buffers usually contain 0.5 % Triton-X100 and 0.1 % Tween-20, 1:100 Solution P, 1:5000 antifoam. A buffer standardly used in our laboratory for complex affinity purification is 20 mM HEPES-KOH pH 7.4, 110 mM KOAc, 100 mM to 300 mM NaCl or 250 mM NaCitrate, 0.5 % Triton-X100, 0.1 % Tween-20, 1:100 Solution P, and 1:5000 antifoam.

7. The SpeedVac must be compatible with organic solvents or else it will corrode over time.

8. *Buffer Removal* should be performed by vacuum aspiration, which will result in no to negligible loss of beads. Removing the buffer by flipping the open tubes upside down on layered Kimwipes while they are placed on the magnet could lead to contamination of the sample.

9. *Determining NaPO4 Amount for AB Mix*: If another antibody/tag combination than IgG/PrA is used, the antibody conjugated to the beads may come at a different concentration, volume or may be available in limiting amounts. Changes in antibody concentration/volume during the conjugation will change the volume of the AB mix. Therefore the concentration of $NaPO_4$ needs to be recalculated as follows. The original total AB Mix volume (IgG + Sodium Phosphate + Ammonium sulfate) is 20 ml per 300 mg of Dynabeads. To determine the required volume of Sodium Phosphate subtract (1) the amount of IgG (or other antibody) being used, and (2) the amount of Ammonium Sulfate (final concentration 1 M) being used from the total reaction mixture volume (20 ml for entire 300 mg bottle). This will leave you with the volume of Sodium phosphate required. In this formula only the final concentration of the Sodium Phosphate varies. While all three buffers change in the volume being used to make the AB Mix, the Ammonium Sulfate should always be at a final concentration of 1 M. If less than 300 mg of Dynabeads are conjugated at a time, then the entire conjugation protocol has to be scaled accordingly, including the volume of AB mix. Amounts of $NaPO_4$ and Ammonium Sulfate have to be recalculated as described above, according to the amount of antibody used.

10. During the washes, make sure the beads are completely resuspended in the wash solutions specified. This resuspension is usually achieved by closing the tube and turning it overend several times until the beads are homogenously dispersed.

11. Take special care in ensuring that the beads are covered by buffer. If they dry out, the efficiency of the affinity purification will not be optimal.

12. Make sure the 50 ml tube is always nearly completely filled with liquid nitrogen during this step. Do not place the syringe too close to the liquid nitrogen or else the cell paste will freeze inside the syringe tip and clog it. If this happens, wait for a moment, and apply pressure on the needle until it unclogs. The yeast "noodle" coming out of the syringe should always be immersed in liquid nitrogen. If there is not enough liquid nitrogen in the tube, fill it immediately, and break the noodles with the cooled spatula if there is no more space in the 50 ml tube.

13. *Handling liquid nitrogen.* Liquid nitrogen temperature is −196 °C. Since this procedure requires handling material prechilled in liquid nitrogen, protect the hand that handles the prechilled material using a cryo-glove. Any material immersed in liquid nitrogen will "boil" vigorously until the material reaches liquid nitrogen temperatures.

14. *Taking care of your sample*: The jar does not need to be completely immersed in liquid nitrogen for it to be appropriately cooled. As a guide, place a metal ball outside of the jar and make sure there is always enough liquid nitrogen to immerse it. Once the jar is closed, it is not recommended to open it unless the stainless steel balls are stuck (no more rattling sound can be heard; *see* **Note 17**). Avoid keeping the jar open for prolonged periods as condensation can occur as a result, and may mix with your yeast cell grindate. The precooled jar can be left at RT for several minutes before it reaches temperatures that will melt your sample. Caution: if the cell grindate starts to look like "ice-cream," the sample is no longer usable and needs to be discarded.

15. *Milling Safety*: The planetary ball mill has a metal holder, which perfectly fits the bottom of the jar to hold it securely in place during the run. The top of the jar is held in place by a clamp that has to be manually secured with a force of ~10 Nm. The movement of the clamp is secured by a red ring, which needs to be pushed up. As the clamp secures the jar, tightening will become more difficult. No more than two turns are necessary at this point to get to ~10 Nm. At the end, the red ring should be down and it may be necessary to turn the clamp back by approximately an eighth of a turn, resulting in a small clicking sound. Be careful not to over-tighten the jar, as it may be difficult to remove afterwards.

16. It is important to ensure the counterbalance has been appropriately set.

17. *Requirements for a full milling cycle.* When the planetary ball mill rotates clockwise, the balls inside the secured jar rotate counter clockwise. This generates a rattling sound, which indicates that the balls are moving. It is normal that the rattling stops for a few seconds and starts again. However, if the rattling is not heard for more than ten consecutive seconds, it indicates that are that the balls are stuck. If this happens, stop the cycle, cool down the metal jar in liquid nitrogen, and cool the tip of a pair of tweezers. Once the jar is completely cooled (to prevent condensation), open the lid and remove/unstuck the balls (tweezers) and place them in liquid nitrogen. Secure the closed jar back on the machine as usual (*see* **Note 15**) and resume the cycle. It is important to listen to the rattling sound of the machine. If the balls are continuously stuck for more than 20 s during a cycle, the cycle should be repeated. It may be necessary to add or remove balls from the jar to achieve the correct ratio of balls to grindated volume.

18. *ssAP*: Before starting, clean your bench and other materials you will use. Wear a clean lab coat and gloves throughout. Any sample contaminated with keratin from hair will appear normal. However, keratin may saturate the MS signal and less of your actual sample will be injected and analyzed as a result. Keratin contaminations will not be recognized if samples are matched against a *S. cerevisiae* database only.

19. *Constant volumes*: If the previous buffer is not completely removed in this step, the final sample volume will not be consistent between samples and between trials. This could affect the amount of sample loaded onto the MS column, which can impact the amount of background observed (*see* **Note 23**).

20. *Trypsin digest*: Trypsin digestion is commonly used for downstream MS analysis because it generates peptides with C-termini basic residues [26]. Moreover, the fragments generated are of a suitable mass range which allows their identification with high specificity [26]. However, because enzymatic digestion can still be incomplete, the data analysis is set to also identify longer peptides with up to two [27] missed cleavage sites. Trypsin digests as described can also be performed overnight followed by a 3-h overday incubation.

21. *Sample treatment*: Many MS samples are alkylated to avoid disulphide bridge formation between cysteines as most programs used to analyze MS spectra do not recognize peptides linked by disulphide bonds. This is not done in our lab as yeast proteins contain overall only ~1.2 % cysteines, and we found alkylation of cysteines not necessary to obtain better spectra [27].

22. *Sample clean up*: Often samples are cleaned prior to loading onto the HPLC column and injection into the mass spectrometer.

Different methods can be used such as MCX (Mixed Cation Exchange), SCX (Strong Cation Exchange), or ZipTips, which, concentrate and purify the samples for sensitive downstream analysis [19]. In our hands, sample clean up using ZipTips is the most successful and most reproducible, maintaining a high ratio of identified unique peptides to low traces of polymer and detergent contamination.

23. *Injection volumes*: To determine the amount of material needed to obtain nonsaturated yet sufficient spectra, first inject 2 μl of your sample. Once the best sample amount has been determined, inject a volume of your sample, which is close to saturation and inject the same volume for your negative controls (untagged strain and PrA-tag expressed alone from an endogenous promoter).

24. *Polymer contamination*: If a sample shows a low number of peptides recognized for the bait protein (epitope-tagged protein), the sample is possibly contaminated with polymers. Polymer contaminated samples generate a peptide elution spectra characterized by high intensity peaks evenly spaced in the m/z axis (Fig. 5). These peptides are not recognized in the

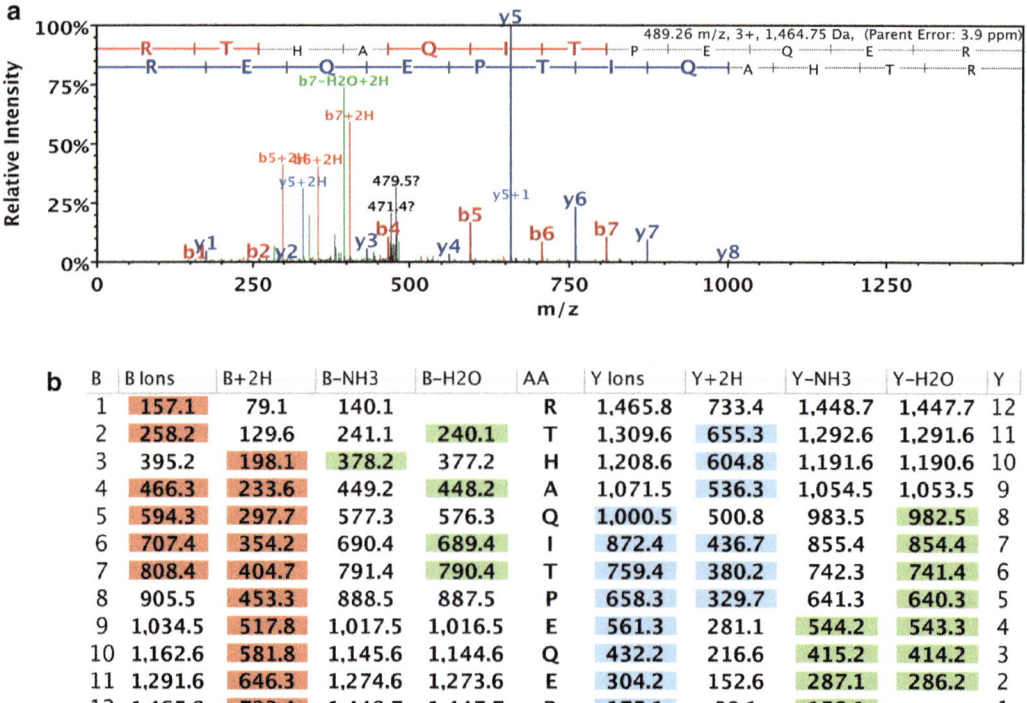

Fig. 5 Peptide Ion Fragmentation. (**a**) Scaffold screenshot of fragmentation pattern for one peptide of the Ssf2 protein. (**b**) Scaffold screenshot of fragment ions experimentally found for this peptide are highlighted in *red, blue,* and *green*

database. If this is the case, the sample should be discarded and the affinity purification should be repeated. To avoid polymer contamination, make sure to use LoBind, acetonitrile washed tubes and tips.

25. *Raw data conversion*: A MM file conversion tool to convert .raw into .mgf or .mzXML files can be found at: http://www.massmatrix.net/mm-cgi/downloads.py or http://source-forge.net/

26. *Common contaminants*: Below is a nonexhaustive list of common contaminants usually found enriched in the control strains: Acs2, Act1, Adh1, Adh2, Adh3, Ado1, Ahp1, Ald6, Arg1, Aro2, Aro3, Aro4, Asn2, Bmh1, Car2, Cdc19, Cys3, Cys4, Ded81, Dps1, Eft2, Eno1, Eno2, Erg6, Erg10, Erg13, Erg20, Fas1, Fas2, Fba1, Frs1, Gpm1, Gus1, Hsp60, Hsp104, Ilv5, Lys1, Lys9, Lys21Met6, Met17, Nap1, Nba1, Pdc1, Pil1, Pfk2, Pgi1, Pgk1, Pma1, Por1, Sah1, Sam1, Sam2, Ser1, Ser33, Shm2, Ssa1, Ssa2, Ssb1, Ssb2, Stm1, Sug1, Tal1, Tdh1, Tdh2, Tdh3, Tef1, Tef2, Thr4, Tkl1,Tpi1, Tsa1, Ura2, Vma2, Vma5, Yad1, Yhb1, and Yhm2. These proteins can be considered true interactors if they are two- to threefold more abundant in the bait samples than in the control strains. Additional research on contaminants can be performed using the www.CRAPome.org website [28].

27. *Matching peptides*: The expected fragmentation of the *b* and *y* ions is known for a given peptide as well as the expected fragmentation of its protonated (+2H), deaminated (−NH$_3$), and dehydrated (−H$_2$O) forms. The algorithm matches experimental fragmentation patterns according to a *best-fit* approach [29] and fragment ions experimentally found for this peptide are highlighted for all experimentally identified peptides in red, blue, and green in Scaffold Viewer (Fig. 3).

Acknowledgments

We thank members of the Oeffinger lab for useful conversations and comments. We also thank Ms. Karen Wei for the development of the liquid N$_2$ decanter depicted in Fig. 1. The Oeffinger laboratory is supported by the Canadian Institutes of Health Research (MOP 106628), the Natural Sciences and Engineering Council of Canada (RGPIN 386315), the Rachel Foundation, the Fonds de recherche Santé Québec, and the Canadian Foundation for Innovation.

References

1. Gavin A, Bosche M, Krause R et al (2002) Functional organization of the yeast proteome by systematic analysis of protein complexes. Nature 415:141–147

2. Goh K-I, Cusick ME, Valle D et al (2007) The human disease network. Proc Natl Acad Sci U S A 104:8685–8690

3. Taylor IW, Linding R, Warde-Farley D et al (2009) Dynamic modularity in protein interaction networks predicts breast cancer outcome. Nat Biotechnol 27:199–204

4. Ho Y, Gruhler A, Heilbut A et al (2002) Systematic identification of protein complexes in Saccharomyces cerevisiae by mass spectrometry. Nature 415:180–183

5. Gavin A, Aloy P, Grandi P et al (2006) Proteome survey reveals modularity of the yeast cell machinery. Nature 440:631–636

6. Krogan N, Cagney G, Yu H et al (2006) Global landscape of protein complexes in the yeast Saccharomyces cerevisiae. Nature 440:637–643

7. Costanzo MC, Hogan JD, Cusick ME et al (2000) The yeast proteome database (YPD) and Caenorhabditis elegans proteome database (WormPD): comprehensive resources for the organization and comparison of model organism protein information. Nucleic Acids Res 28:73–76

8. Goffeau A, Barrell BG, Bussey H et al (1996) Life with 6000 genes. Science 274:546, 563–567

9. Grigoriev A (2003) On the number of protein-protein interactions in the yeast proteome. Nucleic Acids Res 31:4157–4161

10. Collins SR, Kemmeren P, Zhao X-C et al (2007) Toward a comprehensive atlas of the physical interactome of Saccharomyces cerevisiae. Mol Cell Proteomics 6:439–450

11. Johnson ME, Hummer G (2011) Nonspecific binding limits the number of proteins in a cell and shapes their interaction networks. Proc Natl Acad Sci U S A 108:603–608

12. Picotti P, Bodenmiller B, Mueller LN et al (2009) Full dynamic range proteome analysis of S. cerevisiae by targeted proteomics. Cell 38(4):795–806

13. Oeffinger M (2012) Two steps forward-one step back: advances in affinity purification mass spectrometry of macromolecular complexes. Proteomics 12(10):1591–1608

14. Cristea I, Williams R, Chait B, Rout M (2005) Fluorescent proteins as proteomic probes. Mol Cell Proteomics 4:1933–1941

15. Rigaut G, Shevchenko A, Rutz B et al (1999) A generic protein purification method for protein complex characterization and proteome exploration. Nat Biotechnol 17:1030–1032

16. Oeffinger M, Wei KE, Rogers R et al (2007) Comprehensive analysis of diverse ribonucleoprotein complexes. Nat Methods 4:951–956

17. Karlsson R, Jendeberg L, Nilsson B et al (1995) Direct and competitive kinetic analysis of the interaction between human IgG1 and a one domain analogue of protein A. J Immunol Methods 183:43–49

18. López-Ferrer D, Ramos-Fernández A, Martínez-Bartolomé S et al (2006) Quantitative proteomics using $^{16}O/^{18}O$ labeling and linear ion trap mass spectrometry. Proteomics 6(Suppl 1):S4–S11

19. Capelo JL, Carreira R, Diniz M et al (2009) Overview on modern approaches to speed up protein identification workflows relying on enzymatic cleavage and mass spectrometry-based techniques. Anal Chim Acta 650:151–159

20. Belozerov VE, Lin Z-Y, Gingras A-C et al (2012) High-resolution protein interaction map of the Drosophila melanogaster p38 mitogen-activated protein kinases reveals limited functional redundancy. Mol Cell Biol 32:3695–3706

21. Arike L, Peil L (2014) Spectral counting label-free proteomics. Methods Mol Biol 1156:213–222

22. Sambrook J, Russell DW (2001) Molecular cloning: a laboratory manual, 3rd edn. Cold Spring Harbor Laboratory Press, Cold Spring Harbor, NY

23. Bondos SE, Bicknell A (2003) Detection and prevention of protein aggregation before, during, and after purification. Anal Biochem 316:223–231

24. Zhang Y, Cremer PS (2006) Interactions between macromolecules and ions: the Hofmeister series. Curr Opin Chem Biol 10:658–663

25. Damodaran S, Kinsella JE (1983) Dissociation of nucleoprotein complexes by chaotropic salts. FEBS Lett 158:53–57

26. Westermeier R, Naven T (2002) Proteomics in practice: laboratory manual of proteome analysis. Wiley-VCH, Weinheim

27. Miseta A, Csutora P (2000) Relationship between the occurrence of cysteine in proteins and the complexity of organisms. Mol Biol Evol 17:1232–1239

28. Mellacheruvu D, Wright Z, Couzens AL et al (2013) The CRAPome: a contaminant repository for affinity purification-mass spectrometry data. Nat Methods 10:730–736

29. OConnor CD, Hames BD (2008) Proteomics. Scion Publishing Limited, New Delhi

30. Dogruel D, Nelson RW, Williams P (1996) The effects of matrix pH and cation availability on the matrix-assisted laser desorption ionization mass spectrometry of poly(methyl methacrylate). Rapid Commun Mass Spectrom 10:801–804

Chapter 16

Label-Free Quantitative Proteomics in Yeast

Thibaut Léger, Camille Garcia, Mathieu Videlier, and Jean-Michel Camadro

Abstract

Label-free bottom-up shotgun MS-based proteomics is an extremely powerful and simple tool to provide high quality quantitative analyses of the yeast proteome with only microgram amounts of total protein. Although the experimental design of this approach is rather straightforward and does not require the modification of growth conditions, proteins or peptides, several factors must be taken into account to benefit fully from the power of this method. Key factors include the choice of an appropriate method for the preparation of protein extracts, careful evaluation of the instrument design and available analytical capabilities, the choice of the quantification method (intensity-based vs. spectral count), and the proper manipulation of the selected quantification algorithm. The elaboration of this robust workflow for data acquisition, processing, and analysis provides unprecedented insight into the dynamics of the yeast proteome.

Key words Yeast proteomics, Label-free quantification, Mass spectrometry

1 Introduction

Quantitative MS-based proteomics aims to measure protein abundance, and/or variations in protein abundance of proteoforms [1] from a given organism studied under different experimental conditions. Although recent attempts to quantify intact proteins through top-down experiments appear extremely promising [2], the vast majority of quantitative MS-based proteomics is based on bottom-up shotgun or targeted peptide quantification strategies. Recent advances in analytical techniques, both in protein/peptide fractionation with increasingly efficient, high pressure liquid chromatography systems and columns, and in mass spectrometry instrumentation, have made quantitative proteomics a corner stone of modern systems biology. Since the early days of proteomics, yeast, and more specifically the baker's yeast *Saccharomyces cerevisiae*, has been a model organism of choice to develop and validate innovative methods to decipher the complexity of the proteome, and the dynamics of variations in the proteome upon biological or

Frédéric Devaux (ed.), *Yeast Functional Genomics: Methods and Protocols*, Methods in Molecular Biology, vol. 1361, DOI 10.1007/978-1-4939-3079-1_16, © Springer Science+Business Media New York 2016

environmental changes. Recent reports show that it is now possible to obtain a comprehensive view of the complete yeast proteome in a single run of experiments [3–5], which was unheard of only a few years ago. Yeast is a unicellular eukaryote which can grow as a microorganism in complex or synthetic defined media, with a generation time of around 90 min, which allows the easy production of sufficient biomass for biochemical and structural studies. The power of yeast classical and reverse genetics makes it possible to produce virtually any mutant strain to analyze the molecular basis of essential biological processes in great detail. The yeast genome was the first eukaryotic genome to be sequenced, and it is now extremely well annotated, which is an important point to consider when developing proteomic studies. The biological diversity of the single generic denomination, *S. cerevisiae*, can now be analyzed accurately by proteomic techniques. Quantitative proteomics has enabled: the identification of subsets of proteins that are differentially produced by haploid and diploid yeast [6]; the analysis of the effect of aneuploidy on proteome changes and phenotypic variations [7]; and the study of correlations between protein abundance and quantitative traits in a large collection of *S. cerevisiae* strains [4]. Despite the numerous advantages of yeast as a model system, yeast cells are characterized by the presence of a rigid cell wall mainly composed of $\beta1,3$- and $\beta1,6$-glucans, a small amount of chitin, and many different proteins that may bear N- and O-linked glycans and a glycolipid anchor. These components become cross-linked in various ways to form higher-order complex proteins that are heavily reticulated through disulfide bonds (for a recent review, *see* ref. 8). This makes it difficult to prepare cell free extracts under mild conditions, such as hypotonic shock, gentle sonication or Potter homogenization. The only way to get rid of the cell wall and prepare protoplasts is to treat cells with cocktails of specific glycosidases under reducing conditions in isotonic buffers. This is a time consuming process (typically 2 h) where the physiological state of the cells may be severely altered from the original conditions at the time of collection. However, this step may be necessary to prepare highly purified subcellular fractions for proteomics studies.

During the early days of proteomics, MALDI-TOF (and later TOF/TOF) mass spectrometers were extensively used in peptide mapping and identification strategies. Despite its many advantages, the MALDI ionization process does not allow sample to sample comparison of a given ion intensity, which thus precludes the direct utilization of the ion intensity signal for quantification purposes. However, within a single sample, the intensity of chemically identical compounds (peptides) is correlated with their abundance. This led to the development of quantification methods based on the controlled introduction of stable isotopes in proteins and/or peptides, which allow quantitative MS analysis or the quantification of internal reporter ions without affecting the chemical nature of the target compounds. These methods are still very popular because they allow multiplexing of several complex samples, each labeled with a specific isobaric probe, which enables the rapid, quantitative evaluation of protein/peptide abundance in multiple samples (from 2 to 10 depending on the labeling technique). Samples may be labeled by: (1) a chemical reaction on proteins or peptides involving thiol-specific reagents such as ICAT (isotope coded affinity tags) [9, 10] or Iodo-TMT [11], or amine-reactive reagents such as iTRAQ (isobaric Tags for Relative and Absolute Quantification) [12] or TMT (Tandem Mass Tags) [13]; (2) enzymatic labeling, which involves trypsin-catalyzed proteolytic cleavage in ^{16}O to ^{18}O exchange using $^{18}OH_2$ solvent in bottom-up experiments [14, 15]; or (3) metabolic labeling, which involves the incorporation of stable isotope labels by amino acids in cell culture (SILAC) [16], usually with lysine and/or arginine containing combinations of heavy isotopes of carbon or nitrogen. The SILAC strategy is highly efficient in *S. cerevisiae* [17, 18], but it may be inappropriate for some yeast species, due to the possible in vivo metabolic conversion of labeled arginine to proline and other amino acids. For example, Bicho et al. [19] found that arginine conversion in the fission yeast *Schizosaccharomyces pombe* occurs at extremely high levels. The labeling of cells with heavy arginine led to the undesired incorporation of label into essentially all of the proline pool as well as a substantial portion (25–30 %) of glutamate, glutamine, and lysine pools. The resulting isotopic clusters become too complex for the accurate quantification of the cognate peptides. The use of strains with deletions in key enzymes of the arginine conversion pathways may overcome this problem [19].

The intrinsic limitations of these labeling techniques have resulted in the emergence of an alternative quantification method, called the "label-free" approach (reviewed in [20–23]). This method was made possible by remarkable improvements in mass spectrometry instrumentation and the strong correlation between the relative ion intensity and abundance in the electrospray ionization process. It is based on the direct comparison of complex mixtures of native peptides in a series of LC-MS experiments, and the quantification of molecules with similar features, including retention time and mass, over the different LC runs. Peptides are identified from MSMS fragmentation spectra and database searches. Although this approach is well suited for the quantification of variations in the entire proteome, it has also been applied with success to the analysis of sub-proteomes in the nucleus [24], the mitochondrial phosphoproteome [25], and RNA-polymerase complexes [26].

Label-free quantification approaches involve four main steps: the alignment of the LC-MS runs, the detection and quantification of the features, a normalization procedure that enables the comparison of the runs, and the identification of the peptides/ proteins. These approaches can be divided into two distinct groups: (1) signal intensity measurements based on precursor ion spectra and the integration of ion intensities of each peptide detected over its chromatographic elution profile [27, 28]; and (2) spectral counting, which is based on counting the number of peptides assigned to a protein in an MS/MS experiment, which allows protein quantity to be inferred indirectly [29]. Both methods have several advantages. Spectral counting appears to be a more sensitive method for the detection of proteins that undergo changes in abundance, whereas peak area intensity measurements usually yield a more accurate estimate of variations in protein abundance, which are often reported as expression ratios [30]. It is important to gather a sufficient number of experiments to obtain reliable results high statistical power. On a routine basis, we use triplicates of each experimental condition to validate a twofold difference between experimental samples. Replicates also increase the coverage of the proteome (example in Table 1). Peptides are not chemically altered and every sample is analyzed independently from the other samples to be compared; therefore, a large number of different experimental conditions can be explored with this method. However, care should be taken to produce data under the best defined experimental conditions, at all the steps of the procedure, from standardized protein digestion protocols to highly reproducible LC separations and MS acquisition. It is therefore recommended to run LCs in grouped series for a given project to minimize run to run and column to column performance

Table 1
Effect of technical replicates on the identification of proteins (Percolator used and 1 % FDR filter applied)

Mass spectrometer	Gradient time (min)	Species	Runs	Identification number — Mascot (protein groups/unique peptide)
LTQ Orvitrap Velos ETD	240	*Candidaalbicans*	1	2049/7676
			Triplicate	2477/10987
		Saccharomyces cerevisiae	1	2135/7337
			Triplicate	2610/10543
	120	*Candidaglabrata*	1	1598/7097
			Triplicate	1949/10311

variability. The strong relationship between the amount of sample loaded and peptide identification [31] may substantially affect quantitative analyses. It is thus critical to run all LC-MS with the same amount of starting material to ensure reliable comparisons of peptide/protein abundance between samples. Although trypsin is a widely used protease for bottom-up experiments, a LysC/trypsin digestion protocol may be necessary to improve the generation of peptides [32, 33].

The LC-MS data are converted into 2D images and the alignment of different runs consists of optimizing the matches between the spots (associated with an intensity) in the different images (for a review, *see* ref. 34). The quality of the peak picking procedure is critical, especially for peaks with intensity values close to the signal to noise threshold value, or in complex isotope clusters resulting from overlapping peptides, and must be evaluated through rigorous statistical analyses (for a review, *see* ref. 35). A number of software suites are available for label-free quantitative proteomics. One of the most popular is MaxQuant developed by Cox and Mann [36]. Details of the different features of these tools and valuable comparisons can be found in recent studies [37, 38]. This field is still undergoing active research and development and new software, such as Serac [30], IDEAL-Q [39] or LFQuant [40] have been recently developed. Several questions are raised by label-free quantitative proteomics. One issue involves peptides shared by multiple proteins [41], and their use in a quantitative proteomics workflow. Another point of interest involves the adaptation of quantification methods to provide absolute quantifications of proteins, because label-free experiments provide data on the relative abundance of proteins analyzed under different experimental conditions. Modifications of the classical label-free protocols have been used to address this issue, e.g., samples may be spiked with known amounts of labeled standards [42] in intensity-based quantification, or normalized abundance factors [29] may be used in spectral counting methods. An important issue is missing values. Despite care taken to ensure that the analytical procedure is as reproducible as possible, the selection of MS and MSMS precursor ions during data dependent acquisitions is still somewhat stochastic and some ions may be missed in some replicate. This may obviously affect the quality of the quantification procedure.

In the present chapter we discuss several points that are important for the optimization of label-free quantitative proteomics in yeast. We validated the methods described in several quantitative proteomics projects for *S. cerevisiae*, *C. albicans*, and *C. glabrata*. We routinely used the commercial software suite Progenesis-LC QI (http://www.nonlinear.com/progenesis/qi-for-proteomics), which is user-friendly, and complies with the ISO9001 certification of the lab.

2   Materials

1. Reagents for typical yeast growth media.

2. Cell lysis buffer: 40 mM HEPES-KOH pH 7.5, 350 mM NaCl, 10 % glycerol, 0.1 % Tween-20.

3. Protoplast digestion buffer: 0.01 M citrate-phosphate buffer pH 5.8, 1.35 M sorbitol, 1 mM EGTA.

4. Protoplast wash buffer: 0.01 M Tris-maleate buffer pH 6.8, 0.75 M sorbitol, 0.4 M mannitol, 2 mM EGTA.

5. Acid-washed silica beads (0.4–0.6 mm Ø).

6. Teflon/glass Potter homogenizer.

7. One Shot Cell Disrupter (Constant Systems Limited, UK).

8. Trypsin (Proteomic grade).

9. Glutamyl endopeptidase (V8 proteinase).

10. Zymolyase.

11. Low binding microcentrifuge tubes.

12. 4–12 % polyacrylamide gradient gels.

13. Coomassie blue (MS friendly, such as SimplyBlue SafeStain, Invitrogen).

14. BCA protein assay (Pierce).

15. An instrument setup for LC-MS/MSMS data acquisition (*see* **Note 1**).

16. Appropriate software suites for quantification and identification of the peptide/protein content of the samples analyzed (*see* **Note 2**).

3   Experimental Methods

3.1   Preparation of Protein Extracts (See Notes 3 and 4)

On a routine basis, we prefer to prepare protein samples using a non-denaturing extraction process. This allows running both quantitative proteomics experiments and other specific biochemical analysis, such as enzyme activity measurements or protein complexes purification.

3.1.1   Non-denaturing Protein Extraction

1. Grow the cells in the appropriate medium to the expected cell density.

2. Collect the cells by centrifugation for 10 min at $4000 \times g$ at 4 °C.

3. Wash the cell pellet with cold water, and resuspend the cell pellet in the lysis buffer A at a cell density of 0.6 g/ml.

4. To lyse the cells, add to the cell suspension 0.32 ml acid-washed, heat-sterilized silica glass beads and process the cell

suspension by ten cycles consisting of 1 min vortexing at 2500 rpm followed by a 30 s incubation at 4 °C. Decant the beads and pipette the supernatant into a clean tube. Alternatively use a cell disrupter (One Shot, from Constant Systems Limited, UK, or an equivalent device) operating at 2 kbar (30 kPSI), and directly collect the cell lysate from the exit tube of the high pressure chamber.

5. Pellet unbroken cell material by centrifugation at $4000 \times g$ for 20 min at 4 °C. Collect the supernatant and determine the protein concentration by the BCA protein assay (*see* **Note 5**).

6. If the resulting protein solution seems too dilute for further processing, the proteins may be precipitated by the addition of TCA to 20 %, incubated for 3 h at 4 °C, pelleted by centrifugation at $14,000 \times g$ at 4 °C, and washed twice with 500 µl of acetone. The final pellet must be dried in a speed vacuum system.

Occasionally, we may need to prepare purified intact organelles for quantitative studies. To do so, we modify the previous protocol by (1) using isotonic buffers at all the steps of the sample preparation and (2) by removing the cell wall by means of enzymatic digestion of the oligosaccharide components of this structure. This allows the preparation of protoplasts that are further lysed under very gentle conditions.

3.1.2 Protoplast Preparation

1. Grow the cells in the appropriate medium to the expected cell density (*see* **Note 6**).

2. Collect the cells by centrifugation at $4000 \times g$ for 10 min at 4 °C.

3. Wash the cell pellet in 0.01 M Tris-Cl pH 7, 0.5 M KCl and resuspend the cell pellet at a cell density of 0.6 g/ml in the digestion buffer.

4. Add Zymolyase 100T (MP Biomedicals) to a final concentration of 10 mg/ml.

5. Incubate the cell suspension at 30 °C and evaluate protoplast formation by measuring the OD_{600} of a 1/100 dilution of the cells in water.

6. When protoplast formation is nearly complete (>85 %), collect the cells by centrifugation and resuspend them in the wash buffer.

7. The protoplasts are disrupted by gentle homogenization with a Teflon/glass Potter homogenizer.

3.1.3 Optional Clean-Up Method for Protein Extracts Before Processing for Bottom-Up Proteomics Experiments

A simple and efficient way to clean up the samples and remove salts and detergents is to precipitate the proteins with cold acetone.

1. Cool the required volume of acetone to –20 °C.

2. Place protein sample in an acetone-compatible tube, such as an Eppendorf tube with a safe lock cap.

3. Add four times the sample volume of cold (−20 °C) acetone to the tube.

4. Vortex the tube and incubate for 30–60 min at −20 °C.

5. Centrifuge for 10 min at 13,000×g in a refrigerated centrifuge.

6. Remove *very carefully* the supernatant (e.g., by aspiration using a narrow tip for loading gels, after a second pulse centrifugation).

7. Carefully dry the protein pellet by placing the tube in a dry oven at 37 °C for 15 min.

8. Resuspend the pellet in 50 μl of an appropriate solubilization solution (*see* **Note 7**).

3.2 Preparation of Proteolytic Digests for Bottom-Up Proteomics

The input samples for quantitative, label-free proteomics experiments may be protein extracts either in solution, or obtained after separation on SDS-PAGE. In the latter case, a number of additional steps are required to make the in-gel proteins accessible to the protease used.

3.2.1 Trypsin Digestion from Protein Sample Separated on SDS-PAGE

1. Run the gel under standard electrophoretic conditions. It is necessary to load the gel with identical amounts of protein in each lane.

2. Stain the gel with SimplyBlue SafeStain® in ethanol/acetic acid solution. After destaining with the same solvent, take a picture of the gel to keep track of the protein profiles.

3. Soak the gel with 50 mM NH_4HCO_3/50 % acetonitrile for 15 min at 30 °C with agitation.

4. Cut the gel into the required number of bands (3–5 for 1.5 cm short migrations on isocratic gels; 12–15 for long migrations, typically on 4–12 % gradient gels from Invitrogen) to cover the maximum proteome depth.

5. Place the gel pieces into microfuge tubes (Eppendorf type, low protein binding).

6. Wash the gel slices with 50 % acetonitrile/50 mM NH_4HCO_3.

7. Incubate the slices for 15 min at 30 °C.

8. Carefully discard the washing solution by aspiration.

9. Repeat **steps 5–7** twice.

10. Dry the gel slice.

11. Add 100 μl of 100 % acetonitrile.

12. Incubate for 10 min with agitation at room temperature.

13. Discard the supernatant.

14. Dry at 37 °C for 10 min. At this step, it is possible to reduce and alkylate cysteine residues following the steps described below:

15. Add 100 μl of 10 mM dithiothreitol (DTT).

16. Incubate for 45 min at 56 °C with agitation.

17. Discard the excess solution at room temperature.

18. Add 100 μl of 55 mM iodoacetamide (IAA).

19. Incubate at room temperature for 45 min in the dark.

20. Discard the excess of reduction/alkylation solution.

21. Add 100 μl of water (Millipore ultrapure grade).

22. Incubate for 30 min with agitation.

23. Discard the supernatant.

24. Dry the gel slices as described in **steps 10–13**. The samples are now ready for in-gel digestion. The procedure below is given for a trypsin digest (*see* **Note 8**).

25. Prepare a solution of trypsin by dissolving 20 μg of protein (Promega) in 100 μl of 1 mM HCl.

26. Add 20 μl of a 1/20 dilution of the trypsin stock solution to each gel slice in 25 mM NH_4HCO_3 buffer pH 8.0.

27. Incubate for 20 min at 4 °C.

28. Add enough carbonate buffer (20 μl or more) to submerge the piece of gel.

29. Incubate at 37 °C overnight in a dry incubator to prevent evaporation. The resulting peptides are extracted as follows.

30. Transfer the supernatant to a clean, low binding tube.

31. Add 20 μl of 50 % acetonitrile containing 0.1 % formic acid.

32. Incubate for 15 min with agitation.

33. Recover the supernatant and pool it with the supernatant obtained in **step 30**.

34. Remove all solvents by vacuum drying (SpeedVac).

35. Resuspend the peptides in 0.1 % formic acid.

3.2.2 Trypsin Digestion from Protein Samples in Solution

1. Prepare 4–5 μg of total protein in 19 μl of 25 mM ammonium carbonate buffer (NH_4HCO_3).

2. Add 1 μl of the trypsin stock solution prepared as described in Subheading 3.2.1, **step 26**.

3. Incubate at 37 °C overnight in a dry incubator to prevent evaporation.

4. Remove all solvents by vacuum drying (SpeedVac).

5. Resuspend the peptides in 0.1 % formic acid.

3.3 Instrument Setup and Analytical Capability

Background: The relationship between the amount of sample loaded and peptide identification is a crucial factor for the optimization of proteomics experiments [31]. It is therefore critical to evaluate carefully the analytical capability of the available instrument

Table 2
Comparative evaluation of the effect of the instrument setup on the identification of proteins (Percolator used and 1 % FDR filter applied)

Mass Spectrometer	HPLC	Gradient time (min)	Column (cm)	Species	Identification number	
					Mascot (protein groups/unique peptide)	Sequest (protein groups/unique peptide)
LTQ Orvitrap Velos ETD	EasynLC Proxeon	75	10	*Saccharomyces cerevisiae*	972/3912	1111/4884
	RSLC Dionex	120	25	*Saccharomyces cerevisiae*	1234/5245	1402/5948
	U3000	240		*Saccharomyces cerevisiae*	1198/4583	1422/6072
	EasynLC 1000 Proxeon	120	50	*Saccharomyces cerevisiae*	1505/8317	1638/8339
				Candida glabrata	1598/7097	
		240		*Saccharomyces cerevisiae*	2135/7337	
				Candida albicans	2049/7676	2135/7337
Orbitrap Fusion	EasynLC 1000 Proxeon	120	50	*Saccharomyces cerevisiae*	2156/12266	2323/11634

setup. The nature of the chromatographic reversed phase, the length of the separation column (which is correlated with the pressure delivered by the nanoLC pumps) and the duration of the elution gradient substantially affect the detection, identification and quantification of complex proteomes.

As an example, Table 2 shows the differences in the depth of proteome analyses that we obtained with total protein extract from *S. cerevisiae*, *C. albicans*, or *C. glabrata* digested with trypsin (Subheading 3.2.2). All digests of protein extracts were analyzed with an LTQ Velos Orbitrap ETD equipped with a nanoelectrospray ion source.

The following procedure describes our current instrument setup and operating methods. The mass spectrometer is coupled to an EASY-Spray nanoelectrospray ion source and an Easy nano-LC Proxeon 1000 system (all devices are from Thermo Fisher Scientific, San Jose, CA).

1. The chromatographic separation of peptides is performed with the following parameters: Acclaim PepMap100 C18 pre-column (2 cm, 75 μm i.d., 3 μm, 100 Å), Pepmap-RSLC Proxeon C18 column (50 cm, 75 μm i.d., 2 μm, 100 Å), 300 nl/min flow.

2. Typically, 5 µl samples, *containing equal amounts of peptide for quantitative experiments*, are injected onto the pre-concentration column.

3. The chromatographic separation of peptides is obtained with a gradient consisting of 95 % solvent A (water, 0.1 % formic acid) to 35 % solvent B (100 % acetonitrile, 0.1 % formic acid) in 217 min followed by column regeneration for 23 min giving a total run time of 4 h.

4. Peptides are analyzed in the Orbitrap® in full ion scan mode at a resolution of 30,000 (at m/z 400) and with a mass range of m/z 400–1800.

5. Fragments are obtained by collision-induced dissociation (CID) activation with a collisional energy of 40 %, an isolation width of 2 Da, and an activation Q of 0.250 for 10 ms. MS/MS data are acquired in the linear ion trap in a data-dependent mode in which the 20 most intense precursor ions are fragmented, with a dynamic exclusion of 20 s, an exclusion list size of 500 and a repeat duration of 30 s.

6. The maximum ion accumulation times are set to 100 ms for MS acquisition and 50 ms for MS/MS acquisition.

7. All MS/MS data are processed with an in-house Mascot search server (Matrix Science, Boston, MA; version 2.4.1). The mass tolerance is set to 7 ppm for precursor ions and 0.5 Da for fragments. Some analyses use Sequest as a search engine, because this enables results to be compared with particular specifications of instrument performance provided by Thermo Fischer Scientific.

8. The following alterations are used in variable modifications: carbamidomethylation (C), if the sample is reduced and alkylated, and oxidation (M). Phosphorylation (STY), acetylation (K, N-term), and deamidation (N, Q) are usually added for additional analyses of trypsin digests.

9. The maximum number of missed cleavages by trypsin is limited to two for all proteases used.

10. MS/MS data are searched against protein sequence databases, usually the Uniprot database corresponding to the particular yeast species, but also other databases if necessary, with Fasta files retrieved from the Genolevure website (http://www.genolevures.org/) or from the Candida Genome Database website (http://www.candidagenome.org/).

3.4 Workflow to Analyze a Set of Experiments for the Label-Free Quantification of Yeast Proteins

1. Generate .raw files on the mass spectrometer (see above). The typical size of a .raw file corresponding to a 120 min gradient LC-MSMS analysis of a *S. cerevisiae* total protein trypsin digest is 524,610 kbits on the Velos Mass spectrometer.

2. Create a folder to gather all the .raw files from one project together.

3. Import all the .raw files and visualize the Total Ion Current (TIC) for each run (Fig. 1a).

4. Select a reference run (*see* **Note 9**).

5. Run the alignment procedure. This can be either fully automatic or may involve a preliminary manual step consisting in the addition of several vectors across the whole 2D representation of the run (*see* **Note 10**).

6. Review the global quality of alignment by examining the heat map generated for each run (poorly aligned regions or regions with obvious composition differences appear in red, Fig. 1b) and the overall percentage of alignment.

7. If necessary, edit manually new vectors in the poorly processed regions of the runs to increase the percentage of alignment (Fig. 1c, d).

8. Run the peak picking procedure. The default parameters for signal to noise ratio are usually appropriate to obtain high quality peak lists (*see* **Note 11**). This generates a number of features associated with all the ions that have been detected (*see* **Note 12**). At this step, it is possible to run a normalization procedure of all runs versus a reference run (usually the run presenting the smallest number of differences with the other

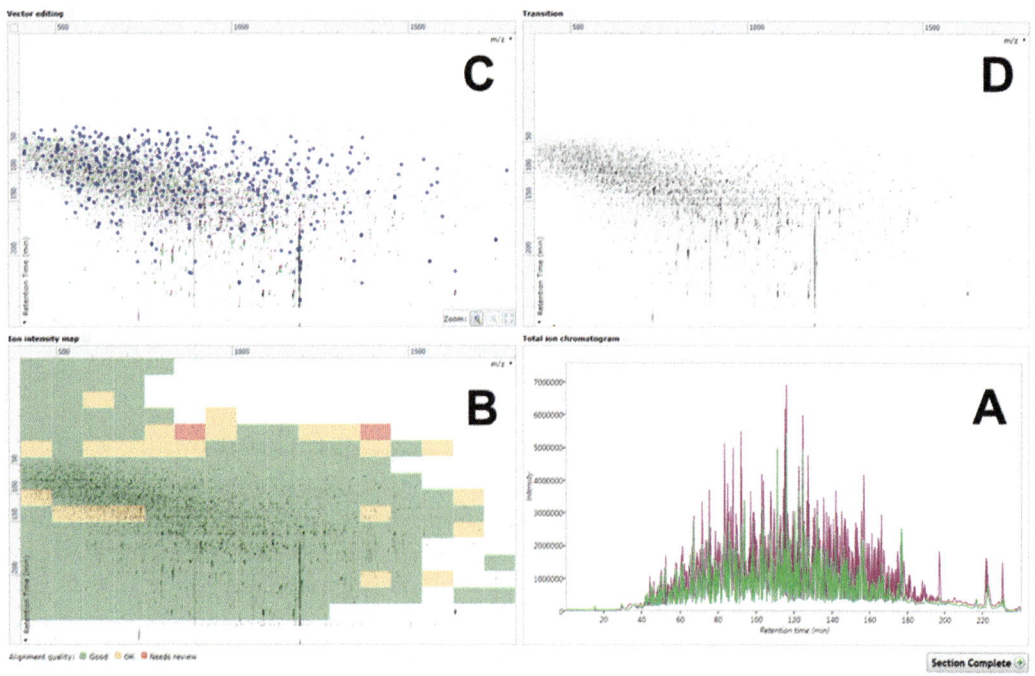

Fig. 1 Screen shot of the alignment procedure for two representative runs: (**a**) Superposition of the reference TIC (*green*) and the TIC from a given experimental condition (*red*). (**b**) Heat map of the automatic alignment procedure, showing well-adjusted areas (*green*), fairly well-adjusted (*yellow*), and poorly adjusted (*red*) ones. (**c** and **d**) vectorization of the 2D image generated from the adjustment of the TIC

runs) either on all features or on sets of features. The user may choose not to use any normalization.

9. Define the experiment design: two types of design are possible: a "between subject design" and a "within subject design". Typically, a "between subject design" is used to compare different mutant strains analyzed under the same experimental conditions, whereas a "within subject" design may apply to the analysis of the same strain analyzed in different growth conditions or in time course experiments.

10. Review peak picking. This step allows the editing of peaks with an abnormal distribution of features (overlapping peptides with intricate isotopic clusters), and generates a list (.txt format) of MS not associated with MSMS that nonetheless present a significant difference between the groups. This list may be used as an inclusion list to identify more peptides, which increases the biological information obtained from the experiments (*see* **Note 13**).

11. Peptide identification. A file aggregating all the features detected in all the runs associated with productive MS/MS is exported as an .mgf file.

12. The mgf file is sent to the mascot server and searched against an appropriate sequence database. A decoy search is performed and the significance threshold is usually fixed at 0.05 (medium stringency) or 0.01 (high stringency).

13. The xml results file is re-imported into the software for matching identifications with the features tables.

14. Refine identification by filtering peptides according to particular parameters such as retention time, mass error, charge, modification or score. We routinely use a filter based on Mascot scores.

15. Resolve conflicts. This step, run manually, may be quite time consuming, but is important for improving the quantification, because only non-conflicting peptides are used in the quantification process. Conflict resolution is based on protein score, mass error and the number of peptides assigned to each protein. Most of the time, it favors the protein with the largest number of peptides (*see* **Note 14**).

16. Review proteins. At this step, it is possible to tag protein entries according to several criteria, such as maximum fold change and ANOVA value, which produces automatic filters for further analysis.

17. Protein statistics. The software integrates a number of built-in statistical tools to analyze the results files, and produce groups of proteins sharing similar expression profiles.

18. Production of a report in .html format including all or selected pieces of information produced during the analysis process (Figs. 2 and 3).

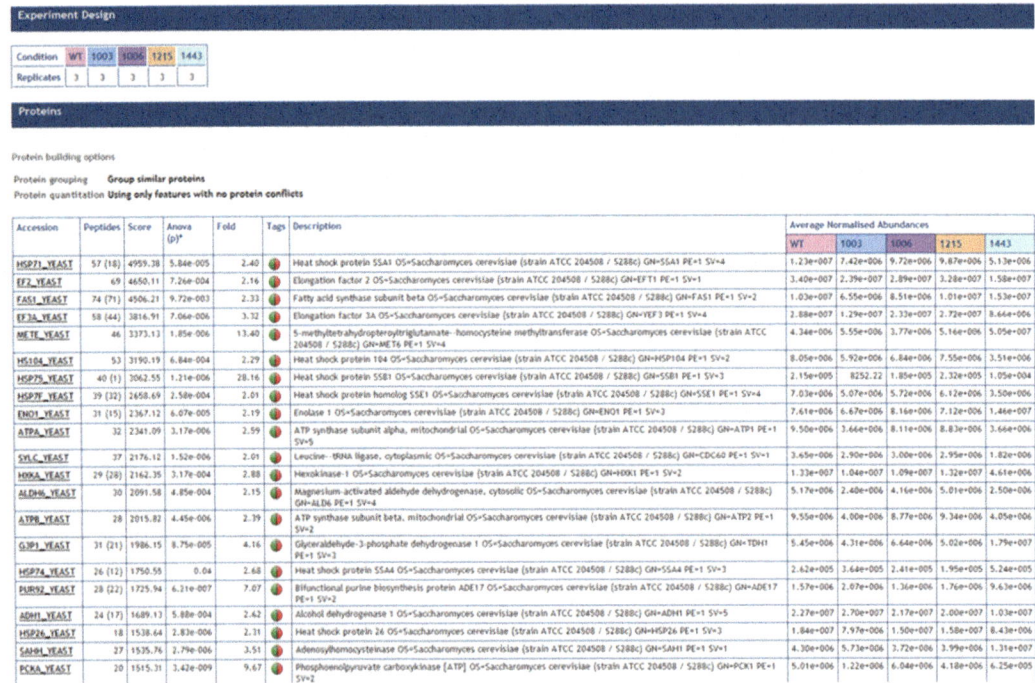

Fig. 2 Typical output of the quantification workflow

4   Notes

1. The instrumental setup in our lab consists of an Orbitrap Fusion, an Orbitrap Q-Exactive Plus, and an LTQ Velos Orbitrap ETD mass spectrometer (data from the latter instrument are presented in Tables 1 and 2), equipped with an Easy-Spray nanoelectrospray ion source. The LC setup consists of an Easy nano-LC Proxeon 1000 system equipped with an Acclaim PepMap100 C18 pre-column and a Pepmap-RSLC Proxeon C18 column. All these devices are from Thermo Fisher Scientific, San Jose, CA.

2. We use Progenesis QI (Nonlinear Dynamics Ltd, Waters, Newcastle, UK) for the quantification steps, and a local Mascot server (Matrix Science, Boston, MA; version 2.4.1) for peptide/protein identification.

3. Three main methods may be used to prepare yeast cell free extracts, depending on the end-purpose of the preparation: (1) non-denaturing protein extraction, with either glass beads to grind the cells, or pressure-based disruption systems, is used if further processing of the samples is required, such as the immuno-precipitation of particular targets or the biochemical

HSP71_YEAST

Heat shock protein SSA1 OS=Saccharomyces cerevisiae (strain ATCC 204508 / S288c) GN=SSA1 PE=1 SV=4
57 peptides

Sequence	Feature	Score	Hits	Mass	Charge	Tags	Conflicts	Modifications	In quantitation	Average Normalised Abundances				
										WT	1003	1006	1215	1443
AEETISWLDSNTTASK	15875	68.76	9	1751.8280	2		0		yes	4.40e+004	1.19e+004	6.04e+004	4.51e+004	1.63e+004
AEETISWLDSNTTASKEEFDDKLK	17803	45.79	10	2756.3027	3		0		yes	8.42e+004	6.55e+004	6.35e+004	5.54e+004	3.15e+004
AEETISWLDSNTTASKEEFDDKLK	5391	21.26	9	2756.3013	4		0		yes	1.31e+005	1.32e+005	9.45e+004	7.09e+004	5.18e+004
ARFEELCADLFR	16543	39.77	9	1468.7137	3		3		no	4.13e+004	3806.71	3.29e+004	5.39e+004	3980.31
ATAGDTHLGGEDFDNR	77	47.92	15	1674.7231	3		2		no	1.27e+006	1.89e+006	1.81e+006	2.01e+006	1.37e+006
ATAGDTHLGGEDFDNR	590	43.99	1	1674.7282	3		2		no	7.64e+005	350.13	550.74	460.50	1581.83
ATAGDTHLGGEDFDNR	618	88.97	14	1674.7243	2		2		no	7.38e+005	1.00e+006	1.05e+006	1.24e+006	7.03e+005
DAGTIAGLNVLR	35799	66.97	15	1198.6676	3		3		no	4887.14	4354.29	4327.48	4115.98	2230.76
DAGTIAGLNVLR	140	77.83	18	1198.6654	2		3		no	1.39e+006	1.04e+006	9.57e+005	1.12e+006	9.87e+005
ELQDIANPIMSK	326	66.28	14	1357.6895	2		0		yes	6.14e+005	7.83e+005	8.40e+005	6.31e+005	4.09e+005
ELQDIANPIMSK	1574	53.55	1	1357.6923	2		0		yes	5.12e+005	8176.03	8557.74	8230.85	1.16e+004
ETAESYLGAK	597	49.52	16	1067.5123	2		1		no	2.28e+005	3.59e+005	3.50e+005	2.64e+005	2.55e+005
FEELCADLFR	1347	51.72	16	1241.5760	2		3		no	2.80e+005	5.45e+004	4.73e+005	3.59e+005	4.40e+004
FELSGPPAPR	872	56.76	43	1182.4										
FKEEDEKESQR	29804	19.16	2	1423.6										
GVPQIEVTFDVDSNGILNVSAVEK	42565	29.11	35	2529.2										
GVPQIEVTFDVDSNGILNVSAVEK	65563	22.21	3	2529.2										
GVPQIEVTFDVDSNGILNVSAVEK	42920	20.31	12	2529.2										
IINEPTAAAIAYGLDK	1470	64.11	17	1658.8										
IINEPTAAAIAYGLDK	246	107.19	16	1658.9										
IINEPTAAAIAYGLDKK	884	44.61	1	1786.9										
IINEPTAAAIAYGLDKK	4329	104.46	9	1786.9										
IINEPTAAAIAYGLDKK	176	45.62	7	1786.9										
ITITNDKGR	28620	17.65	5	1016.5										
ITITNDKGR	490	48.04	2	1017.5										
KAEETISWLDSNTTASK	14275	87.70	1	1879.9										
KAEETISWLDSNTTASK	3657	97.28	14	1879.9										
KAEETISWLDSNTTASK	5277	53.84	1	1879.9										
KAEETISWLDSNTTASK	969	57.96	14	1879.9										
KAEETISWLDSNTTASKEEFDDK	37427	71.30	1	2643.2										
KAEETISWLDSNTTASKEEFDDK	13225	37.36	12	2643.2										
KSEIFSTYADNQPGVLIQVFEGER	4625	64.17	21	2726.										
KSEIFSTYADNQPGVLIQVFEGER	73659	16.22	1	2726.										

Accession HSP71_YEAST

escription Heat shock protein SSA1 OS=Saccharomyces cerevisiae (strain ATCC 204508 / S288c) GN=SSA1 PE=1 SV=4

Peptides 57 (18)

Score 4959.38

Anova 5.84e-005

Fold 2.40

● Anova p-value ≤ 0.05
● Max fold change ≥ 2

WT	1003	1006	1215	1443

Fig. 3 Typical output of the quantification features obtained for a given protein (here HSP71) and graphical representation of the measured variations in abundance

enrichment of subcellular components, (2) protoplasts are prepared if highly purified intact subcellular organelles are to be used as a starting material for quantitative proteomics, or (3) denaturing protein extraction, such as alkaline extraction followed by acid precipitation of total proteins, is used if the sample is intended for deep proteomic analysis, with the largest possible proteome coverage. In the first and second procedures, the buffers may contain detergents, protease and/or phosphatase inhibitors and all chemicals required to preserve the integrity of the particular sets of proteins to be analyzed. These compounds have to be removed before processing the samples for bottom-up proteomics experiments. A simple procedure is described in Subheading 3.1.3.

4. It may sometimes be useful to access the deep proteome of the yeast cells using a fast preparation method. The denaturing protein extraction by alkaline extraction/acid precipitation of total proteins is optimal for this purpose. The following

protocol summarizes the different steps of the procedure (adapted from [43]):

(a) Grow the cells in the appropriate medium to the expected cell density.

(b) Centrifuge one volume of culture corresponding to 1.5 A_{600}.

(c) Remove most of the supernatant, but keep about 0.5 ml.

(d) Add 50 μl of 1.85 M NaOH + 2 % 2-mercaptoethanol and incubate for 10 min on ice.

(e) Add 50 μl of 50 % TCA and incubate for 10 min on ice.

(f) Centrifuge 3 × 5 min at 15,300 × g (in an Eppendorf type refrigerated centrifuge). Remove *very carefully* the supernatant (e.g., by aspiration using a narrow tip for loading gels, after a second pulse centrifugation).

(g) Resuspend the pellet in 50 μl of an appropriate solubilization solution (*see* **Note 7**).

5. Although the clarification step does not affect the abundance of most yeast proteins, the overall proteome coverage seems higher in non-clarified samples than in clarified samples [3, 44], and non-clarified samples are enriched in proteins from the GO terms "Membrane protein" and "Nucleus" [44].

6. The efficiency of protoplast preparation is highly dependent on the growth phase of the yeast cells. Cells in the late stationary phase of growth are very resistant to Zymolyase action. The efficiency is also dependent on the yeast species used. 2 h incubation with Zymolyase is usually sufficient to produce protoplasts with an efficiency of more than 80 % with most strains of *S. cerevisiae*; however, an incubation time of up to 12 h may be required to reach similar proportions of protoplasts with the yeast form of *C. albicans* collected in mid-exponential phase.

7. The following solubilization solution may be used to load the sample onto a SDS-PAGE gel: 2 vol Sample Buffer 2× (100 mM Tris–HCl pH 6.8, 4 mM EDTA, 4 % SDS, 20 % glycerol, Bromophenol Blue)/1 vol Tris Base 1 M/2 % 2-mercaptoethanol. The sample is heated for 10 min at 37 °C, and may then be kept for several months at –20 °C. To digest the proteins in solution with trypsin, resuspend the pellet in 50 mM ammonium acetate buffer in an Eppendorf tube, and immerse the tube in a sonication bath to facilitate the solubilization of the proteins.

8. In some experiments, it may be interesting to quantify peptides produced by a proteolytic enzyme other than trypsin. The glutamyl endopeptidase (V8-proteinase, EC3.4.21.19) is an enzyme of choice owing to its cleavage specificity. The protocol

is very similar to that described for trypsin, with the following differences: V8-proteinase stock solution is prepared by dissolving the lyophilized protein (Promega) in 0.6 M urea. The incubation buffer is made of 0.1 M potassium/sodium phosphate buffer pH 7.5.

9. The software is able to select automatically one reference file, usually the file allowing the maximum matches between the runs included in the project.

10. Manual editing of alignment vectors, although slightly time consuming, facilitates and speeds up the subsequent automatic alignment.

11. It is not necessary to include all the runs for peak picking. At this step, runs presenting outlier specificities may be excluded from the analysis. The features will be added to theses runs after completion of the peak picking procedure.

12. In our lab, we filter the features to save only ions with 2+, 3+, and 4+ charge states.

13. At this step, it is possible to run multivariate statistical analysis on peptide features. However in our hands, this step does not provide any appreciable benefit to the quantification process, but skipping it may increase the number of conflicting peptides.

14. Conflicting peptides are mainly peptides with the same mass but different composition or peptides belonging to a protein group.

References

1. Smith LM, Kelleher NL (2013) Proteoform: a single term describing protein complexity. Nat Methods 10(3):186 187

2. Ntai I, Kim K, Fellers RT, Skinner OS, Smith AD IV, Early BP, Savaryn JP, LeDuc RD, Thomas PM, Kelleher NL (2014) Applying label-free quantitation to top down proteomics. Anal Chem 86(10):4961–4968

3. Hebert AS, Richards AL, Bailey DJ, Ulbrich A, Coughlin EE, Westphall MS, Coon JJ (2014) The one hour yeast proteome. Mol Cell Proteomics 13(1):339–347

4. Picotti P, Clement-Ziza M, Lam H, Campbell DS, Schmidt A, Deutsch EW, Rost H, Sun Z, Rinner O, Reiter L, Shen Q, Michaelson JJ, Frei A, Alberti S, Kusebauch U, Wollscheid B, Moritz RL, Beyer A, Aebersold R (2013) A complete mass-spectrometric map of the yeast proteome applied to quantitative trait analysis. Nature 494(7436):266–270

5. Nagaraj N, Kulak NA, Cox J, Neuhauser N, Mayr K, Hoerning O, Vorm O, Mann M (2012) System wide perturbation analysis with nearly complete coverage of the yeast proteome by single-shot ultra HPLC runs on a bench top Orbitrap. Mol Cell Proteomics 11(3):M111 013722

6. de Godoy LM, Olsen JV, Cox J, Nielsen ML, Hubner NC, Frohlich F, Walther TC, Mann M (2008) Comprehensive mass-spectrometry-based proteome quantification of haploid versus diploid yeast. Nature 455(7217):1251–1254

7. Pavelka N, Rancati G, Zhu J, Bradford WD, Saraf A, Florens L, Sanderson BW, Hattem GL, Li R (2010) Aneuploidy confers quantitative proteome changes and phenotypic variation in budding yeast. Nature 468(7321):321–325

8. Orlean P (2012) Architecture and biosynthesis of the *Saccharomyces cerevisiae* cell wall. Genetics 192(3):775–818

9. Griffin TJ, Gygi SP, Rist B, Aebersold R, Loboda A, Jilkine A, Ens W, Standing KG (2001) Quantitative proteomic analysis using a MALDI quadrupole time-of-flight mass spectrometer. Anal Chem 73(5):978–986

10. Gygi SP, Rist B, Gerber SA, Turecek F, Gelb MH, Aebersold R (1999) Quantitative analysis of complex protein mixtures using isotope-coded affinity tags. Nat Biotechnol 17(10):994–999

11. Pan KT, Chen YY, Pu TH, Chao YS, Yang CY, Bomgarden RD, Rogers JC, Meng TC, Khoo KH (2014) Mass spectrometry-based quantitative proteomics for dissecting multiplexed redox cysteine modifications in nitric oxide-protected cardiomyocyte under hypoxia. Antioxid Redox Signal 20(9):1365–1381

12. Ross PL, Huang YN, Marchese JN, Williamson B, Parker K, Hattan S, Khainovski N, Pillai S, Dey S, Daniels S, Purkayastha S, Juhasz P, Martin S, Bartlet-Jones M, He F, Jacobson A, Pappin DJ (2004) Multiplexed protein quantitation in *Saccharomyces cerevisiae* using amine-reactive isobaric tagging reagents. Mol Cell Proteomics 3(12):1154–1169

13. Thompson A, Schafer J, Kuhn K, Kienle S, Schwarz J, Schmidt G, Neumann T, Johnstone R, Mohammed AK, Hamon C (2003) Tandem mass tags: a novel quantification strategy for comparative analysis of complex protein mixtures by MS/MS. Anal Chem 75(8):1895–1904

14. Yao X, Afonso C, Fenselau C (2003) Dissection of proteolytic ^{18}O labeling: endoprotease-catalyzed ^{16}O-to-^{18}O exchange of truncated peptide substrates. J Proteome Res 2(2):147–152

15. Heller M, Mattou H, Menzel C, Yao X (2003) Trypsin catalyzed ^{16}O-to-^{18}O exchange for comparative proteomics: tandem mass spectrometry comparison using MALDI-TOF, ESI-QTOF, and ESI-ion trap mass spectrometers. J Am Soc Mass Spectrom 14(7):704–718

16. Ong SE, Blagoev B, Kratchmarova I, Kristensen DB, Steen H, Pandey A, Mann M (2002) Stable isotope labeling by amino acids in cell culture, SILAC, as a simple and accurate approach to expression proteomics. Mol Cell Proteomics 1(5):376–386

17. de Godoy LM, Olsen JV, de Souza GA, Li G, Mortensen P, Mann M (2006) Status of complete proteome analysis by mass spectrometry: SILAC labeled yeast as a model system. Genome Biol 7(6):R50

18. Dilworth DJ, Saleem RA, Rogers RS, Mirzaei H, Boyle J, Aitchison JD (2010) QTIPS: a novel method of unsupervised determination of isotopic amino acid distribution in SILAC experiments. J Am Soc Mass Spectrom 21(8):1417–1422

19. Bicho CC, de Lima Alves F, Chen ZA, Rappsilber J, Sawin KE (2010) A genetic engineering solution to the "arginine conversion problem" in stable isotope labeling by amino acids in cell culture (SILAC). Mol Cell Proteomics 9(7):1567–1577

20. Bantscheff M, Schirle M, Sweetman G, Rick J, Kuster B (2007) Quantitative mass spectrometry in proteomics: a critical review. Anal Bioanal Chem 389(4):1017–1031

21. Bantscheff M, Lemeer S, Savitski MM, Kuster B (2012) Quantitative mass spectrometry in proteomics: critical review update from 2007 to the present. Anal Bioanal Chem 404(4):939–965

22. Zhu W, Smith JW, Huang CM (2010) Mass spectrometry-based label-free quantitative proteomics. J Biomed Biotechnol 2010:840518

23. Neilson KA, Ali NA, Muralidharan S, Mirzaei M, Mariani M, Assadourian G, Lee A, van Sluyter SC, Haynes PA (2011) Less label, more free: approaches in label-free quantitative mass spectrometry. Proteomics 11(4):535–553

24. Mosley AL, Florens L, Wen Z, Washburn MP (2009) A label free quantitative proteomic analysis of the *Saccharomyces cerevisiae* nucleus. J Proteomics 72(1):110–120

25. Renvoise M, Bonhomme L, Davanture M, Valot B, Zivy M, Lemaire C (2014) Quantitative variations of the mitochondrial proteome and phosphoproteome during fermentative and respiratory growth in *Saccharomyces cerevisiae*. J Proteomics 106:140–150

26. Mosley AL, Sardiu ME, Pattenden SG, Workman JL, Florens L, Washburn MP (2011) Highly reproducible label free quantitative proteomic analysis of RNA polymerase complexes. Mol Cell Proteomics 10(2):M110 000687

27. Bondarenko PV, Chelius D, Shaler TA (2002) Identification and relative quantitation of protein mixtures by enzymatic digestion followed by capillary reversed-phase liquid chromatography-tandem mass spectrometry. Anal Chem 74(18):4741–4749

28. Tu C, Li J, Sheng Q, Zhang M, Qu J (2014) Systematic assessment of survey scan and MS2-based abundance strategies for label-free quantitative proteomics using high-resolution MS data. J Proteome Res 13(4):2069–2079

29. Neilson KA, Keighley T, Pascovici D, Cooke B, Haynes PA (2013) Label-free quantitative shotgun proteomics using normalized spectral

abundance factors. Methods Mol Biol 1002:205–222

30. Old WM, Meyer-Arendt K, Aveline-Wolf L, Pierce KG, Mendoza A, Sevinsky JR, Resing KA, Ahn NG (2005) Comparison of label-free methods for quantifying human proteins by shotgun proteomics. Mol Cell Proteomics 4(10):1487–1502

31. Liu K, Zhang J, Wang J, Zhao L, Peng X, Jia W, Ying W, Zhu Y, Xie H, He F, Qian X (2009) Relationship between sample loading amount and peptide identification and its effects on quantitative proteomics. Anal Chem 81(4):1307–1314

32. Glatter T, Ludwig C, Ahrne E, Aebersold R, Heck AJ, Schmidt A (2012) Large-scale quantitative assessment of different in-solution protein digestion protocols reveals superior cleavage efficiency of tandem Lys-C/trypsin proteolysis over trypsin digestion. J Proteome Res 11(11):5145–5156

33. Wisniewski JR, Zougman A, Nagaraj N, Mann M (2009) Universal sample preparation method for proteome analysis. Nat Methods 6(5):359–362

34. Vandenbogaert M, Li-Thiao-Te S, Kaltenbach HM, Zhang R, Aittokallio T, Schwikowski B (2008) Alignment of LC-MS images, with applications to biomarker discovery and protein identification. Proteomics 8(4):650–672

35. Podwojski K, Eisenacher M, Kohl M, Turewicz M, Meyer HE, Rahnenfuhrer J, Stephan C (2010) Peek a peak: a glance at statistics for quantitative label-free proteomics. Expert Rev Proteomics 7(2):249–261

36. Cox J, Mann M (2008) MaxQuant enables high peptide identification rates, individualized p.p.b.-range mass accuracies and proteome-wide protein quantification. Nat Biotechnol 26(12):1367–1372

37. Nahnsen S, Bielow C, Reinert K, Kohlbacher O (2013) Tools for label-free peptide quantification. Mol Cell Proteomics 12(3):549–556

38. Sandin M, Teleman J, Malmstrom J, Levander F (2014) Data processing methods and quality control strategies for label-free LC-MS protein quantification. Biochim Biophys Acta 1844(1 Pt A):29–41

39. Tsou CC, Tsai CF, Tsui YH, Sudhir PR, Wang YT, Chen YJ, Chen JY, Sung TY, Hsu WL (2010) IDEAL-Q, an automated tool for label-free quantitation analysis using an efficient peptide alignment approach and spectral data validation. Mol Cell Proteomics 9(1):131–144

40. Zhang W, Zhang J, Xu C, Li N, Liu H, Ma J, Zhu Y, Xie H (2012) LFQuant: a label-free fast quantitative analysis tool for high-resolution LC-MS/MS proteomics data. Proteomics 12(23–24):3475–3484

41. Zhang Y, Wen Z, Washburn MP, Florens L (2010) Refinements to label free proteome quantitation: how to deal with peptides shared by multiple proteins. Anal Chem 82(6):2272–2281

42. Ahrne E, Molzahn L, Glatter T, Schmidt A (2013) Critical assessment of proteome-wide label-free absolute abundance estimation strategies. Proteomics 13(17):2567–2578

43. Horvath A, Riezman H (1994) Rapid protein extraction from *Saccharomyces cerevisiae*. Yeast (Chichester, England) 10(10):1305–1310

44. Kulak NA, Pichler G, Paron I, Nagaraj N, Mann M (2014) Minimal, encapsulated proteomic-sample processing applied to copy-number estimation in eukaryotic cells. Nat Methods 11(3):319–324

Chapter 17

Profiling of Yeast Lipids by Shotgun Lipidomics

Christian Klose and Kirill Tarasov

Abstract

Lipidomics is a rapidly growing technology for identification and quantification of a variety of cellular lipid molecules. Following the successful development and application of functional genomic technologies in yeast *Saccharomyces cerevisiae*, we witness a recent expansion of lipidomics applications in this model organism. The applications include detailed characterization of the yeast lipidome as well as screening for perturbed lipid phenotypes across hundreds of yeast gene deletion mutants. In this chapter, we describe sample handling, mass spectrometry, and bioinformatics methods developed for yeast lipidomics studies.

Key words Yeast, Lipidomics, Mass spectrometry, Lipids

1 Introduction

The yeast *Saccharomyces cerevisiae* is a widely used model organism for the study of eukaryotic cell biology, biochemistry, and metabolism. More specifically, studies in yeast have revealed important insights into the regulation of storage lipid metabolism, sphingolipid homeostasis, cell cycle-dependent fine tuning of lipid biosynthesis as well as sphingolipid–sterol interactions within the cellular membranes [1–6]. Generalization of these findings is facilitated by the similarity of the yeast lipid composition with those of higher organisms such as mammals. The major glycerophospholipid and glycerolipid classes present in mammalian systems can be found in yeast as well: phosphatidic acid (PA), phosphatidylcholine (PC), phosphatidylethanolamine (PE), phosphatidylinositol (PI), phosphatidylserine (PS), diacylglycerol (DAG), phosphatidylglycerol (PG), and their respective lyso-derivatives, and cardiolipin (CL). Triacylglycerols (TAG) and sterol esters (SE) serve as storage lipids. The fatty acid composition of these lipid classes is fairly simple. The major fatty acids are palmitic (C16:0), palmitoleic (C16:1), and oleic acid (C18:1) [7]. Given the commonalities with respect to storage lipids and glycerophospholipids, yeast cells are special with respect to their sterol and sphingolipid composition. Firstly,

Frédéric Devaux (ed.), *Yeast Functional Genomics: Methods and Protocols*, Methods in Molecular Biology, vol. 1361,
DOI 10.1007/978-1-4939-3079-1_17, © Springer Science+Business Media New York 2016

the major yeast sterol is ergosterol (instead of cholesterol found in mammalian organisms). Secondly, the yeast sphingolipids (SP) consist of inositolphosphorylceramide (IPC), mannosyl-inositol phosphorylceramide (MIPC) and mannosyl-di-(inositolphosphoryl) ceramide (M(IP)2C) with a phytoceramide backbone. A third peculiarity of the yeast lipidome is the presence of a C26 very long chain fatty acid in the phytoceramide backbone, which is mostly alpha-hydroxylated. Taken together, the yeast lipidome consists of several hundred lipid molecules and is hence not much less complex than its mammalian counterparts. The recent advances in mass spectrometry-based lipid analysis have enabled a fast, comprehensive and quantitative assessment of cellular lipidomes within a single experiment and constitute a means for an advanced, systems level understanding of lipid-related metabolic processes, pathways and networks [8]. The aim of this chapter is, to provide yeast biologists with a protocol for conducting lipidomics experiments. We hope to give enough details to facilitate adaptation of lipidomics by a broader scientific audience, some of which may sound trivial for analytic chemists. However, it is clear that yeast biologists will have to resort to collaboration with an experienced mass spectrometrist in order to perform the analysis. In addition, an experienced statistician/data analyst is almost unavoidable to be able to handle and analyze large data sets (>50 samples).

The experimental protocols provided here are based on the Shotgun Lipidomics methodology as described and exploited in refs. [6, 9–12, 13]. Shotgun Lipidomics means the direct infusion of a lipid extract into the mass spectrometer, without chromatographic separation of the lipid classes [14, 15]. This reduces analysis time and simplifies data analysis because the temporal dimension (i.e., elution time) is not required for lipid identification. Since the protocols have evolved and improved over the years, we present here a robust and reproducible approach that should enable everyone equipped with the necessary instrumentation to obtain reasonable data from a lipidomic experiment.

In the present chapter, we provide a protocol for low throughput yeast lipidomics because this method may be used for a broad range of applications. It is suitable for processing 20–50 samples per day by means of manual lipid extraction. Robotic extraction can increase the throughput to about 100 samples per day (*see* **Note 1**).

2   Materials

2.1   Chemicals

1. ABC: 150 mM ammonium bicarbonate (NH_4HCO_3) pH 8.

2. MSmix MA: 0.05 % methylamine in chloroform–methanol 1:5 (v/v).

3. MSmix AA: 7.5 mM ammonium acetate in chloroform–methanol–propanol 1:2:4 (v/v).

4. C/M 15:1: chloroform–methanol 15:1 (v/v).

5. C/M 2:1: chloroform–methanol 2:1 (v/v).

6. Acetylchloride/chlorofom 1:5 (v/v).
All chemicals should be of LC-MS grade.

2.2 Internal Standard Mix

Internal standards used are listed in Table 1.

2.3 Software

1. LipidXplorer: (https://wiki.mpi-cbg.de/wiki/lipidx/index.php/Main_Page).

2. Alex Software: (www.msLipidomics.info).

3. MSConvert: (http://proteowizard.sourceforge.net).

4. Tableau Software (http://www.tableau.com/).

5. Orange Data Mining (http://orange.biolab.si/).

6. SAS Enterprise Guide (SAS Institute).

3 Methods

3.1 Culturing of Yeast Cells for Lipidomics Analysis

Yeast cells should be grown under conditions that will obviously depend on the objective of the given experiment. Growth temperature, media and supplements depend, among others, on the experimental design, strain background and genotype. Therefore, no specific guidelines for culturing conditions will be given, only a few general remarks that might help to obtain reproducible results (*see* **Note 2**):

1. Inoculate experimental cultures to an OD600 = 0.2.

2. Culture volume should be at least 20 ml in a flask that allows for proper shaking of the culture (for 20 ml cultures we usually use 100 ml flasks).

3. The growth curves of the investigated strains under the experimental conditions should be determined prior to running the lipidomics experiment. Decide for a growth stage that suits your experimental needs.

The last point is of particular importance. It is to avoid artifacts in the lipid compositions that are simply due to differences in growth rates (*see* **Note 2**).

3.2 Harvest and Lysis

Perform all steps at 4 °C or on ice and try to minimize the time it takes to freeze or extract the lysate (*see* **Note 3**).

1. After growth, spin down yeast cells at 5000 × g for 3 min.

2. Wash cells twice in 1 volume of ABC and spin down for 3 min at 5000 × g.

Table 1
Lipid classes and the corresponding internal standards

Lipid class	Internal standard name	Typical amount	Extraction phase	Ionization	Adduct	Mass range	Commercially available
Cer	Cer 35:1:2 (18:1;2/17:0;0)	20	15:1	Positive	H+	500–1000	Yes
DAG	DAG 34:0 (17:0/17:0)	30	15:1	Positive	NH4+	500–1000	Yes
Erg	Stigmastatrienol	150	15:1	positive	H+	400–500	No
TAG	TAG 51:0 (17:0/17:0/17:0)	60	15:1	Positive	NH4+	500–1000	Yes
LPC	LPC 17:1	10	15:1	Negative	Ac–	400–650	Yes
LPE	LPE 17:1	20	15:1	Negative	H–	400–650	Yes
PC	PC 31:1 (17:0/14:1)	50	15:1	Negative	Ac–	500–1200	Yes
PE	PE 31:1 (17:0/14:1)	50	15:1	Negative	H–	500–1200	Yes
PG	PG 31:1(17:0/14:1)	40	15:1	Negative	H–	500–1200	Yes
CL	CL 61:1 (15:0(3)-16:1)	40	2:1	Negative	2H–	500–1200	Yes
IPC	IPC 44:0;2	50	2:1	Negative	H–	500–1200	No
LPA	LPA 17:0	20	2:1	Negative	H–	400–650	Yes
LPI	LPI 17:1	20	2:1	Negative	H–	400-650	Yes
LPS	LPS 17:1	20	2:1	Negative	H–	400–650	Yes
M(IP)2C	M(IP)2C 44:0;2	70	2:1	Negative	2H–	500–1200	No
MIPC	MIPC 44:0;2	70	2:1	Negative	H–	500–1200	No
PA	PA 31:1 (17:0/14:1)	40	2:1	Negative	H–	500–1200	Yes
PI	PI 31:1 (17:0/14:1)	60	2:1	Negative	H–	500–1200	Yes
PS	PS 31:1 (17:0/14:1)	40	2:1	Negative	H–	500–1200	Yes

For some lipid classes, internal standards are not commercially available and have thus to be synthesized or purified on demand. *H+* protonated ion, *NH4+* ammonium adduct, *Ac–* acetate adduct, *H–* deprotonated ion, *2H–* doubly deprotonated ion

3. After the last wash, the pellet can be snap-frozen in liquid nitrogen and stored at –80 °C or processed immediately.

4. Resuspend (thaw before if applicable) pellet in a volume of ABC to reach a concentration of 20 OD600 nm units/ml.

5. Transfer 1 ml of suspension to fresh 2 ml eppendorf tube.

6. Lyse cells with 200 μl of glass or zirconia beads (0.5 mm diameter) using vortex cell disruptor, cell lyser or similar equipment for 10 min, 4 °C.

7. Lysed cells (including the glass beads) can be snap-frozen in liquid nitrogen and stored at –80 °C or processed immediately.

3.3 Internal Standard Mix

A key feature of quantitative mass spectrometry-based lipidomics is the inclusion of lipid class-specific internal standards. There are to be added *prior* to lipid extraction. In that way, they allow for the normalization of differences in extraction efficiency and ionization behavior of the different lipid classes. Alternative methods of lipid quantification based on relative normalization are discussed below.

Suitable internal standards fulfill the following criteria (*see* **Note 4**):

1. They should be lipid class-specific, i.e., have the same headgroup as the lipids to be quantified (the endogenous lipids).

2. They should have a molecular mass different from the endogenous lipids. This is usually achieved by using lipids labeled with stable isotopes (deuterium or C13) or containing "nonnatural" (combinations of) fatty acids.

3. They should have a molecular mass so similar to the endogenous lipids that they can be measured in the same mass range.

4. They should be added in amounts that ensure that their signal is well above noise and comparable to the intensity of the lipids to be analyzed. The exact value depends on the type of mass spectrometer used and should be specified by appropriate dynamic range experiments.

Table 1 summarizes suggestions for lipid standards and their amounts to be used in a typical lipidomics experiment for yeast samples. Additionally, it contains information regarding ionization mode of the lipid classes and the adducts these classes form.

For most lipid classes, internal lipid standards are commercially available. However, there is a subset of yeast-specific lipids, for which no standards are available on the market. These are the major yeast sterol ergosterol, the corresponding ergosterol esters, and the inositol-containing complex yeast sphingolipids. The internal standard lipid for ergosterol, stigmastatrienol, can be obtained by de-acetylation of commercially available stigmastatrienyl-acetate as described in ref. [9]. The complex yeast sphingolipids can be purified from crude yeast lipid extracts by preparative thin layer chromatography or liquid chromatography as described in refs. [9, 16, 17]. Here it is important to note that the complex yeast sphingolipids should be purified from a *sur2scs7Δ* strain. This strain produces sphingolipids with 1 free hydroxyl group only, while a wild-type strain produces mainly phytocer`amide-based sphingolipids with an alpha-OH at the fatty acid moiety. The lack of these free OH groups in the *sur2scs7Δ* strain provides the required mass shift that makes those lipids suitable as internal standards. However, care must be taken when working with mutants that result in differences in the hydroxylation pattern of sphingolipids. In that case, alternative ways of normalization have to be considered.

3.4 Two-Step Lipid Extraction

1. Take volume of yeast lysate that corresponds to 0.2 ODu of yeast cells.

2. Add ABC to a final volume of 200 μl.

3. Add Internal Standard mix.

4. Mix 10 min, 1400 rpm, 4 °C.

5. Add 1000 μl C/M 15:1.

6. Extract for 120 min at 1400 rpm, 4 °C.

7. Spin 3 min, $3000 \times g$.

8. Collect organic (lower) phase (ca. 800 μl).

9. Split extract: ca. 300 μl for acetylation and rest for normal MS.

10. Re-extract aqueous phase with 1000 μl C/M 2:1 for 120 min at 1400 rpm, 4 °C.

11. Spin 3 min, $3000 \times g$.

12. Collect organic (lower) phase (ca. 800 μl).

13. Dry the lipids in a vacuum concentrator ("speed vac") or desiccator.

14. C/M 15:1 extract resuspended in 100 μl MS mix AA.

15. C/M 2:1 extract resuspended in 100 μl MS mix MA.

After resuspension, the extracts are ready for direct infusion into the mass spectrometer (see below).

3.5 Sterol Derivatization

While most of the yeast lipid classes can be analyzed by mass spectrometry directly from the extract, ergosterol requires derivatization in order to achieve acceptable signal-to-noise ratios. There are two principal methods for sterol derivatization for mass spectrometric analysis: sulfation [18] and acetylation [19]. For us, acetylation has proven more reliable and robust and does not require the determination of response factors for ergosterol vs. its internal standard stigmastatrienol [9]. In addition, due to its simplicity in terms of handling, it is highly suitable for a high sample throughput.

For the chemical acetylation of sterol:

1. Dry 1/3 of the (ca. 300 μl) volume of the 15:1 extract in a vacuum concentrator.

2. Resuspend the dried lipid film in acetylchloride/chloroform 1:5 (v/v).

3. Incubate for 1 h at room temperature.

4. Dry the reaction mix in an desiccator (do not use a vacuum concentrator as acetylchloride is a very harsh chemical that will cause corrosion of metallic surfaces!).

5. Resuspend the dried lipids in 40 μl of MS mix AA by shaking for 3 min at 1400 rpm.

The suspension is ready to use for infusion into the mass spectrometer in positive ion mode.

3.6 Sample Infusion and Mass Spectrometry

Sample infusion and mass spectrometric acquisition described in the next subsections are interlinked processes, which are performed simultaneously.

3.6.1 Sample Infusion with a TriVersa NanoMate

In our laboratories, sample infusion in a shotgun lipidomics experiment is usually carried out with a TriVersa NanoMate nano-ESI source (Advion). It features the following advantages:

1. Automated sample infusion in a 96-well format.

2. Disposable tips, thus no sample carry over and cross contamination.

3. Robustness: spray failure rates well below 5 %.

4. The polarity switch option allows for the acquisition of both positive and negative ion mode spectra within one and the same infusion event [20].

5. Flow rates in the nanoliter range ensure high sensitivity and low background signals.

Essentially two settings are required for the execution of a lipidomics experiment of the kind described here: one for positive and one for negative ion mode. The settings shown in Table 2 are a good starting point and should yield stable spray. However, further optimizations may be required in order to achieve optimal performance of individual Nanomate devices.

The parameters shown here are the "core settings". Many other parameters are available whose values/settings depend on the actual needs and preference of the experimenter. In principle, any other ESI source should work as well.

3.6.2 Mass Spectrometry with LTQ Orbitrap or Q-Exactive

A straightforward and efficient way of analyzing yeast lipid extracts quantitatively is high-resolution mass spectrometry. In our laboratories we usually use Orbitrap technology in the form of LTQ

Table 2
Settings for sample infusion with TriVersa NanoMate

125.5 pt	Negative ionization	Positive ionization
Backpressure (psi)	0.5	1.25
Ionization voltage (kV)	0.95	0.95
Sample volume (μl)	10	10

Orbitrap, LTQ Velos or Q-Exactive mass spectrometers from Thermo Fisher Scientific. These devices provide resolutions >=100,000 (FWHM) and therefore reduce potential overlaps of isobaric lipid species substantially. Due to the rather simple lipid composition of yeast cells, it is usually sufficient to acquire spectra in full MS mode, i.e., without fragmentation of the lipid molecules. This is a timesaving approach both because it reduces MS time and time for lipid identification in MS/MS spectra, which is computationally demanding. Therefore, our suggestion is to confine oneself to full MS analysis for initial screening for lipid phenotypes. A more detailed structural analysis of interesting hits may follow in the framework of a secondary screen. Mass spectra should be acquired [13]:

1. Using the highest possible target mass resolution (e.g., 100,000 or higher).

2. With an automated gain control value of 1e6 (this value refers to the number of ions of injected into the Orbitrap mass analyzer per scan).

3. And a maximum injection time of 250 ms.

These settings are applicable for both positive and negative ion mode. Here, they may serve as a starting point for establishing an optimized acquisition method. Together with the information provided in Table 1, these settings should enable an experienced mass spectrometrist to not only set up a method on an Orbitrap device, but also to implement appropriate methods for any other type of mass spectrometer. Alternatives to full high-resolution FT-MS scans are, among others, data-dependent acquisitions (DDA), precursor ion scans (PIS), neutral loss scans (NLS) or multiple reaction monitoring (MRM). These approaches deliver more detailed information as they include the fragmentation of lipid molecules. The advantages, however, of high-resolution full MS are: short acquisition times (less than 5 min per ion mode and extract) and rather simple data structure: spectra can be exported as two-dimensional peak lists and subjected to straightforward lipid identification algorithms.

3.7 Raw Data Processing, Lipid Identification and Quantification

3.7.1 Raw Data Handling

1. Results of lipidomics experiments are first gathered as raw tables of molecular masses and peak intensities (peak heights and areas) in each sample. Vendors of the mass spectrometers provide the software for extracting such raw data for further processing (XCalibur for .raw files of ThermoFisher instruments and Analyst for .wiff files of AppliedBiosystems Sciex instruments). Alternatively, the vendor specific output files can be converted into an open source format (mzXML, mzML) by the ProteoWizard MSConvert software (http://proteowizard.sourceforge.net/).

2. Next, raw data from the specific vendor's analytical software are exported as text, csv, or Excel files for further processing.

3.7.2 Lipid Identification

1. Lipid identification is performed by matching m/z values of detected ion peaks obtained in MS and MS/MS scans with theoretical lipid masses (see "Comprehensive Classification System for Lipids" developed by LIPID MAPS Consortium [21] and LipidHome database [22]) with the help of pregenerated target lists (Alex software [23]) or using a molecular query language (LipidXplorer [24]). Alex software is able to process proprietary mass spectral data files while LipidXplorer handles data in the mzXML format. Details about various software packages for lipidomic data handling are available in refs. [25] and [26].

2. An important parameter for the lipid identification is an m/z tolerance window. This value is dependent on mass resolution of a particular instrument. For example, data acquired with FT MS scan with a target resolution at 100,000 is typically processed with an m/z tolerance window set to ±0.0020 amu.

3. Potential calibration drifts can be compensated using Alex software by a lock mass adjustment calculated as an m/z calibration offset based on defined ubiquitous mass ions that were well characterized in previous experiments. Even though the uniform adjustment for all samples is possible, an automatic sample specific approach yields more accurate results, because the effect of a calibration drift might be not constant across the samples. We previously demonstrated that the addition of the automated sample specific lock mass adjustment step to the data processing routine improves the accuracy of the identification of endogenous lipid species by decreasing the lipid detection mass error from 1 to 0.4 ppm.

4. Commonly, lipidomics data are normalized using one internal standard (stable-isotope labeled or an unnatural lipid molecule) per lipid class. Therefore, it is important to supply the lipid class information together with a lipid name of the identified peak. The lipid class information is used to define the appropriate internal standard for the normalization.

5. Defining additional information about lipid molecules at this stage, will help to interpret the results and perform relative normalization (described below). These parameters are referred to as lipidomics features and represent a condensed, yet fully informative, overview of the lipidome [11, 23]. The features include lipid categories consisting of several lipid classes (e.g., glycerophospholipid, sphingolipid, glycerolipid, sterol lipid, energy storage, membrane lipids), structural attributes (e.g., number of double bonds, fatty acid chain length), chemical formula, mono-isotopic mass, and isotope information.

3.7.3 Quantification   Quantification of lipid species can be absolute or relative. As described below, different normalization methods allow to analyze lipidomic changes from different angles. Therefore, it is advantageous to apply multiple normalizations on the same data and compare the results. However, if no internal standards were used for the experiment only relative quantification is possible.

1. In absolute quantification, lipid concentrations are calculated as endogenous lipid peak intensities divided by the peak intensities of the appropriate internal standards and adjusted for the amount of the spiked internal standard. The concentration is expressed as mol/ml of volume or mol/mg of dry weight or mol/mg of protein.

2. A relative concentration is calculated as the amount of a particular lipid divided by the sum of the amounts of all detected lipids. The resulting value is expressed as molar percentage. Molar percentage can else be calculated relative to a particular lipid class. This kind of normalization better reflects qualitative, class specific changes. Furthermore, such normalization is beneficial for minor lipid classes that would otherwise be greatly affected by quantitative changes in major, highly abundant lipid classes (e.g., TAGs and PCs).

3. Recently, intensity percentage normalization has been introduced that does not use internal standards at all. Instead, relative values are expressed as ratios of a peak intensity normalized to total peak intensities in a particular mass spectrometry scan [13]. The normalization based on the intensity percentages can be beneficial for a rapid profiling in a screening of a large number of samples to help identify samples with most prominent changes in lipid profiles that can be further validated with experiments involving absolute quantification.

4. Lipid profiles can be also presented by summing up the corresponding values of lipids with particular lipidomics features such as fatty acid length or number of double bonds as outlined above. Again, these parameters may be normalized to total lipid content or lipid class amount.

3.8 Quality Control and Data Filtering

3.8.1 Overall Signal Quality

The quality of experimental results can be greatly influenced by the overall spraying stability and instrument performance during analysis of a particular sample (*see* **Note 5**).

1. Plot total lipid intensities or concentrations of all lipids detected in a particular analytical mode for identifying samples with poor spraying. It is important to note, that low total intensities could be also due to biological properties of a particular sample.

2. Plot total lipid intensities vs. number of detected lipids in a sample for identifying technical outliers that should be removed from a dataset or reanalyzed (for example, samples with less than 70 % of detected lipids and intensities lower than 15 % of the average).

3.8.2 Background Filter

In lipidomic experiments, it is common to include blank controls that contain only buffers or buffers with internal standards.

1. Peak intensities found in these control samples, except for the internal standards if they were added, are subtracted from peak intensities of the endogenous lipids identified in the other samples.

2. It is also common to remove molecules from further analysis if corresponding peaks are highly abundant in the blank samples (e.g., peak intensity is 25 % or higher than a corresponding peak in a not blank sample).

3.8.3 Noise Filter

Similarly, peaks with intensities lower than a certain threshold value empirically defined are filtered out. This value may be defined as a multiple of the spectral noise. The threshold depends on sample properties (i.e., matrix effects) and on the instrumentation used and should be defined by the determination of the limits of quantification (LOQ) for each lipid class.

3.8.4 Coefficient of Variation and Outliers

Lipid abundances (absolute or relative) can be evaluated for reproducibility of the quantification.

1. The variation of lipid quantification should be addressed by calculating a coefficient of variation (CV), which is a ratio of a standard deviation of a lipid abundance and a mean lipid abundance multiplied by 100 to express the value as percentage. The CV value is calculated for each lipid in a set of replicated quality control samples (*see* **Note 6**).

2. Lipid species with high CV values should be filtered out. A CV of up to 25 % may be acceptable (*see* **Note 7**). In analyses that include internal standards, the CV of the quality control samples should not exceed 15 %. CV values in the experimental samples will be higher than in controls because the resulting values also

reflect the biological differences between the samples. However, experimental errors can be identified by screening for high CV values within a set of biological replicates.

3. Identify outliers by calculating differences of lipid abundances in a particular sample from the corresponding mean or median values. Thresholds for outlier removal are determined empirically and could be selected to reject lipids with threefold higher or lower abundances than mean or median values of that lipid in a set of biological replicates.

3.8.5 Filtering Based on Reproducibility of Lipid Detection

Counting how many times a lipid species is detected in a group of samples or within a certain number of technical replicates is another quality control measurement of a reliable lipid detection. Lipids that are detected in only one out of three replicates or in less than 75 % of a large number of replicates or samples of a particular group are generally excluded from further analyses (*see* **Note 8**).

3.8.6 Batch Corrections

Development of high-throughput lipidomics workflows allows for screening lipidomic profiles in hundreds of samples. In such screens, it is important to evaluate the sample quality and quantification between different batches of samples and samples from the same batch injected at different times. Nonbiological systematic differences between samples could occur because of use of different batches of chemicals and standards, sample degradation, and accumulation of contaminants during a long analytical run.

1. The presence of a batch bias is investigated by comparing means and medians of lipid species abundances between the plates in quality control samples. Graphical representation of lipid abundances detected in ordered samples between different batches are very illustrative for identifying the bias. Furthermore, statistical testing for significance of difference between lipid species abundances means (*t*-test) and medians (Mann–Whitney *U*-test) can be employed to detect the batch effects automatically.

2. Systematic differences between batches should be corrected before further statistical data analyses. A simple method for correcting for a batch effect is Ratio-Based Calibration Method, which is based on multiplication of an abundance value of a lipid in a sample from a particular batch by a factor calculated as a ratio of a mean or median abundance of that lipid in control samples of that batch to a mean or median calculated for the whole set of the controls [27]. We routinely use this method because of its simplicity and ease of calculations and interpretation. It is important to perform the correction separately for each lipid and not on total lipid intensities or internal standard signals because batch effects affect differently particular lipid species. It should be noted that a variety of other

simple as well as computationally more sophisticated methods exist for correcting for the batch effect [28]. In order to decide which one is suitable, it is beneficial to apply several and compare the results to choose a method that works better in a particular situation.

3.8.7 Data Processing Tools

The speed and quality of data analysis of novel lipidomics workflows can be facilitated by employing modern visual programming and visual analytics software systems [13, 23]. Visual programming systems, such as SAS Enterprise Guide (SAS institute) and Orange Data Mining [29] provide intuitive wizards for creating independent data analysis tasks that can be connected into executable visual data processing workflows. A major advantage of such systems is the availability of powerful data processing tools that can be used by researchers without programming expertise. In addition, such systems keep transparency of data processing steps that can be easily shared between researches and allow to perform modifications to particular steps without the need for reprogramming other parts of the system.

Visual analytics platform, such as Tableau Software (http://www.tableau.com) and Tibco Spotfire (http://spotfire.tibco.com/) provide an opportunity to create highly customized and interactive graphics that are best suited for particular datasets without the need for programming. A broad selection of available graphical data representations are particularly useful in discovery experiments where data exploration is the main task. Graphical views can be easily linked with each other, which allows connecting various statistical results with raw experimental data (e.g., raw intensities, used internal standards, sample amounts).

4 Notes

1. Scaling up lipidomics experiments is dependent on automation of cell growth and sample extraction procedures. Described sample injection method relies on TriVersa NanoMate capable of handling a plate with 96 samples automatically.

2. Lipidomics provides for an unprecedented precision, sensitivity and coverage in the quantitative analysis of lipids. Hence, mass spectrometry will detect even subtle changes in lipid compositions in a reproducible way. This in turn will yield differences between samples that are statistically significant but may not be of any biological importance. Therefore, it is crucial to tightly control for experimental and culture conditions.

   It has been shown that culture conditions can have a tremendous effect on the lipid composition of yeast cells [11]. This flexibility of the yeast lipidome is not confined to certain lipid classes, but may affect any lipid class.

Even subtle differences in growth phases can strongly affect the lipid composition, e.g., a prominent increase in TAGs. Given these peculiarities of the yeast lipidome, it may be advisable to determine the lipid composition of the mutant strains under investigation under different growth conditions and growth phases. A weak phenotype in one condition might be stronger in another. The failure to accumulate TAGs at later growth phases might be a phenotype not detected in logarithmic phase and therefore provides for an illustrative example for the necessity of determining lipidomes under different conditions or growth phases. An unsupervised clustering analysis should be used to reveal groups of samples with such systematic differences.

3. The extraction of a volume lysate that corresponds to 0.2 ODu of yeast cells should result in about 3000–5000 pmol of lipids per extraction. Increasing the sample amount in order to achieve a broader coverage by extending the spectrum to low abundant lipid classes is usually not successful as ion suppression and matrix effects will actually reduce sensitivity and coverage.

4. Internal standards should be validated gravimetrically, by phosphorus assay [30] for phospholipids and/or against an independently validated standard, if applicable.

5. Order of quality control procedures. For absolute normalization is does not really matter in which order the filtering steps are performed. However, the relative quantification molar percentages are dependent on all lipid species in the dataset. Therefore, it is important to complete all the filtering steps before making the calculations of molar percentages. The CV values typically are lower for normalized values than for raw intensity values because lipid normalization to internal standard or other lipid compensates for analytical variation, such as spraying instability.

6. We distinguish between biological and technical replicates. Biological replicates control for biological variation among the samples. Technical replicates control for technical variation. In the case of an entire lipidomics experiment, this is the variation introduced by sample preparation (cell lysis and extraction), preparation of internal lipid standards for normalization, and measurement of the sample by means of mass spectrometry. Extracting and measuring the same biological sample multiple times therefore helps assessing technical variation.

A meaningful reference sample should be an aliquoted yeast lysate that is extracted and measured multiple times (at least twice) together with the actual set of samples. Calculation of the coefficient of variation will provide a measure for technical variation within each run. When normalizing to lipid class-specific internal standards, the technical variation should be around 10 % (based on lipid class abundances).

In our experience, technical replicates for all samples in an experiment are not required as long as: technical variation (i.e., precision) of the lipidomics workflow was assessed during method development; and reference samples are included in the sample set. Typically, most of the variation is introduced by biological variability. Therefore, biological triplicates are recommended.

7. High CV values are typical for low abundant lipids. Plots of CV values vs. lipid abundance are helpful to verify if this is the case. Such plots will identify high abundant lipids that are not reliably detected and highlight a threshold of a reliable quantification. These parameters are important for optimization of the analytical methods.

8. For data filtering based on reproducibility of lipid detection, lipids should be counted in specific groups of samples because some sample types might not express particular lipids that could be a valuable biological observation.

Acknowledgements

The authors would like to thank Kai Simons for critically reading the manuscript. C.K. acknowledges fruitful discussions with Julio L. Sampaio and Michal A. Surma.

References

1. Breslow DK, Collins SR, Bodenmiller B et al (2010) Orm family proteins mediate sphingolipid homeostasis. Nature 463(7284):1048–1053

2. De Smet CH, Vittone E, Scherer M et al (2012) The yeast acyltransferase Sct1p regulates fatty acid desaturation by competing with the desaturase Ole1p. Mol Biol Cell 23(7): 1146–1156

3. Guan XL, Souza CM, Pichler H et al (2009) Functional interactions between sphingolipids and sterols in biological membranes regulating cell physiology. Mol Biol Cell 20(7):2083–2095

4. Kohlwein SD (2010) Triacylglycerol homeostasis: insights from yeast. J Biol Chem 285(21):15663–15667

5. Kurat CF, Wolinski H, Petschnigg J et al (2009) Cdk1/Cdc28-dependent activation of the major triacylglycerol lipase Tgl4 in yeast links lipolysis to cell-cycle progression. Mol Cell 33(1):53–63

6. Surma MA, Klose C, Peng D et al (2013) A lipid E-MAP identifies Ubx2 as a critical regulator of lipid saturation and lipid bilayer stress. Mol Cell 51(4):519–530

7. Daum G, Lees ND, Bard M et al (1998) Biochemistry, cell biology and molecular biology of lipids of Saccharomyces cerevisiae. Yeast 14(16):1471–1510

8. Han X, Yang K, Gross RW (2012) Multidimensional mass spectrometry-based shotgun lipidomics and novel strategies for lipidomic analyses. Mass Spectrom Rev 31(1):134–178

9. Ejsing CS, Sampaio JL, Surendranath V et al (2009) Global analysis of the yeast lipidome by quantitative shotgun mass spectrometry. Proc Natl Acad Sci U S A 106(7):2136–2141

10. Klemm RW, Ejsing CS, Surma MA et al (2009) Segregation of sphingolipids and sterols during formation of secretory vesicles at the trans-Golgi network. J Cell Biol 185(4):601–612

11. Klose C, Surma MA, Gerl MJ et al (2012) Flexibility of a eukaryotic lipidome—insights from yeast lipidomics. PLoS One 7(4), e35063

12. Surma MA, Klose C, Klemm RW et al (2011) Generic sorting of raft lipids into secretory vesicles in yeast. Traffic 12(9):1139–1147

13. Tarasov K, Stefanko A, Casanovas A et al (2014) High-content screening of yeast mutant libraries by shotgun lipidomics. Mol Biosyst 10(6):1364–1376

14. Shevchenko A, Simons K (2010) Lipidomics: coming to grips with lipid diversity. Nat Rev Mol Cell Biol 11(8):593–598

15. Wang M, Han X (2014) Multidimensional mass spectrometry-based shotgun lipidomics. Methods Mol Biol 1198:203–220

16. Ejsing CS, Moehring T, Bahr U et al (2006) Collision-induced dissociation pathways of yeast sphingolipids and their molecular profiling in total lipid extracts: a study by quadrupole TOF and linear ion trap-orbitrap mass spectrometry. J Mass Spectrom 41(3): 372–389

17. Klose C, Ejsing CS, Garcia-Saez AJ et al (2010) Yeast lipids can phase-separate into micrometer-scale membrane domains. J Biol Chem 285(39):30224–30232

18. Sandhoff R, Brugger B, Jeckel D et al (1999) Determination of cholesterol at the low picomole level by nano-electrospray ionization tandem mass spectrometry. J Lipid Res 40(1): 126–132

19. Liebisch G, Binder M, Schifferer R et al (2006) High throughput quantification of cholesterol and cholesteryl ester by electrospray ionization tandem mass spectrometry (ESI-MS/MS). Biochim Biophys Acta 1761(1):121–128

20. Schuhmann K, Almeida R, Baumert M et al (2012) Shotgun lipidomics on a LTQ Orbitrap mass spectrometer by successive switching between acquisition polarity modes. J Mass Spectrom 47(1):96–104

21. Fahy E, Subramaniam S, Brown HA et al (2005) A comprehensive classification system for lipids. J Lipid Res 46(5):839–861

22. Foster JM, Moreno P, Fabregat A et al (2013) LipidHome: a database of theoretical lipids optimized for high throughput mass spectrometry lipidomics. PLoS One 8(5), e61951

23. Husen P, Tarasov K, Katafiasz M et al (2013) Analysis of lipid experiments (ALEX): a software framework for analysis of high-resolution shotgun lipidomics data. PLoS One 8(11), e79736

24. Herzog R, Schwudke D, Schuhmann K et al (2011) A novel informatics concept for high-throughput shotgun lipidomics based on the molecular fragmentation query language. Genome Biol 12(1):R8

25. Ejsing CS, Husen P, Tarasov K (2012) Lipid informatics: from a mass spectrum to interactomics. Lipidomics. Wiley, Weinheim, pp 147–174. doi:10.1002/9783527655946.ch8

26. Fahy E, Cotter D, Byrnes R et al (2007) Bioinformatics for lipidomics. Methods Enzymol 432:247–273

27. Kamleh MA, Ebbels TM, Spagou K et al (2012) Optimizing the use of quality control samples for signal drift correction in large-scale urine metabolic profiling studies. Anal Chem 84(6):2670–2677

28. Wang SY, Kuo CH, Tseng YJ (2013) Batch Normalizer: a fast total abundance regression calibration method to simultaneously adjust batch and injection order effects in liquid chromatography/time-of-flight mass spectrometry-based metabolomics data and comparison with current calibration methods. Anal Chem 85(2):1037–1046

29. Demšar J, Curk T, Erjavec A (2013) Orange: data mining toolbox in python. J Mach Learn Res 14:4

30. Bartlett GR (1959) Phosphorus assay in column chromatography. J Biol Chem 234(3): 466–468

Chapter 18

Identification of Links Between Cellular Pathways by Genetic Interaction Mapping (GIM)

Christophe Malabat and Cosmin Saveanu

Abstract

The yeast systematic deletion collection offered the basis for a number of different strategies that establish functional links between genes by analyzing the phenotype of cells that combine two different deletions or mutations. A distinguishing feature of the collection is the presence of molecular barcodes at each deleted locus, which can be used to quantify the presence and abundance of cells bearing a given allele in a complex mix. As a result, a large number of mutants can be tested in batch cultures, replacing tedious manipulation of thousands of individual strains with a barcode microarray readout. Barcode-based genetic screens like Genetic Interaction Mapping (GIM) thus require little investment in terms of specific equipment, are fast to perform, and allow precise measurements of double mutant growth rates for both aggravating (synthetic sick) and alleviating (epistatic) effects. We describe here protocols for preparing the pools of haploid double mutant *S. cerevisiae* cells, testing their composition with barcode microarrays, and analyzing the results to extract useful functional information.

Key words Genetic screen, Gene deletion, Growth rate, Barcode microarray, Data analysis

1 Introduction

Tests of the effect on growth rate for a large number of combinations of gene deletions became possible in the last decade through technical advances that arose after the completion of *S. cerevisiae* genome sequencing. An essential development in this respect has been the creation of a collection of systematic deletion for all the predicted yeast genes [1, 2]. Each cell of this collection bears specific short 20 nucleotide long sequences that uniquely identify the affected locus. These "molecular barcodes" allow the analysis of thousands of strains in parallel, to estimate growth of any single strain in a complex mixed population. The second most important advance in the field has been the invention of methods that would allow efficient production of thousands of double mutant strains. Two strategies were employed to this end. One, called SLAM (Synthetic Lethality Analysis with Microarrays), was based on

Frédéric Devaux (ed.), *Yeast Functional Genomics: Methods and Protocols*, Methods in Molecular Biology, vol. 1361,
DOI 10.1007/978-1-4939-3079-1_18, © Springer Science+Business Media New York 2016

massive transformation of cells from the collection with DNA fragments that ensure the replacement of a gene of interest with a selection cassette [3, 4]. The other, which was employed in several laboratories under various names, is based on efficient mass mating [5] of haploid yeast cells, bringing together two mutations of interest. The development of a method for efficient selection of double mutant haploid strains issued through meiosis from these diploids was crucial to the ability to obtain and test a large number of mutation combinations in parallel. Both Synthetic Genetic Array analysis (*SGA*, 6, 7) and Genetic Interaction Mapping (*GIM*, 8) use this approach. Quantitative tests of the extent by which two concomitant mutations affect growth rate in comparison with each single mutation present in a yeast strain, were pioneered in Epistatic Miniarray (E-MAP) studies [9], and used in SGA [10] and GIM [8] screens. The GIM method, presented here, combines the efficient generation of haploid double mutant strains through mating and sporulation with a direct quantitative test of strains growth based on DNA microarrays.

Genetic screens of the type described above can be extended to other collections of mutant strains. For example, substituting the normal 3′ UTR region of a gene with a long 3′ UTR destabilize mRNA through nonsense-mediated mRNA decay (*NMD*, 11). The reduction of the amounts of proteins allowed testing of phenotypes for some of the yeast essential genes [9]. Our recent analysis of a collection of such strains indicate, however, that this strategy is biased and cannot guarantee that the obtained strains show a phenotype, since NMD shows preference for mRNAs with short coding sequences [12]. The study of phenotypes for essential gene mutants can also be done with collections of thermosensitive yeast strains [10].

The details provided in the present chapter should be helpful to set up and perform reproducible GIM screens and identify either direct genetic interactions or specific interaction profiles that indicate which cellular pathways are required for the cells to adapt to the absence of a given gene. Our experience with large-scale quantitative data indicates that crucial to the identification of patterns or interesting correlation is easy visual data representation. Several software solutions answer this question including Mayday [13], geWorkBench [14], or Perseus [15].

In the end, GIM screen results are only as good as the capacity of the experimenter to deduce useful information from them. In most of the cases, the screens are only the starting point for projects that need years of focused work until completion. Thus, such screens are most of the time a discovery tool that provides first hints for focused analyses in functional genomics.

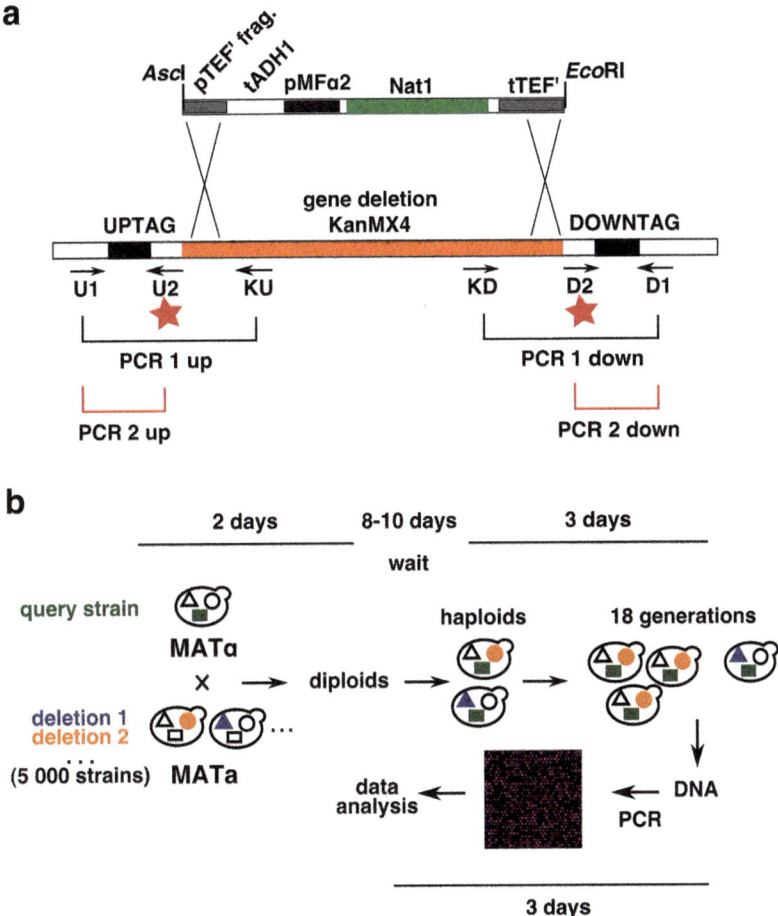

Fig. 1 (a) The haploid-specific nourseothricin resistance cassette (generated by digestion of pGID3 with *AscI* and *Eco*RI) replaces the KanMX4 cassette from a *MATα* strain by homologous recombination with fragments of pTEF′ and tTEF′. The barcodes from the KanMX4 strains are amplified in two steps. The second step includes fluorescent labeled oligonucleotides (indicated by *red star*). (b) Timing for a typical GIM screen includes 2 days for the obtention of diploid strains and about 1 week for the selection, processing and microarray analysis of haploid double mutants. Eight to ten additional days are required for the sporulation step

The preparation of the query strain as well as the relative timing of a GIM screen is indicated in Fig. 1. We perform between 4 and 16 parallel screens in parallel and the limiting step is the selection and culture of the pool of haploid double mutants.

2   Materials

2.1 Growth Media, Antibiotics and Molecular Biology Reagents

1. YPD medium: 2 % D-glucose, 1 % Difco yeast extract, 1 % Difco Bacto peptone. Autoclave at 110 °C for 20 min.

2. GNA medium: 5 % D-glucose, 3 % Difco nutrient broth, 1 % Difco yeast extract. Autoclave at 110 °C for 20 min. Prepare fresh before use.

3. Sporulation medium: 1 % potassium acetate (10 % stock), 0.005 % zinc acetate (0.5 % stock), uracil (2 mg/100 ml, 0.2 % stock), histidine (2 mg/100 ml, 0.4 % stock), and leucine (6 mg/100 ml, 1.2 % stock in dH₂O). Resuspend the stocks in dH₂O and filter-sterilize.

4. Antibiotics: 0.2 mg/ml hygromycin B (stock 50 mg/ml); 20 µg/ml nourseothricin, clonNAT (Werner Bioagents, ref 5.1000, 200 mg/ml stock); 0.2 mg/ml G418 sulfate (100 mg/ml stock).

5. Synthetic complete medium without uracil: 6.7 g/L Yeast Nitrogen base without amino acids, with ammonium sulfate, 2 % glucose, and 0.2 % amino acids mix. The mix of amino acids is composed of alanine, arginine, aspartic acid, asparagine, cysteine, glutamic acid, glycine, histidine, isoleucine, leucine, lysine, methionine, phenylalanine, proline, serine, threonine, tryptophan, tyrosine, and valine, in equal weight proportion, with the exception of leucine, for which twice as much is added. In the same mix, add for 1 g of amino acid 0.5 g of adenine. Autoclave or filter-sterilize. For plates 2 % agar is used in the final medium composition.

6. Culture plates (96 deep wells, 2 ml), 96-long pin replicator.

7. Gas permeable and aluminum seal for 96-well plates.

8. NEB4 buffer: 50 mM potassium acetate, 20 mM Tris-acetate, pH 7.9 at 25 °C, 10 mM magnesium acetate, 1 mM dithiothreitol.

9. *Asc*I and *Eco*RI restriction enzymes. pGID3 plasmid.

2.2 Microarray Reagents

1. Oligonucleotides: **U1** 5′-GAT GTC CAC GAG GTC TCT; **KU** 5′-AAG AAG AAC CTC AGT GGC; **D1** 5′-CGG TGT CGG TCT CGT AG; **KD** 5′-GGA TCT TGC CAT CCT ATG; **U2block** 5′-CGT ACG CTG CAG GTC GAC; **D2block** 5′-ATC GAT GAA TTC GAG CTC; **U2-Cy3/5** 5′-Cy3/5-GTC GAC CTG CAG CGT ACG; **D2-Cy3/5** 5′-Cy3/5-CGA GCT CGA ATT CAT CGA T.

2. Custom Agilent microarray (8×15k version for barcodes, Agilent-026035   Scer_barcode_v2_200911,   platform GPL18088 at the GEO database, www.ncbi.nlm.nih.gov/geo/) (*see* **Note 1**).

3. Extraction buffer: 2 % Triton X-100, 1 % SDS, 0.1 M NaCl, 10 mM Tris–HCl pH 8, 1 mM EDTA.

4. TE buffer: 10 mM Tris–HCl pH 8, 1 mM EDTA.

5. Acid washed glass beads of 0.4–0.6 mm diameter.

6. Phenol–chloroform–isoamyl alcohol mix (25:24:1, pH 8).

7. Linear acrylamide, 5 mg/ml.

8. DIG Easy Hyb solution (Roche Applied Science).

9. 20× SSPE buffer: 3 M NaCl, 20 mM EDTA, 0.2 M NaH_2PO_4, pH 7.4.

10. Agilent hybridization chamber and gasket slides for 8×15k microarrays.

11. Agilent stabilization and drying solution, required to protect Cy5 fluorescence from fading under the action of ambient O_3.

12. Scanner (GenePix 4000B) and scanning software (GenePix Pro 6, Molecular Devices, LLC).

13. System for normalization and data analysis (R from cran.r-project.org/ with the ggplot2 package installed).

3 Methods

3.1 Obtention of a Pool of MATa Mutant Cells from the Collection

This is probably one of the most time consuming steps when initiating the first series of GIM screens in a laboratory. It requires manipulation of 74 plates with 96 wells under sterile conditions (*see* **Note 2**). The pools of mutants can be obtained in large amounts and aliquoted, to be used by several laboratories, as a common resource.

1. Thaw several plates (usually no more than 12, for ease of manipulation) from the collection on the bench until the medium is half-molten. Carefully pick cells from the plate by using a 96-pin replicator. Pay attention that no droplets of medium are transported from one well to another. Transfer the cells to two different 96-well plates that are already filled with YPD medium. A standard 96-well plate will serve as a duplicate of the original collection, while the second one, with deep wells that contain up to 2 ml of YPD, will constitute the culture of individual strains to constitute the pool.

2. Add a fresh aluminum sticking cover on the original plate and return it to –80 °C (*see* **Note 3**).

3. Leave the cells in the new plates to grow for 48–72 h at 30 °C, under a gas permeable seal. No continuous agitation is required.

4. Verify that the cells have grown in all the wells that contained cells in the original plate and are recorded as such in the spreadsheet that accompanies the collection. Transfer the cells to a single flask by using a multichannel pipetting device. Vortex briefly, centrifuge at 3200×g for 10 min at room temperature. Resuspend the cells in YPD containing 25 % (w/v) glycerol and take an OD_{600nm} reading. Store the pool of cells in aliquots of different sizes that will be directly used in the GIM screens.

3.2 Generation of a "Query" Mutant Strain for the Screen

1. Thaw one or several haploid BY4742 strains from the collection that carry the G418 resistance cassette as a replacement of the coding sequence for the genes of interest. Carry out the maximum number of validations required to be sure the strain lacks the gene of interest and shows no additional mutations (*see* **Note 4**).

2. Prepare enough pGID3 plasmid to be able to transform as many strains as you wish (1 μg digested plasmid/strain). For a single strain, place 1 μg pGID3 plasmid in 10 μl of NEB4 buffer containing 5 U of *Asc*I and 5 U of *Eco*RI. Digest for 1 h at 37 °C. Test digestion efficiency by electrophoresis on 1 % agarose gel. Two bands corresponding to the digestion fragments (2066 and 2484 nt) should be obvious.

3. Prepare for each strain 10 ml of culture in YPD medium, leave to grow overnight. In the next morning, the culture should be saturated (OD_{600nm} between 10 and 15). Dilute culture in 50 ml fresh YPD to an OD_{600nm} of 0.15. Proceed with a DMSO enhanced yeast transformation protocol [16] with 1 μg of digested pGID3. Spread transformed cells on YPD plates containing nourseothricin to select for the recombination event. Typical yields are 50–500 colonies, visible after 2 days of plate incubation at 30 °C. Streak single colony cells on YPD plates containing G418 and on YPD plates containing nourseothricin. Store cells that have lost G418 resistance and became resistant to nourseothricin at −80 °C.

3.3 Obtention of a Pool of Double Mutant Strains

1. Transform previously obtained strains, that bear the pMFα2-NatR cassette as a replacement for the coding sequence of the gene of interest, with 0.3 μg pGID1 plasmid, containing two different markers: URA3 and hygromycin resistance (*see* **Note 5**). Select transformants on SC-URA plates (optionally, these strains can be also stored at −80 °C, to speed up the time until a screen starts).

2. The evening before starting the screen, prepare a pre-culture of the query strain in 10 ml YPD medium containing hygromycin. The next morning, dilute the cells to an OD_{600nm} of 0.2 in 25 ml GNA + hygromycin medium and leave to grow until the OD_{600nm} reaches 0.8. Thaw an aliquot of the pool of mutants in 25 ml GNA medium and incubate for 30 min at 30 °C with agitation; the expected OD_{600nm} is 0.4 (*see* **Note 6**).

3. Combine the two cultures in a 50 ml tube, centrifuge 10 min. at $3200 \times g$ at room temperature, resuspend the cells in 0.5 ml GNA medium and spread the suspension on a 90 mm diameter GNA plate. Leave at 30 °C for 5 h to allow mating.

4. Recover (scrape) the cells from the plate in 1 ml GNA medium containing hygromycin and G418 and resuspend them in

100 ml of the same medium. Incubate for 18 h, usually overnight, at 30 °C with agitation.

5. Take an OD_{600nm} reading for the overnight culture (dilute 1/50 in GNA medium) and recover an amount of culture that, when diluted in 150 ml would yield an OD_{600nm} of 0.8 (typically 6–10 ml). Centrifuge at room temperature in a 50 ml Falcon tube for 10 min at $3200 \times g$.

6. Thoroughly resuspend the pellet in 25 ml sporulation medium. Centrifuge a second time and resuspend the cells in another 25 ml sporulation medium (*see* **Note 7**). Resuspend the cells in 150 ml sporulation medium in a 1 l culture flask.

7. Leave the cells at 25 °C for 5 days with continuous agitation and shift the cultures to 30 °C for an additional 3 days. At the end of this sporulation period check by light microscopy that specific tetrads are a sizable proportion of the cells.

8. Recover the cells from 100 ml of sporulation flask by centrifugation at $3200 \times g$, 10 min, RT. Resuspend the pellet in 1 l YPD in a large 2 l culture flask and incubate with agitation at 30 °C for 5.5 h.

9. Add antibiotics (G418 and nourseothricin) to the YPD medium and the cells and leave with agitation at 30 °C.

10. Take repeated readings of OD_{600nm} until the values reach a value of 2. Dilute the cells in 1 l of fresh YPD medium with antibiotics (G418 and nourseothricin) to a OD_{600nm} of 0.1. Estimate the number of generations from the slope of the growth curve of the newly diluted culture to know at which moment the average number of generations is 18 (in general 45–60 h, *see* **Note 8**).

11. Recover the equivalent of 10 OD_{600nm} culture for each screen and the corresponding reference pool of cells. Pellet the cells by centrifugation at $3200 \times g$, 10 min, RT, and resuspend in 1 ml ice cold H_2O, then split in two aliquots in 1.5 ml Eppendorf tubes. Pellet the cells by 20 s of $13,000 \times g$ centrifugation at 4 °C, remove supernatant and freeze the cells at –80 °C until further processing.

3.4 Growth Speed Estimation from Barcode Microarray Data

1. Resuspend the cell pellet recovered in the previous step in 0.2 ml extraction buffer and add approximately 0.2 ml acid-washed glass beads (0.4–0.6 mm diameter, Sigma) and 0.2 ml phenol–chloroform–isoamyl alcohol mix (25:24:1, pH 8). Vortex vigorously for 7 min under a chemical hood.

2. Add 0.2 ml TE buffer and vortex for 5 s. Centrifuge at $13,000 \times g$ for 10 min, at RT. Transfer the supernatant in a 1.5 ml Eppendorf tube that already contains 1 ml pure ethanol. Mix by turning the tube upside down several times.

3. Centrifuge at $13,000 \times g$, at RT, for 5 min. Wash the white small pellet that sticks to the bottom of the tube with 0.5 ml of 70 % ethanol. Centrifuge at $13,000 \times g$ for 5 min at RT. Throw the supernatant, leave the pellet at room temperature to dry out, between 5 and 10 min and resuspend the dry pellet in 30 µl of TE (optionally add 10 ng/ml RNase A).

4. Use 1 µl of the previously prepared DNA extract as template in each of two PCR reactions that are set up using oligonucleotides U1-KU (uptag PCR 1, UPCR1) and D1-KD (downtag PCR 1, DPCR1) (*see* Fig. 1a and **Note 9**). Set up standard 50 µl reactions for each PCR reaction (25 cycles, 94 °C denaturation 30 s, 50 °C annealing 30 s, 72 °C elongation 30 s). Verify the size and abundance of PCR products by electrophoresis on a 1 % agarose gel.

5. Prepare four different PCR reactions for each comparison between a screen using a query mutation and a screen performed under identical conditions with a reference mutation (*see* **Note 10**). Set up standard 50 µl PCR reaction with 0.5 µl UPCR1 or DPCR1 in the following combinations with oligonucleotides: UPCR1 query with oligonucleotides U1 and U2-Cy3, UPCR1 reference with oligonucleotides U1 and U2-Cy5, DPCR1 query with oligonucleotides D1 and D2-Cy3, DPCR1 reference with oligonucleotides D1 and D2-Cy5. Use 15 cycles of amplification (94 °C 15 s, 55 °C 15 s, 72 °C 15 s). Test the reactions by electrophoresis using a 3 % agarose gel and a loading buffer that only contains xylene cyanol. The expected size of PCR products is 60 nt.

6. Mix the four PCR products obtained in the previous step. Add a premix containing 5 µl each of oligonucleotides U1, D1, U2block, and D2block (100 µM each, *see* **Note 11**) and 2 µl linear polyacrylamide (5 mg/ml), as carrier. Add 22.5 µl sodium acetate (3 M, pH = 5.2), mix well and precipitate with 550 µl pure ethanol. Leave at –20 °C for at least 1 h.

7. Centrifuge 30 min at $15,000 \times g$, 4 °C. Pay attention to the purple pellet when removing the supernatant. From now on, avoid direct exposure of the pellet or solution to light. Wash the pellet once with 70 % ethanol. Recentrifuge 5 min at $15,000 \times g$, 4 °C. Remove the ethanol and leave the tube on the bench for 5 min to dry.

8. Add 50 µl DIG Easy Hyb buffer to the pellet. Mix by repeated pipetting. Dilute 20 µl of the sample with 100 µl DIG Easy Hyb and keep the rest at –20 °C, for the case when the hybridization needs to be repeated. Heat for 2 min at 95 °C and then switch to ice for 5 min. Keep the sample at room temperature until hybridization.

9. Add the cover slide, with the gaskets, to the lower part of the hybridization chamber. Pipet carefully 47 µl of sample in the

middle of a gasket square. Carefully layer the microarray slide over the sample, oligonucleotide side down (can be detected by breathing on the sides of the slide) and close the hybridization chamber. Pay attention that a single air bubble is formed in the chamber. Place the chamber on a rotating wheel (*see* **Note 12**) in a 24 °C incubator for at least 12 h.

10. Prepare several dilutions of SSPE buffer (6×, 2×, 0.2×) containing 0.05 % Triton X-100. Preheat 50 ml SSPE 6×, 0.05 % Triton X-100 at 30 °C. Open the chamber and transfer the two slides together in a Falcon tube with the wash 6× SSPE buffer. Remove the slide with the gaskets and transfer the microarray slide to a second bath of 6× SSPE, 0.05 % Triton X-100. Leave for 5 min with occasional agitation, then move the slide to the 2× SSPE solution, 0.2× SSPE solution and finally to a 0.2× SSPE solution containing no Triton X-100. Incubate at each step for 5 min (RT).

11. After the last wash, slowly remove the slide from the liquid, dry its lower edge on a piece of Whatman paper, and plunge in stabilization solution for 10 s (under a chemical hood). Slowly remove the slide from the liquid so that no droplets remain; droplets can also be removed by a flux of dry air.

12. Scan the slide on a GenePix 4000B or equivalent scanner that is capable of 5 μm resolution. Analyze the scan images using an associated file that describes the position and annotation of each spot (**gal** file for GenePix Pro 6 or 7). The results are exported in the **gpr** format, a text file that contains columns separated by the **tab** character.

3.5 Microarray Data Analysis

1. We provide here an example of R session commands used to normalize data post-acquisition and to be able to do diagnostic plots. The input text file is the **gpr** result of image analysis with GenePix Pro.

2. Define a function allowing to extract the ORF names from the name of the probes (to be changed depending on how probes are identified). An example of identifier is YDR439W-U, where U shows that the barcode is the one located in the upstream position of the KanMX4 cassette. The **"namesextract"** function uses regular expressions to identify and grab ORF information from the identifier.

```
namesextract <- function(IDs) {

    orfnms1 <- gsub("(Y......-[ABC]?).*", "\\1", IDs)

    orfnms2 <- gsub("-$", "", orfnms1)

    return(orfnms2)

}
```

3. Read the tab delimited file in a data frame, choose columns and create a column with the ORF name extracted from the barcode identifiers. Replace the file name with your actual **gpr** file name. A **setwd**( ) command indicating the "working directory" is useful at the beginning of the R session.

```
gprfname <- "252603510022_8_G100526_12vsownmix.gpr"

gprdata <- read.delim(gprfname, skip = 35, header = TRUE, sep = "\t",

stringsAsFactors = F)

gprdata$ORF <- namesextract(gprdata$ID)
```

4. Find the columns that will be used in the normalization process by using either string identity (==) or regular expressions:

```
header = names(gprdata)

block_c = which(header == "Block")

col_c = which(header == "Column")

row_c = which(header == "Row")

id_c = which(header == "ID")

name_c = which(header == "Name")

genename_c = which(header == "GeneName")

ORF_c = which(header == "ORF")

median_ratios_c = which(header == "Median.of.Ratios..635.532.")

SNR_c = grep("SNR", header)

F_median_c = grep("^F[0-9]{3}.Median$", header)

B_median_c = grep("^B[0-9]{3}.Median$", header)
```

5. Extract only a few columns of interest and do simple filtering and calculations for **M**, the log transformed ratio of red and green signal, *log2(Red/Green)* and **A**, the average intensity of the two signals *0.5*log2(Red*Green)*. **A** is a log-average, the log value for the geometric mean of the two signals. The plot of ratios against average intensity can reveal intensity dependent artifacts that need correction.

```
F_median <- gprdata[F_median_c]

B_median <- gprdata[B_median_c]

FB_median <- F_median - B_median

FB_median[FB_median < 1] <- 1   #adjust missing values to 1

names(FB_median) <- c("signal_R", "signal_G")

to_norm = data.frame(gprdata[, c(ORF_c, block_c, col_c, row_c, id_c,

name_c,genename_c, SNR_c, median_ratios_c)], FB_median)
```

6. Filter and calculate values that have low signal:

```
minSNR = 5  #minimum acceptable signal to noise ratio

goodones <- which(to_norm$signal_R > 1 & to_norm$signal_G > 1)

filterSNRidx <- which(to_norm$SNR.635 > minSNR & to_norm$SNR.532 >

minSNR)
```

7. Calculate M and A only on the signals that are worth it, others leave as **NA** (not available):

```
goodx <- intersect(goodones, filterSNRidx)

to_norm$M <- NA

to_norm$A <- NA

to_norm$M[goodx] <-

log2(to_norm$signal_R[goodx]/to_norm$signal_G[goodx])

to_norm$A[goodx] <- 0.5 * log2(to_norm$signal_R[goodx] *

to_norm$signal_G[goodx])
```

8. Split the data by group of barcodes (separate for up and down, since coming from two separate PCR reactions).

```
ups <- grep("-U", to_norm$ID)

downs <- grep("-D", to_norm$ID)

to_norm$UD <- NA

to_norm$UD[ups] <- "U"

to_norm$UD[downs] <- "D"

to_norm$UD <- as.factor(to_norm$UD)
```

9. Create a diagnostic plot for the two classes of barcodes (Fig. 2a):

```
library(ggplot2)

ggplot(to_norm[goodx, ], aes(x = A, y = M, color = UD)) + facet_grid(UD ~ .)

+ theme_bw() + geom_point(alpha = 0.33) + geom_smooth(method = "loess",

size = 0.5, color = "black", alpha = 0.5) + ylim(-3, 6)
```

10. Find the row numbers for cases when signal higher than background:

```
indexU <- intersect(goodx, which(to_norm$UD == "U"))

indexD <- intersect(goodx, which(to_norm$UD == "D"))
```

11. Simple normalization, by median subtraction, will not work well in this case, since the dependence between signal intensity and ratio changes is not linear. Lowess, or loess is a more versatile method. First, define the function, based on the internal **"loess"** function:

```
loessnorm <- function(M, A) {

  loessfit <- loess(M ~ A, span = 0.8) #span is adjustable

  normM <- M - predict(loessfit)

  return(normM)

}
```

12. Apply **loess** normalization to the two classes of signal (U/D):

```
normU <- loessnorm(to_norm$M[indexU], to_norm$A[indexU])

normD <- loessnorm(to_norm$M[indexD], to_norm$A[indexD])

to_norm$normML <- NA  #create a new column

to_norm$normML[indexU] <- normU

to_norm$normML[indexD] <- normD
```

13. Verify that the normalization changed the distribution of the values (Fig. 2b):

```
ggplot(to_norm[goodx, ], aes(x = A, y = normML, color = UD)) +

facet_grid(UD ~ .) + theme_bw() + geom_point(alpha = 0.33) +

geom_smooth(method = "loess", size = 0.5, color = "black") +

ggtitle("Normalized") + ylim(-3, 6)
```

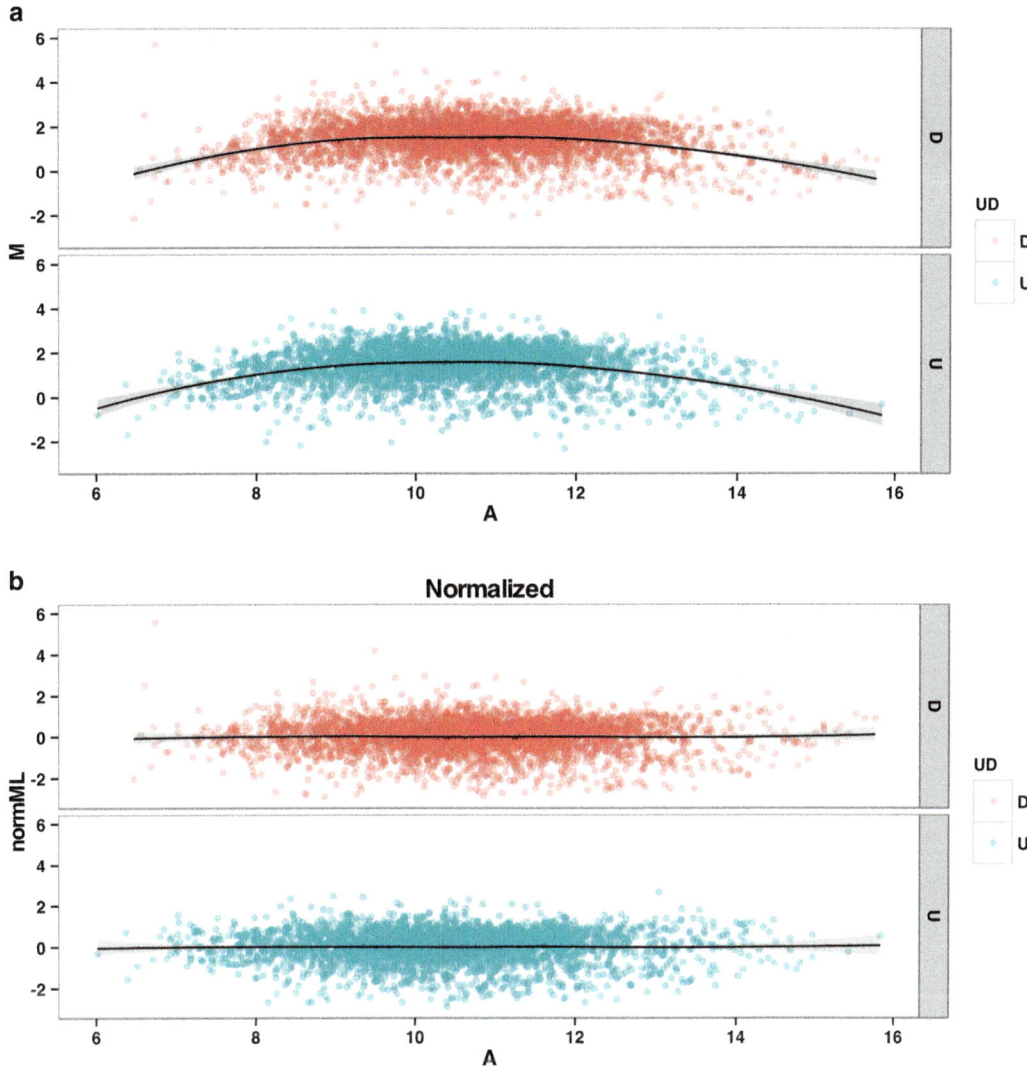

Fig. 2 (**a**) Diagnostic MA plot to estimate variations in the observed signal ratios (*M*, log$_2$ transformed) with the intensity of the signal (*A*, average of signals for the two fluorescence channels, log$_2$ transformed). (**b**) MA plot after **loess** normalization

14. Recalculate the green and red signal values from the normalized ratios (if needed):

    to_norm$norm_R <- **sqrt**(2^(2 * to_norm$A + to_norm$normML))

    to_norm$norm_G <- **sqrt**(2^(2 * to_norm$A - to_norm$normML))

15. Combine the results for each UP and DOWN values for each ORF and combine the obtained UP and DOWN medians for each ORF.

```
aggregated_UD <- aggregate.data.frame(to_norm[, c("normML", "A")], by =

list(ORF = to_norm$ORF, UD = to_norm$UD), median, na.rm = T)

aggregated_ORF <- aggregate.data.frame(aggregated_UD[, c("normML",

"A")], by = list(ORF = aggregated_UD$ORF),median, na.rm = T)
```

16. Read a list of genomic positions for the different features of SGD (text file, tab delimited) and add this information to the normalized table of values:

```
genopos <- read.delim("ordre_geno.txt", stringsAsFactors = F)

aggregated_ORFgp <- merge.data.frame(aggregated_ORF, genopos, by.x =

"ORF",

by.y = "orf")
```

17. Estimate the presence of the "exclusion region" in the position that corresponds to the gene of interest, in this case the deleted gene is the 4057th feature in the list (Fig. 3, *see* **Note 13**):

```
ggplot(aggregated_ORFgp, aes(x = ordre_geno, y = -normML)) + theme_bw()

+ geom_point(alpha = 0.5) + geom_vline(xintercept = 4057, alpha = 0.5, col =

"blue")
```

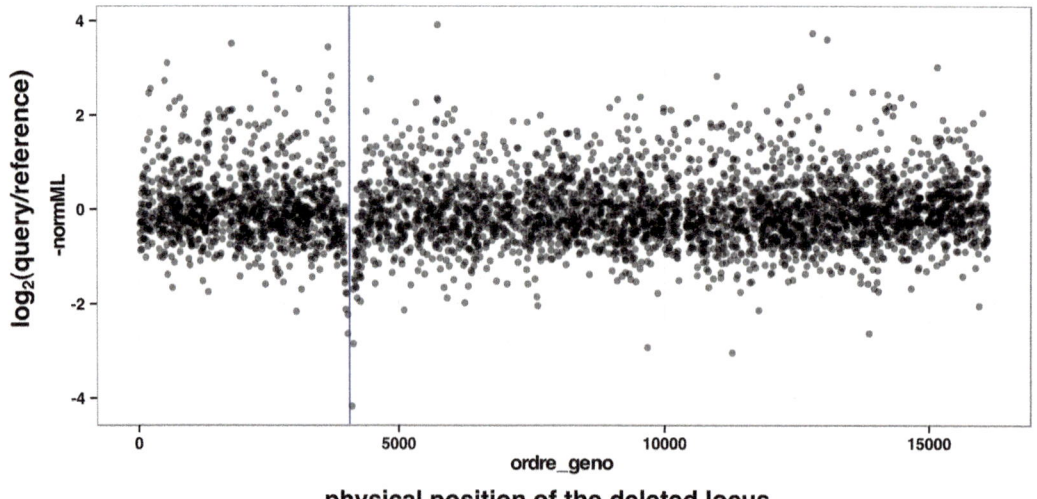

Fig. 3 Diagnostic plot showing measured double deletion strains growth as a function of the physical position of the tested loci along the yeast chromosomes. The characteristic exclusion peak is marked with a *blue vertical line* and corresponds to the position of the query gene locus

18. The various computations can be better organized in a series of functions to be sequentially applied to the data. Interactive R sessions are best done with RStudio (www.rstudio.com/). For further data processing or analysis, the results can be exported to a comma separated text file:

```
write.csv(aggregated_ORFgp, "aggregated_ORFgp.csv", row.names = F)
```

3.6 Visualization of Series of GIM Results

1. To be able to visualize the data from several GIM screens at the same time, build a tabulated file that has a format accepted by Cluster 3 (http://bonsai.hgc.jp/~mdehoon/software/cluster/). It includes a first column with unique identifiers for the genes or mutations followed by an optional, column with annotations (gene name, SGD description line) labeled "NAME" and columns corresponding to normalized numerical values (usually *log2* transformed) for different GIM screens. Cluster 3 has a graphical user interface that allows to keep only rows for which values were obtained in at least 70 % of the experiments. Proceed with hierarchical clustering and test several methods. We obtained good functionally relevant grouping of data with 'Correlation (uncentered)' as the similarity metric and with 'Average linkage' as clustering technique.

2. Open the clustering result (.cdt file) with Java TreeView (http://jtreeview.sourceforge.net/) and use either the tabular color-coded view of the cluster or scatterplots (Analysis>Scatterplot) to identify interesting correlations between genetic interaction profiles or similarity in response of the tested mutants as measured in several different GIM screens.

3. More involved data analysis depends on the integration of data from several different sources (SGA, GIM, protein-protein interactions) and gene ontology (GO) enrichment analysis in groups of related genes (by using, for example, the generic GO term finder, http://go.princeton.edu/cgi-bin/GOTermFinder).

4   Notes

1. Agilent microarrays for barcodes, unlike those used for transcriptome studies, can be stripped and reused, since hybridization is done on a relatively short, 20 nt, region (instead of 60 nt, as usual for expression microarrays). If slides have been previously treated with the Agilent stabilization and drying solution, they should be kept for 4–5 h in a Tris–HCl 10 mM, EDTA 1 mM, pH 8 solution at 45 °C to remove the protective coating. Stripping is performed by incubating the slides twice

for 5 min in a solution of 1 % SDS, 5 mM Tris–HCl pH 8, 0.5 mM EDTA that was brought to 95 °C. The stripped slides should be re-scanned to verify that no signal is present after the procedure.

2. The collection of yeast mutants has been distributed to many laboratories in the yeast community under the form of 96-well plates. Single strains should be checked thoroughly for the absence of the deleted gene and for the presence of the KanMX4 cassette at the appropriate locus. Cross-contamination is not a problem when working with pools of mutants, where we can follow a deleted locus via the molecular barcodes, with the exception of slow growing mutants that could be "invaded" readily by faster growing contaminating strains.

3. Thawing plates leads to viability loss; however since there is a large excess of viable cells in a typical well of the collection plates, re-freezing plates allows an economy of plates and manipulation time to be made.

4. Many viable yeast deletion strains have been previously shown to accumulate secondary mutations that alleviate the growth defect of the original strain [17]. The extent of this phenomenon is remarkable, as it was estimated that more than half of the strains from the collection are heterogeneous mixtures of cells bearing different genome variants as a result of the original gene deletion [18].

5. The presence of two markers on the plasmid used for diploid selection eases selection of transformed yeast cells on SC-URA plates, while culture and selection of diploids is done in rich medium with hygromycin.

6. For a single screen we use an estimated number of 3×10^8 mutant cells. Since only a fraction of the cells stored at −80 °C is viable and the mating efficiency is variable, we estimate the number of diploids obtained to be around 5×10^7, which corresponds to around 10,000 heterozygous double mutants per locus, assuming an uniform distribution for the number of cells for each mutant in the pool of cells. These relatively large numbers tend to minimize between-screen variation. Another step in which the number of cells is critical is sporulation, since this step is particularly inefficient in cells derived from S288c strain [19].

7. If some of the very rich GNA medium gets carried in the sporulation medium, an important decrease in the sporulation efficiency can be observed. Thorough wash of diploid cells is thus mandatory for the efficiency of this step.

8. Growth rate estimates are useful to ensure that the query screen pool of double mutants and the reference pool of double mutants grow for a comparable number of generations.

If, for example, the doubling time of a pool of mutants is 2.5 h, cells should be recovered after 45 h from the time the antibiotics have been added to the culture. Such timing can be difficult to follow, and an automatic system for cell culture can be set up in such a way that several screens are performed in parallel. The system involves the use of a turbidostat, for which readings of absorbance (proportional with the turbidity and the total number of cells) are done periodically, and defined amounts of fresh media are automatically added to ensure continuous dilution of the culture.

9. We use two separate PCR reactions for barcode amplification and labeling to avoid the massive expansion of the barcode regions from the query strain deletion that is present in all the cells of the population. The KU and KD oligonucleotides are complementary to regions of the KanMX4 cassette that have no equivalent in the MFα2-NAT cassette. While we used to gel purify the products of the first PCR reaction, a simple dilution before the second PCR reaction is sufficient to yield excellent hybridization results. It is important, especially for the second PCR reaction, to set up control PCR reactions, without template DNA. Contamination with PCR products from previous screens is a common occurrence.

10. In general we use the deletion of YEL068C as a reference strain. We observed no specific effects of this deletion on cell growth in previous GIM screens or under various stress conditions [20]. We have also used successfully mixes of genomic DNA obtained from 16 different GIM screens as a base for comparison with a given gene deletion screen.

11. Oligonucleotides used at this step in great excess hybridize with the universal sequences found on both sides of the barcode region, so that only that region remains available for hybridization with the complementary regions on the barcode microarray.

12. In the absence of a specific rotating system for Agilent hybridization chambers, we adapted a rotating wheel by carving slots of the right size in a polystyrene foam pad. The hybridization chambers are placed and secured with elastic bands and the rotating wheel is left in a 24 °C incubator overnight.

13. The observed peak, centered on the position of the query strain deletion, correspond to gene deletions that are physically close on the chromosome. If the two markers are on the same chromosome, to find both of them in a single cell, a meiotic recombination event must occur. It is possible to use simple probabilistic rules and the Haldane transformation to correct the observed values and estimate genetic linkage distances, as explained in the supplementary information of the initial

GIM paper [8]. Briefly, we can estimate the recombination frequency $r = 1/(1 + \text{query/reference})$, where the ratio query/reference is the one obtained in the GIM screen after normalization. Haldane genetic distance is found as $\ln(1 - 2r)/2$ and shows a linear correlation with the known physical distance between the markers (Fig. S1E in the mentioned supplementary information). Once a fit has been made, we correct the observed values for the effect of genetic linkage. The same method has been recently used to estimate changes in crossing over rates at different chromosomal positions, based on the results of large numbers of SGA screens [21]. Peaks can be also inspected visually, for single GIM screens and the data having obvious biases can be removed from the final table.

Acknowledgments

The protocol presented here was set up initially by Laurence Decourty, Gwenael Badis-Breard and Alain Jacquier, data management and analysis were performed by Christophe Malabat, and large scale developments were made possible by the ANR funding of the GENO-GIM project (ANR-2008-JCJC-0019-01).

References

1. Giaever G, Chu AM, Ni L et al (2002) Functional profiling of the Saccharomyces cerevisiae genome. Nature 418:387–391

2. Winzeler EA, Shoemaker DD, Astromoff A et al (1999) Functional characterization of the S. cerevisiae genome by gene deletion and parallel analysis. Science 285:901–906

3. Ooi SL, Shoemaker DD, Boeke JD (2003) DNA helicase gene interaction network defined using synthetic lethality analyzed by microarray. Nat Genet 35:277–286

4. Pan X, Yuan DS, Xiang D et al (2004) A robust toolkit for functional profiling of the yeast genome. Mol Cell 16:487–496

5. Lindegren CC, Lindegren G (1943) A new method for hybridizing yeast. Proc Natl Acad Sci U S A 29:306–308

6. Tong AH, Evangelista M, Parsons AB et al (2001) Systematic genetic analysis with ordered arrays of yeast deletion mutants. Science 294:2364–2368

7. Tong AHY, Lesage G, Bader GD et al (2004) Global mapping of the yeast genetic interaction network. Science 303:808–813

8. Decourty L, Saveanu C, Zemam K et al (2008) Linking functionally related genes by sensitive and quantitative characterization of genetic interaction profiles. Proc Natl Acad Sci U S A 105:5821–5826

9. Schuldiner M, Collins SR, Thompson NJ et al (2005) Exploration of the function and organization of the yeast early secretory pathway through an epistatic miniarray profile. Cell 123:507–519

10. Costanzo M, Baryshnikova A, Bellay J et al (2010) The genetic landscape of a cell. Science 327:425–431

11. Muhlrad D, Parker R (1999) Aberrant mRNAs with extended 3′ UTRs are substrates for rapid degradation by mRNA surveillance. RNA 5:1299–1307

12. Decourty L, Doyen A, Malabat C et al (2014) Long open reading frame transcripts escape nonsense-mediated mRNA decay in yeast. Cell Rep 6:593–598

13. Battke F, Symons S, Nieselt K (2010) Mayday—integrative analytics for expression data. BMC Bioinformatics 11:121

14. Floratos A, Smith K, Ji Z et al (2010) geWorkbench: an open source platform for integrative genomics. Bioinformatics 26:1779–1780

15. Cox J, Mann M (2012) 1D and 2D annotation enrichment: a statistical method integrating quantitative proteomics with complementary

high-throughput data. BMC Bioinformatics 13:S12

16. Hill J, Donald KA, Griffiths DE et al (1991) DMSO-enhanced whole cell yeast transformation. Nucleic Acids Res 19:5791

17. Hughes TR, Marton MJ, Jones AR et al (2000) Functional discovery via a compendium of expression profiles. Cell 102:109–126

18. Teng X, Dayhoff-Brannigan M, Cheng W-C et al (2013) Genome-wide consequences of deleting any single gene. Mol Cell 52: 485–494

19. Ben-Ari G, Zenvirth D, Sherman A et al (2006) Four linked genes participate in controlling sporulation efficiency in budding yeast. PLoS Genet 2, e195

20. Hillenmeyer ME, Fung E, Wildenhain J et al (2008) The chemical genomic portrait of yeast: uncovering a phenotype for all genes. Science 320:362–365

21. Baryshnikova A, VanderSluis B, Costanzo M et al (2013) Global linkage map connects meiotic centromere function to chromosome size in budding yeast. G3 3:1741–1751

Chapter 19

On the Mapping of Epistatic Genetic Interactions in Natural Isolates: Combining Classical Genetics and Genomics

Jing Hou and Joseph Schacherer

Abstract

Genetic variation within species is the substrate of evolution. Epistasis, which designates the non-additive interaction between loci affecting a specific phenotype, could be one of the possible outcomes of genetic diversity. Dissecting the basis of such interactions is of current interest in different fields of biology, from exploring the gene regulatory network, to complex disease genetics, to the onset of reproductive isolation and speciation. We present here a general workflow to identify epistatic interactions between independently evolving loci in natural populations of the yeast *Saccharomyces cerevisiae*. The idea is to exploit the genetic diversity present in the species by evaluating a large number of crosses and analyzing the phenotypic distribution in the offspring. For a cross of interest, both parental strains would have a similar phenotypic value, whereas the resulting offspring would have a bimodal distribution of the phenotype, possibly indicating the presence of epistasis. Classical segregation analysis of the tetrads uncovers the penetrance and complexity of the interaction. In addition, this segregation could serve as the guidelines for choosing appropriate mapping strategies to narrow down the genomic regions involved. Depending on the segregation patterns observed, we propose different mapping strategies based on bulk segregant analysis or consecutive backcrosses followed by high-throughput genome sequencing. Our method is generally applicable to all systems with a haplodiplobiontic life cycle and allows high resolution mapping of interacting loci that govern various DNA polymorphisms from single nucleotide mutations to large-scale structural variations.

Key words Genetic interaction, Bimodal trait, Segregation, Bulk segregant analysis, Introgression, Next-generation sequencing

1 Introduction

With the advent of next-generation sequencing technologies, we are currently entering an era where whole genome data among individuals of a same species are routinely generated. This has led to an unprecedented understanding of the amount of genetic variation within a species; however, how these variants interact and affect a given phenotype remains a challenging question. The yeast *Saccharomyces cerevisiae* has played an emerging role in deciphering the genetic architecture of many complex phenotypes [1–7].

Frédéric Devaux (ed.), *Yeast Functional Genomics: Methods and Protocols*, Methods in Molecular Biology, vol. 1361,
DOI 10.1007/978-1-4939-3079-1_19, © Springer Science+Business Media New York 2016

Nevertheless, identification of the causative variants is often biased toward variants with a large effect, which only explains a fraction of the phenotype observed [8–11]. It has become increasingly evident that non-additive effects of complex genetic interactions are likely one of the major sources of missing heritability, a problem that stems the difficulty in understanding the basis of complex traits, including many human diseases [12, 13].

Yeast models present a powerful toolset to identify genetic interactions. For example, development of the synthetic genetic array (SGA) in *S. cerevisiae* enabled systematic construction of double deletion mutants from ordered arrays of single mutants, which allowed for genome-wide profiling of the synthetic genetic interaction networks [14–16]. These advances provided deep insights into the functional connections of genes at an organismal level. However, much is still unknown about how different types of genetic variations, other than deletions, would interact and impact the phenotypic diversity in a non-laboratory setting.

Natural isolates of *S. cerevisiae* are universally isolated from many ecological (soil, tree exudate, immunocompromised patients, for example) and geographical (Europe, Africa, America, Asia) niches and constitute a rich repertoire of genetic diversity [17]. Genetic variations acquired in their natural context including sequence differences, regulatory changes and structural variations, could potentially lead to non-additive interactions when tested in another genetic background. We define such interactions here as epistasis, where novel combinations of alleles sampled from different genetic backgrounds result in unexpected phenotypic deviation in the offspring. Identifying the molecular bases of such interactions is valuable to better understand the phenotypic consequences of genetic differentiation in natural isolates of yeast.

When working with natural variation, no prior knowledge is available concerning which loci or genes could potentially interact. Therefore, sampling a large number of crosses is essential to find out which combination of genetic backgrounds is of particular interest for the phenotype studied. For any given cross, analysis of the phenotypic segregation in the F1 segregants is key to determine whether the observed phenotype is under epistatic genetic control. For example, for a qualitative trait, the phenotype of the F1 segregants is categorical, i.e., presence *vs.* absence of the trait or viable *vs.* nonviable of the segregant. In this case, any phenotypic class in the offspring that is unexpected from the additive effect of the parental phenotypes could indicate the presence of an epistatic interaction. This also holds true for quantitative traits, in which epistasis is implied when a fraction of the segregants present phenotypic values that are strongly deviated from the population mean, resulting in a bimodal distribution of the trait (Fig. 1).

Once a case of potential interaction is identified, the goal is to pinpoint the genomic regions involved in the phenotype of interest.

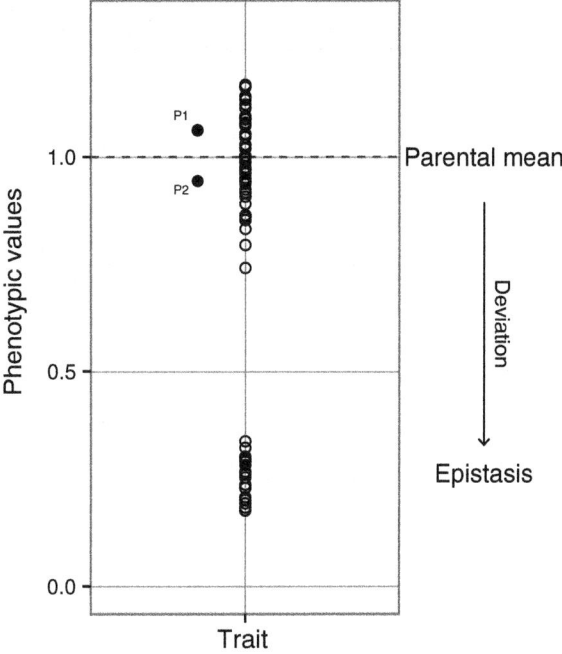

Fig. 1 Phenotypic distribution of traits with epistatic control. Each *dot* corresponds to the phenotypic value for one offspring for a given trait. *Dashed line* represents the mean phenotypic value between the parents, which are symbolized by *black dots*. In the presence of epistasis, offspring with phenotype deviation from the parental mean could be observed

Concerning the choice of mapping strategies, bulk segregant analysis following by high-throughput sequencing (BSA-seq) has become a common measure for many yeast geneticists [1, 18–20]. This strategy generally consists of dividing the segregating population into phenotypically distinct populations. These populations, or bulks, are then subjected to whole genome sequencing and the causative loci could be identified by looking at regions with biased allele frequencies. Apart from the fact that BSA-seq is precise and cost-effective, this method also provides information about the sequence depth across the genome, which is essential to identify causative loci governing structural changes such as copy number variation and translocation.

Nevertheless, in many of the experimental designs using BSA-seq, only segregants with extreme phenotypes are considered, which results in a major challenge for the statistical power to detect causative loci when strong epistasis is present [10]. To this end, classical segregation analysis of the phenotypic distribution in the offspring is important for efficient mapping. For any cross of interest, hybrid diploids are sporulated to generate offspring in the form of tetrads. Each tetrad contains meiotic product of haploid spores, where each spore represents only one allelic combination of any pair of parental loci. Analyzing the phenotypic distribution in the tetrads is therefore essential to understand the basis and complexity of the observed interaction, and incorporating this information into the design of the mapping method significantly increase the power of detection.

Here, we provide a complete workflow starting from the identification of potential epistatic interactions using classical genetic analysis to the strategies for mapping the loci involved using high throughput sequencing methods. In the following section, we discuss the basic theoretical interaction models involving two independent loci and their possible phenotypic penetrance. We then illustrate the segregation patterns of the phenotype by genetic analysis of the tetrad type distribution and describe two mapping strategies, namely bulk segregant analysis and successive backcrosses, designed to precisely locate the loci involved according to different instance of tetrad type distribution. In Subheadings 2 and 3, we present in detail the protocol for the required experimental procedures and downstream bioinformatic analysis.

Consider a basic interaction model involving two unlinked loci A and B. Locus A has two alleles **A** and **a**; and locus B has two alleles **B** and **b**. Suppose we have two parental strains P1 and P2 with genotypes **Ab** and **aB**, respectively. Given that allele **a** and allele **b** interact recessively with a phenotypic effect of ε, the penetrance of loci A and B in the offspring from the cross between P1 and P2 could be summarized by Fig. 2a. In this case, both parental combinations **Ab** and **aB** have the same phenotypic value as well as the recombinant genotype **AB**, whereas **ab** shows an epistatic effect resulting in a phenotypic value of $1 - \varepsilon$. If we admit that the allele frequencies of **A**, **B**, **a**, and **b** are equal in the offspring, which is expected from the cross between P1 and P2, the overall frequency of the parental phenotype will be 75 % (Fig. 2a).

What does this model imply in a biological sense? There are several possibilities. For instance, the Dobzhansky-Müller incompatibility predicts that genes evolving independently in different lineages could accumulate mutations which have no effect in its original genetic background but cause negative epistasis when tested together in the hybrid [21]. Take for example the present case, suppose that P1 and P2 are two lineages which descended from an ancestral population P with a genotype **AB**. P1 and P2 independently acquired mutations **a** and **b**, where **a** is compatible with **B** and **b** is compatible with **A**; however, the combination **ab** has never been evolutionarily tested and is incompatible. It is worth noting that this model also applies to non-genic scenarios, such as asymmetrical resolution after genome duplication and the presence of non-equivalent reciprocal translocations. Example of such cases will be further discussed later in the text.

In the case of two loci interactions, another model is possible when the loci involved interact dominantly. Consider two independent

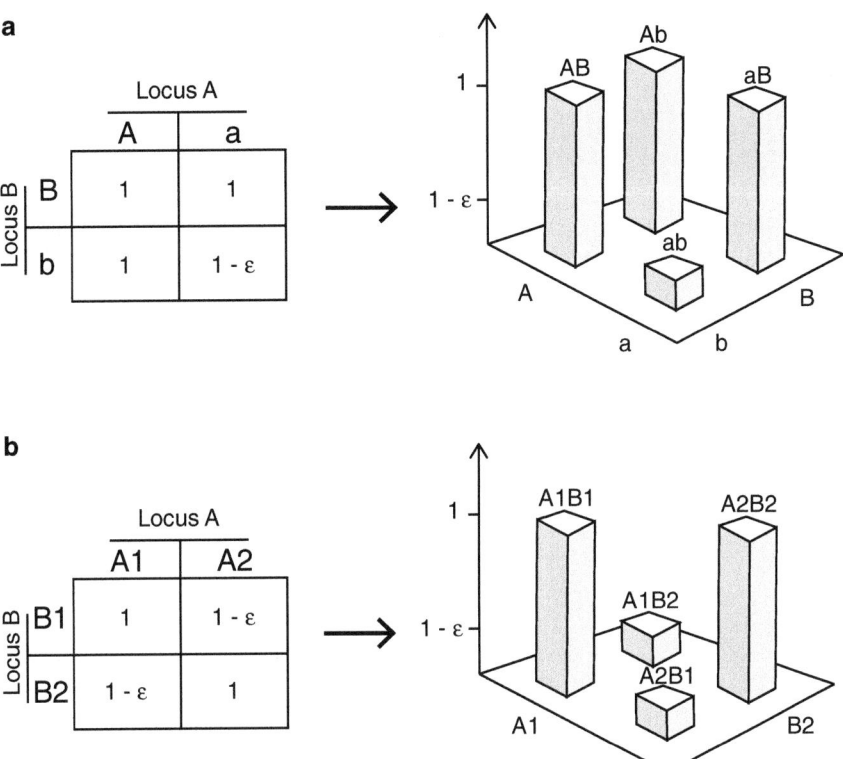

Fig. 2 Possible penetrance of two loci interactions and the resulting genotype-phenotype map. The penetrance table of two loci interacting recessively (**a**) and dominantly (**b**) are presented. *Right panels* represent the corresponding phenotypic outcome for each possible genotypic combination

loci A and B each with two alleles denoted **A1/A2** and **B1/B2**. Suppose we have two parental strains P1 and P2 with the genotypes **A1B1** and **A2B2**. If A and B interact dominantly, any non-parental genotype combination would have an epistatic effect ε. The penetrance of loci A and B in the offspring could then be summarized in Fig. 2b, where the parental combinations **A1B1** and **A2B2** have the same phenotypic value, and the recombinant genotypes **A1B2** and **A2B1** have a phenotype value of $1 - \varepsilon$. Again, as equal allelic frequencies of **A1**, **A2**, **B1**, and **B2** are expected, the overall frequency of the parental phenotype in the offspring will be 50 % (Fig. 2b).

In this scenario, A and B could represent coevolving genes in the same complex or pathway, in which **A1B1** and **A2B2** form a "lock-and-key" type of interaction in independent lineages [22]. Recombined genotypes **A1B2** and **A2B1** are therefore non-functional. Nevertheless, the presence of large-scale chromosomal rearrangements such as inversions and reciprocal translocations will also result in the same penetrance.

For both models considering the segregation of two loci for a given cross, the resulting tetrads could be assigned to different types

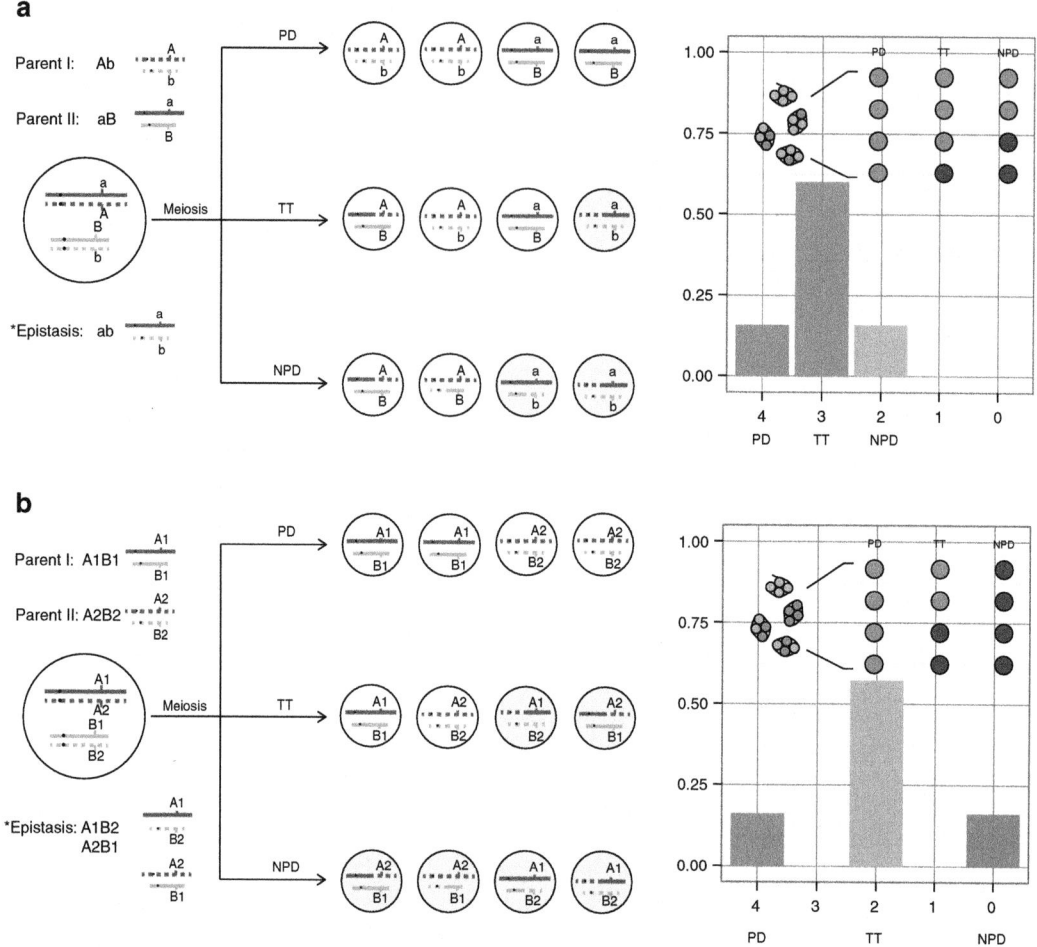

Fig. 3 Phenotypic segregation pattern and tetrad type distribution in the offspring. Segregation in the F1 offspring for two loci recessive (**a**) and dominant **b**) interactions are presented. The phenotypic segregation in each type of tetrad PD, TT and NPD are shown on the *left panel*, and the distribution of the tetrad types are shown on the *right panel* supposing that the two loci involved are independent

according to the allelic recombination. There are three possible types of tetrads: parental ditype or PD contains only parental alleles, non-parental ditype or NPD contains only recombined alleles, and tetratype or TT contains all four possible allelic combinations. As all spores in a tetrad are haploids from a single meiosis event, the phenotype distribution in the tetrad could thus reflect directly the type of interaction of the loci in question.

For example, in the scenario with a recessive interaction between two loci (Fig. 2a), parental genotypes **Ab** and **aB** as well as the recombined genotype **AB** have the same phenotype P_{parent}, whereas the allelic combination of **ab** shows epistasis resulting in a different phenotype $P_{epistasis}$. Given the possible genotypes in the PD, TT and NPD tetrads, the distribution of P_{parent} would be PD:TT:NPD = 4:3:2 (Fig. 3a). Assuming that the loci A and B are unlinked, equal number of PD and NPD could be observed and the overall ratio of $P_{epistasis}/P_{parent}$ in the offspring will be 1:3 (Fig. 3a).

Alternatively, in the case of a dominant interaction (Fig. 2b), parental genotypes **A1B1** and **A2B2** have the phenotype P_{parent}, whereas any non-parental allelic combination **A1B2** or **A2B1** lead to the phenotype $P_{epistasis}$. As a result, the distribution of P_{parent} in different tetrad types would be PD:TT:NPD = 4:2:0 (Fig. 3b). Additionally, when no linkage is assumed, the number of PD and NPD would be equivalent and the overall ratio of $P_{epistasis}/P_{parent}$ in the offspring will be 1:1 (Fig. 3b).

By evaluating the segregation pattern of the phenotype of interest in the tetrads, we can easily infer the type of interaction that we are dealing with. Careful analysis of the segregation is essential to map the genomic regions governing the causative loci, especially when the epistatic effect is strong and affects the viability of the offspring. In fact, depending on the dominance or recessivity of the interaction, different mapping strategies could be employed, namely bulk segregant analysis and successive back-crossing, for example.

Since its first implementation in yeast [23], bulk segregant analysis strategy has become increasingly popular among yeast geneticists. The principle of the strategy is to group segregants from a mapping cross according to their phenotypes, and then genotype this pool of segregant all together [24]. Genomic regions containing the causative loci will have a skewed allele frequency whereas the rest of the genome will have an equal proportion of alleles from each parent.

While traditional design in BSA-seq for mapping quantitative trait loci (QTL) usually focuses on pools of segregants with upper and lower extreme phenotypes, the same design is less applicable when mapping epistatic interactions. Take for example the case of a two loci interaction (Fig. 4). If the loci involved interact recessively, the lower phenotypic group $P_{episitasis}$ contains segregants with only one possible genotype **ab**. Sequencing of this pool will efficiently localize the causative loci, as the allele frequency of loci A will be biased toward the allelic version of **a** (**A/a** = 0/1) and the allele frequency of loci B will be biased toward the allelic version of **b** (**B/b** = 0/1) (Fig. 4b). However, when the epistatic effect is strong enough to affect the viability of the segregants, mapping using the lower phenotypic group will simply be impossible. In this case, only the upper phenotypic group P_{parent} could be used, which contains equal proportion of segregants with the genotype **Ab**, **aB**, and **AB**. As a result, only a small variation of allele frequency at both loci could be observed (**A/a** = 0.67/0.33, **B/b** = 0.67/0.33) (Fig. 4a), and the power of detecting these loci will be extremely limited due to the presence of experimental noise (that is, random allele frequency variation at unassociated loci).

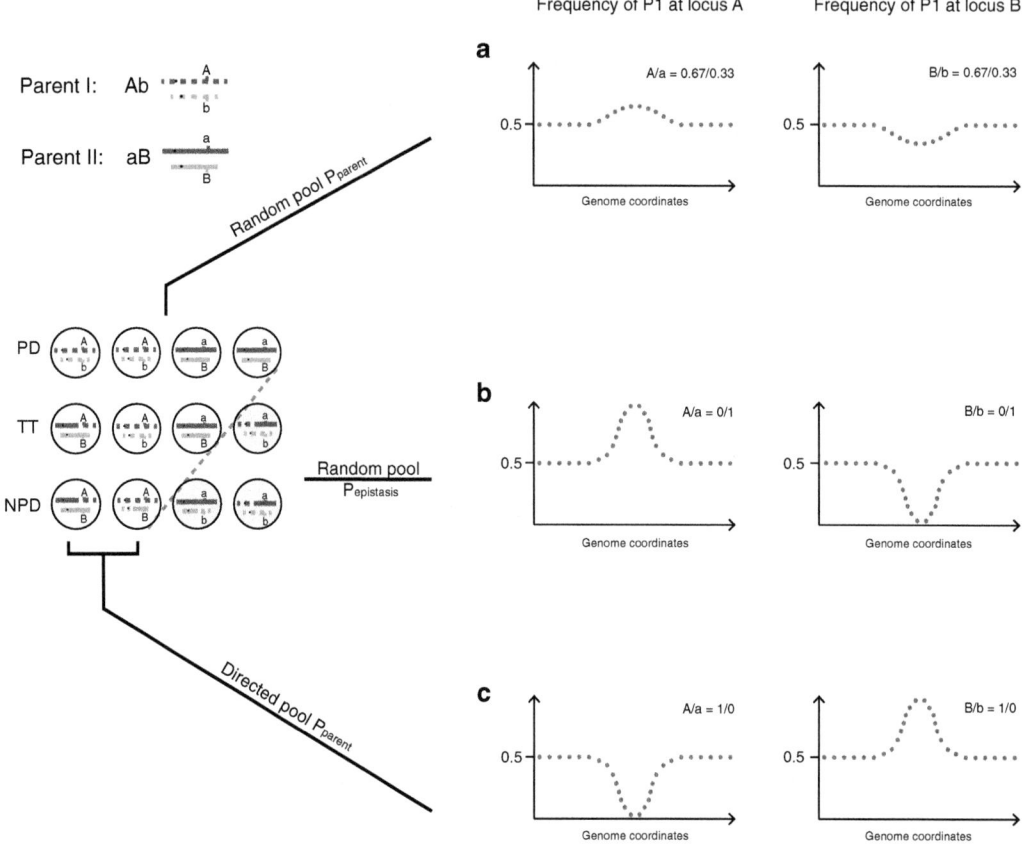

Fig. 4 Comparison of random and segregation-directed bulk segregant analysis strategy in mapping a two loci recessive interaction. Allele frequencies of (**a**) random F1 pool of P$_{parent}$ segregants; (**b**) random F1 pool of P$_{epistasis}$ segregants and (**c**) segregation directed pool of P$_{parent}$ segregants in NPD tetrads are represented. The allele frequency equal to 1 means only the allele from parent P1 is present

For efficient mapping of the aforementioned scenario, the segregation pattern of the phenotype has to be taken account of. Suppose that the combination **ab** cause a lethal phenotype, then the distribution of viable segregant in the tetrads will be PD:TT:NPD = 4:3:2. Knowing that the lethal combination **ab** is absent in the NPD tetrads, the mapping could be achieved by pooling segregants from independent NPD tetrads. In this case, as all segregants in this pool will only have the genotype **AB**, the allele frequency at A locus will bias toward **A** ($A/a = 1/0$) and the allele frequency at B locus will bias toward **B** ($B/b = 1/0$) (Fig. 4c).

By incorporating the phenotypic segregation, BSA-seq could be extremely powerful in mapping potential genetic interactions, which is not limited to interactions between genes, but also applicable to structural variations. Recently, we applied this strategy in the study of intraspecific reproductive isolation in *S. cerevisiae* [25]. After crossing a large number of natural isolates with the reference strain S288c, several cases of reduced offspring viability were identified, of which eight crosses showed a segregation of PD:TT:NPD = 4:3:2, indicating a potential two loci interaction. According to the segregation pattern, the lethal genotype combination was likely absent in viable spores from the NPD tetrads. To map the genomic regions involved, 50 independent segregants in NPD tetrads were selected and the genome of the pool was sequenced. The allele frequencies of S288c at each polymorphic position were scored, and two regions with significantly skewed allele frequencies were identified (Fig. 6a). Additionally, variations in sequence coverage were also observed, which ultimately led to the identification of a non-equivalent reciprocal translocation, explaining the observed cases of reduced offspring viability (Fig. 6a).

Nevertheless, a major limit of this method is that it relies on the ability of selecting a pool of segregant with biased genotype. For example, in the case of a dominant interaction between two loci, upper and lower phenotypic groups as well as different types of tetrads will always have the same frequencies of each allele. The application of BSA-seq is simply powerless in this scenario and another mapping strategy is required.

Introgression of alleles with major phenotypic effects by consecutively backcrossing one strain to another is not new, especially in organisms such as yeast where backcrossing is timely effective. However, the use of introgression in mapping epistatic interaction is not yet common. The concept here is to treat the segregation pattern as a phenotype itself, and simultaneously introduce all interacting loci into a single genetic background. The identification of the causative loci is then possible by sequencing only one backcrossed segregant and looking for introgressed regions. Even though this strategy is somewhat more labor intensive, it allows for efficient mapping of dominant interactions, which compensate the major short coming of BSA-seq.

Take for example a dominant interaction between two loci (Fig. 2b). Parental strains P1 and P2 have the genotype **A1B1** and **A2B2**, which result in a phenotypic value of P_{parent}. Any recombined genotype in the offspring **A1B2** and **A2B1** cause an epistatic effect ε, which result in a phenotypic value of $P_{epstasis}$. Therefore, the distribution of P_{parent} in different tetrad types would be PD:TT:NPD = 4:2:0 (Fig. 3b). To map these loci, the idea is to introduce both alleles **A1** and **B1** into the genetic background of the parental strain P2 (Fig. 5). To do so, one PD tetrad in the generation F1 is selected, and all four spores from this tetrad are backcrossed with P2. For all four crosses, the segregation pattern of the phenotype is scored again. Since PD tetrads contain only segregants with the parental genotype **A1B1** and **A2B2**, half of these segregants (**A1B1**) with retain the epistatic segregation, whereas the other half of the segregants (**A2B2**) will show no phenotypic effect when backcrossed to P2. Then, one segregant that retained

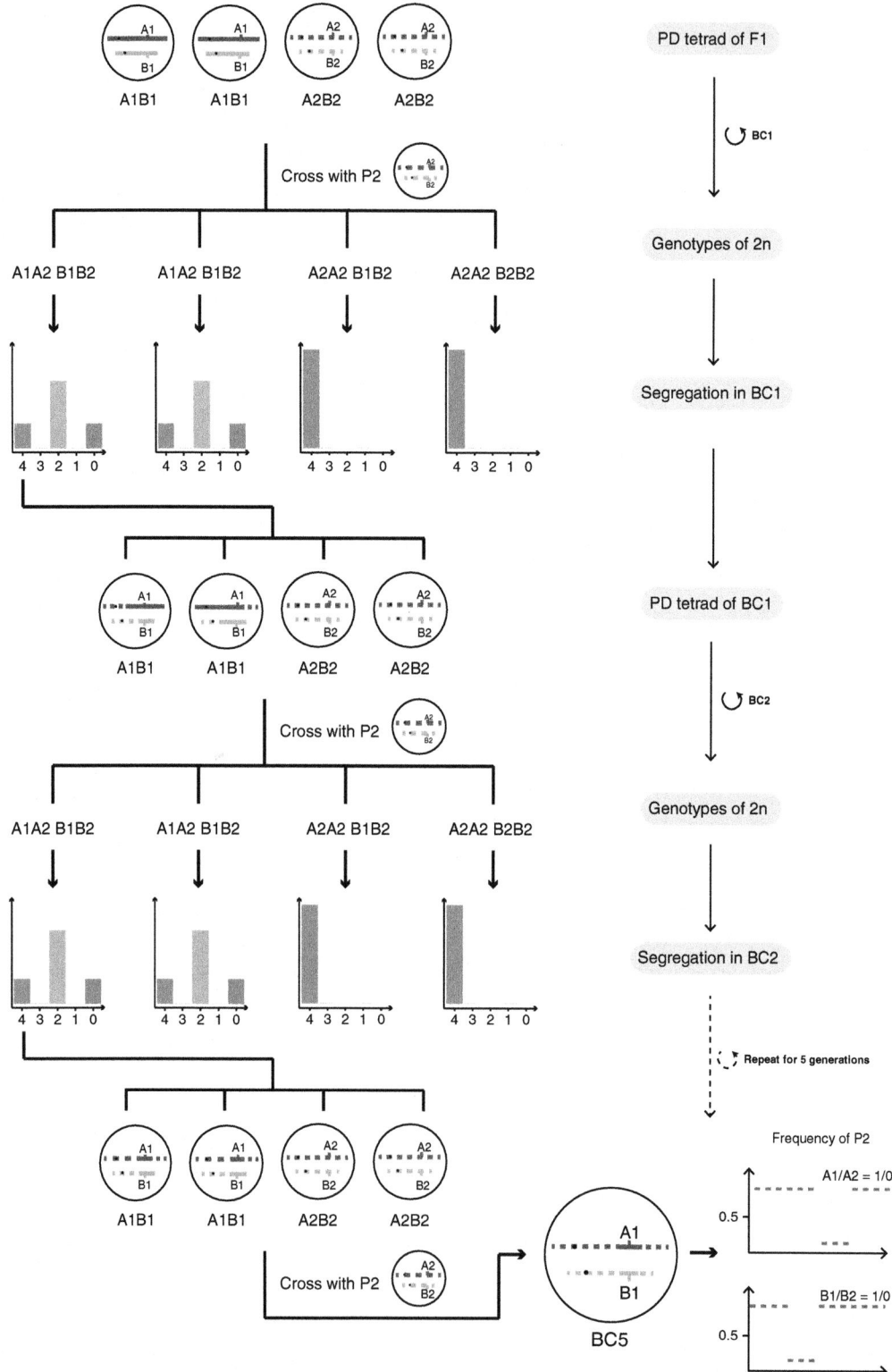

Fig. 5 Overview of double introgression of interacting loci using successive backcrossing. For each generation the tetrad type segregation is treated as a phenotype. One full PD tetrad from F1 is selected and backcrossed to the parental strain P2 and the segregation pattern is scored. One segregant that retained the segregation pattern is then selected and five subsequent backcrosses are performed. The resulting BC5 segregant will be enriched for the P2 genome except for regions involved in the interaction

the 4:2:0 segregation is selected, and again one PD tetrad is taken to perform another round of backcross. By repeating this procedure for several generations, the genome of the backcrossed segregant will be highly enriched by the allele of P2, except for the regions containing the causative loci (Fig. 5).

By sequencing the backcrossed segregant, loci involved can be easily located by looking for regions in the P2 genome that came from P1. Combined with some karyotype analysis, this method is also useful in mapping large-scale chromosomal rearrangements such as reciprocal translocation. An example is given in the same study of reproductive isolation, where cases of reduced offspring viability showing a segregation of PD:TT:NPD = 4:2:0 were mapped using backcrosses [25]. To do so, the incompatible strain was successively backcrossed with the reference strain S288c for five generations. For each round of backcross, one segregant that retained the 4:2:0 segregation was selected for the subsequent cross. When the five generations of backcrosses were complete, one final segregant was chosen for whole genome sequencing. Using this strategy, the causative loci were introgressed into regions spanning ~100 kb intervals. Further examination of the mapped regions revealed the presence of several transposable elements, which ultimately allowed for the identification of a large-scale reciprocal translocation responsible for the observed phenotype of reproductive isolation (Fig. 6b).

2 Materials

2.1 Media

1. Standard YPD media 1 % yeast extract, 2 % peptone, and 2 % glucose, is used for common strain growth and maintenance.

2. Sporulation media 1 % potassium acetate, 2 % agar.

3. High grade Difco YPD agar for dissection plates to ensure transparency.

4. 20 % Zymolyase solution (v/v, 5 mg/ml) was used to digest the ascus of the tetrad.

5. When phenotyping on a specific media is required, simply prepare them with proper instructions.

2.2 Dissection Microscope

A dissection microscope is necessary to obtain segregants in a complete tetrad. In our lab, we have a MSM 400 from Singer instruments which has a computer controlled motorized platform.

2.3 Phenotyping Device

When the phenotype of interest is categorical, phenotyping can be achieved by replicating cell culture on a solid media with a replicator or with the drop test. However, if the phenotype is quantitative, some high throughput phenotyping device may come in handy.

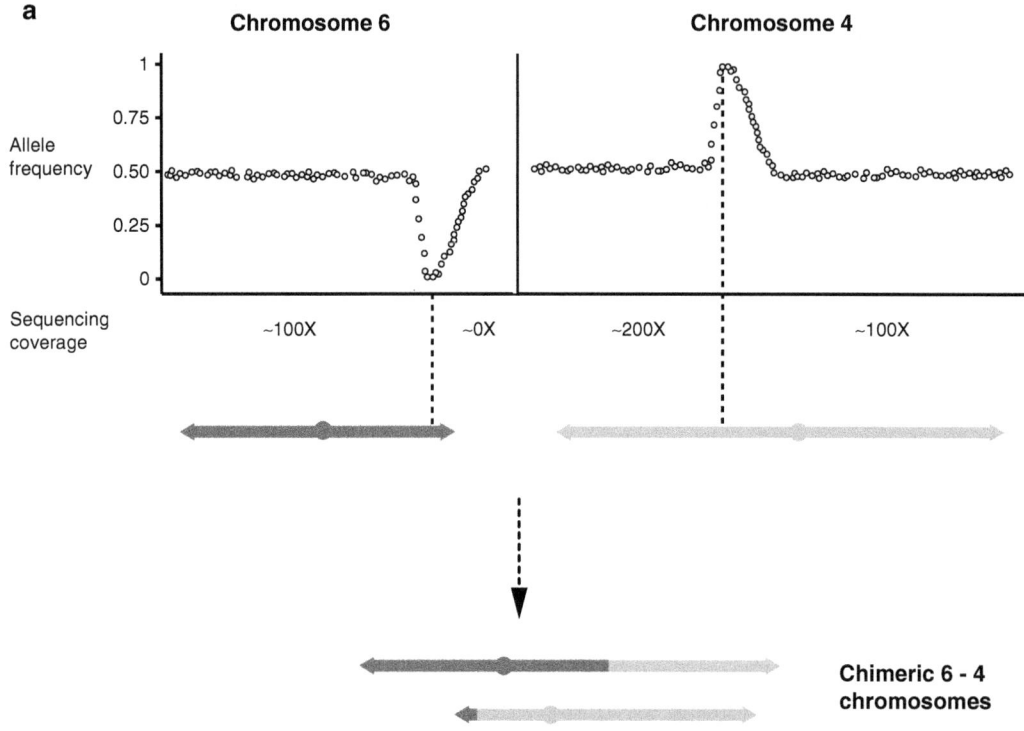

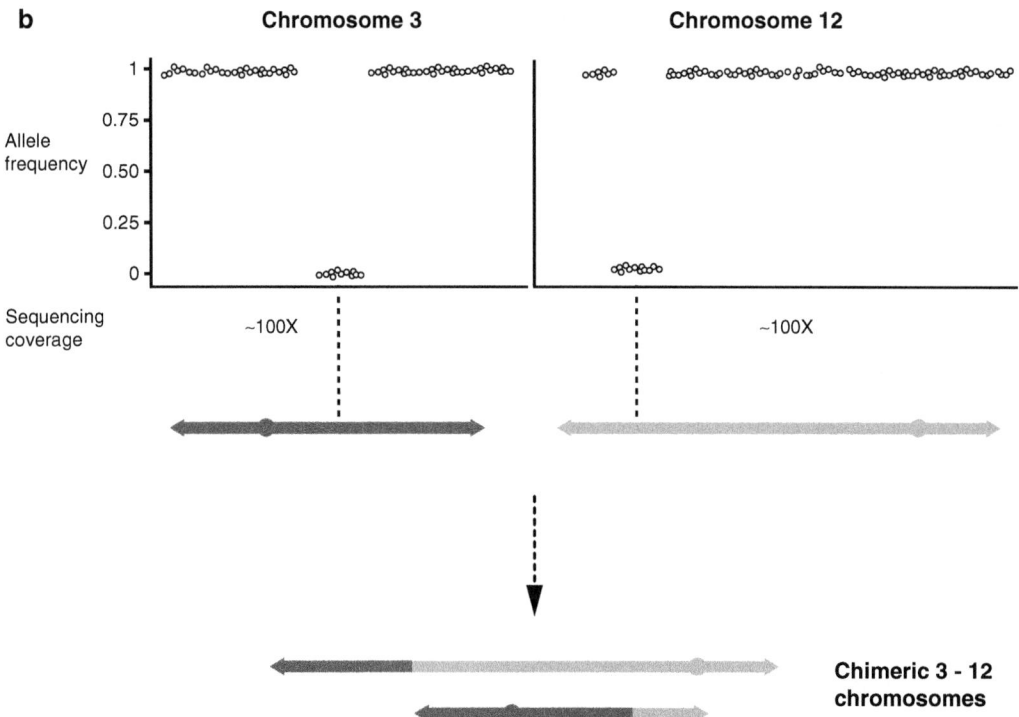

Fig. 6 Example mapping results using segregation directed BSA-seq and backcrossing. Schematic of the allele frequency variation and coverage results using segregation directed BSA-seq (**a**) and successive backcrosses (**b**) are presented. Mapped regions correspond to junctions of two translocations affecting 25 % (**a**) or 50 % of the offspring viability (**b**) in a heterozygous diploid background. Figures are modified from Hou et al. [25]

For example, the replication robot RoTor allows quantitative measurement of normalized growth in multiple conditions on solid plates; or a microplate reader such as the Tecan Infinite series which monitors strain growth in microcultures.

2.4 DNA Extraction

We used the Qiagen Genomic-tip kit to prepare sequencing samples. However, any kit that meets the required quantity and purity specified by the sequencing platform can be used.

3 Methods

3.1 Preparation of the Pool of Segregant

For the bulk segregant analysis to be efficient, selected spores should be from independent tetrads. In our experience a pool size of ~50 segregants is good enough to map the interacting loci. Additionally, equal representation of each segregant in the pool is important to ensure no additional bias is introduced in the sequence.

1. Grow each segregant in a separate tube with 5 ml YPD overnight.

2. The next day, measure the cell density of each culture, then take approximately equal amounts of cell from each culture and put them in a flask. Always make a replicate for this step in case of insufficient DNA extractions. Empirically, for each segregant, a volume sufficient for an O.D reading at 600 nm of 1.5 (that is, approximately 1 ml of an overnight culture) would be enough for a bulk with 50 segregants. The quantity of cells should be adjusted according to method used in DNA extraction to ensure maximum efficiency.

3. Centrifuge the pooled culture ($5000 \times g$ for 10 min). If desired, the cell pellet can be frozen to coordinate with other DNA extractions. Briefly, after culture elimination of centrifuged cell pellet, the cells can be left in Falcon tube with caps and store at $-20\ °C$.

4. DNA extraction. 10–20 µg of high quality DNA is generally required by most Illumina HiSeq platforms. We use Qiagen Genomic-tip kit with 100G columns, which ensures high purity DNA with minimum fragmentation.

3.2 Successive Backcrossing

For most *S. cerevisiae* strains, the turnover of one generation of backcrosses takes at least 4–5 days. In our case we used segregation as a phenotype; therefore, analyzing a sufficient number of tetrads at each generation is essential to confirm the right segregation.

1. Streak out the strains to be crossed. It is essential to perform crosses with fresh colonies. We generally grow them over night on YPD at $30\ °C$.

2. Mix equal amounts of cells from each mating type. In our case, we crossed all four spores from the same tetrad with a parental strain. As the mating type of the spores is unknown, the parental strain used for backcrossing should present both isogenic mating types and should both be crossed with the segregants.

3. Grow the mixtures over night at 30 °C. Zygotes should start to form after 2–3 h. Although, we found newly formed zygotes tend to form abnormally shaped tetrads, which are hard to dissect. Thus it is important to wait sufficient amount of time so that the zygotes could divide and develop before putting them on sporulation media.

4. The next day, put the mix on sporulation media. It will take 1–2 days for the tetrads to appear.

5. Once sporulation is complete, digest the ascus wall with zymolyase and dissect at least ten tetrads per cross.

6. The colonies take 48 h to grow. Once colonies appear on the plate, score the phenotype of each segregant and identify the cross that retained the phenotypic segregation. These segregants are the first backcrossed generation BC1.

7. Take 4 spores from a PD tetrad in BC1. Repeat the procedure until a fifth generation of backcrossed segregant is obtained.

8. Extract the DNA and send for sequencing.

3.3 Bioinformatics Analysis

For both mapping strategies, downstream bioinformatic analysis remains more or less the same. Here we give a general pipeline for Illumina HiSeq data.

1. We use Illumina HiSeq 2000 with paired-end libraries, 101 bp per read and a coverage of 50× per genome. Samples can be multiplexed to reduce sequencing cost.

2. Once the sequencing is completed, clean reads are obtained by removing paired end adaptors, short and low quality reads. This step could have been already completed depending on the sequencing platform used.

3. Quality controlled reads are then aligned to the genome of interest using BWA [26] with "-n 5 -o 2" options. These options are stringent enough for most reads to align specifically, but still allow the mapping of reads that contains polymorphic sites.

4. SNP calling is then performed using SAMtools [27]. First, the alignment file .sam from BWA is converted to .bam using the "samtools view -bT" command. Then the .bam file is sorted ("samtools sort") and indexed (samtools index), which allows for variant calling using the command "samtools pileup -c -f". At each SNP variant position, the allele frequency of the reference genome is scored as the frequency of reads carrying the variant divided by the sequencing coverage at that position.

5. Allele frequency and coverage information at polymorphic positions are extracted from the variant calling file .pileup using "samtools varFilter". Analyze the allele frequency variation or coverage by simply plotting them against their chromosome coordinates. In the case of BSA-seq, most genomes would have an allele frequency ~0.5 and biased allele frequencies near 1 or 0 for the causative loci. In the case of introgression, only introgressed regions would be polymorphic compared to the reference, which would have an allele frequency near 0.

4   Notes

1. In the context of using natural isolates for crossing, stable haploid parental strains are preferred. In *S. cerevisiae*, most wild isolates are diploids; however, haploid derivatives could be easily obtained by deleting the *HO* gene responsible for the mating type switch [28]. Moreover, an extensive panel of monosporic stable haploid derivatives of wild isolates was used in numerous published studies, which are available upon request [29–31]. It is also worth noting that for a large number of these strains, whole genome data have already been generated and are publically available [32].

2. It is implied that no interaction preventing the crossing or sporulation capacity between the parental pairs should be present. All crosses should have a reasonable good sporulation rate for the method to be valid.

References

1. Liti G, Louis EJ (2012) Advances in quantitative trait analysis in yeast. PLoS Genet 8(8): e1002912

2. Steinmetz LM et al (2002) Dissecting the architecture of a quantitative trait locus in yeast. Nature 416(6878):326–330

3. Deutschbauer AM, Davis RW (2005) Quantitative trait loci mapped to single-nucleotide resolution in yeast. Nat Genet 37(12):1333–1340

4. Nogami S, Ohya Y, Yvert G (2007) Genetic complexity and quantitative trait loci mapping of yeast morphological traits. PLoS Genet 3(2):e31

5. Cubillos FA et al (2013) High-resolution mapping of complex traits with a four-parent advanced intercross yeast population. Genetics 195(3):1141–1155

6. Brion C et al (2013) Differential adaptation to multi-stressed conditions of wine fermentation revealed by variations in yeast regulatory networks. BMC Genomics 14:681

7. Albert FW et al (2014) Genetics of single-cell protein abundance variation in large yeast populations. Nature 506(7489):494–497

8. Sinha H et al (2008) Sequential elimination of major-effect contributors identifies additional quantitative trait loci conditioning high-temperature growth in yeast. Genetics 180(3): 1661–1670

9. Smith EN, Kruglyak L (2008) Gene-environment interaction in yeast gene expression. PLoS Biol 6(4):e83

10. Magwene PM, Willis JH, Kelly JK (2011) The statistics of bulk segregant analysis using next generation sequencing. PLoS Comput Biol 7(11):e1002255

11. Yang Y et al (2013) QTL analysis of high thermotolerance with superior and downgraded parental yeast strains reveals new minor QTLs and converges on novel causative alleles involved in RNA processing. PLoS Genet 9(8):e1003693

12. Baryshnikova A et al (2013) Genetic interaction networks: toward an understanding of

heritability. Annu Rev Genomics Hum Genet 14:111–133

13. Carlborg O, Haley CS (2004) Epistasis: too often neglected in complex trait studies? Nat Rev Genet 5(8):618–625

14. Tong AH et al (2004) Global mapping of the yeast genetic interaction network. Science 303(5659):808–813

15. Giaever G et al (2002) Functional profiling of the Saccharomyces cerevisiae genome. Nature 418(6896):387–391

16. Costanzo M et al (2010) The genetic landscape of a cell. Science 327(5964):425–431

17. Cromie GA et al (2013) Genomic sequence diversity and population structure of Saccharomyces cerevisiae assessed by RAD-seq. G3 (Bethesda) 3(12):2163–2171

18. Wilkening S et al (2014) An evaluation of high-throughput approaches to QTL mapping in Saccharomyces cerevisiae. Genetics 196(3): 853–865

19. Duveau F et al (2014) Mapping small effect mutations in Saccharomyces cerevisiae: impacts of experimental design and mutational properties. G3 (Bethesda) 4(7):1205–1216

20. Ehrenreich IM et al (2010) Dissection of genetically complex traits with extremely large pools of yeast segregants. Nature 464(7291): 1039–1042

21. Dobzhansky T (1937) Genetics and the origin of species. Columbia biological series. New York: Columbia Univ. Press. xvi, 364 p

22. Morrison JL et al (2006) A lock-and-key model for protein-protein interactions. Bioinformatics 22(16):2012–2019

23. Brauer MJ et al (2006) Mapping novel traits by array-assisted bulk segregant analysis in Saccharomyces cerevisiae. Genetics 173(3) :1813–1816

24. Dunham MJ (2012) Two flavors of bulk segregant analysis in yeast. Methods Mol Biol 871: 41–54

25. Hou J et al (2014) Chromosomal rearrangements as a major mechanism in the onset of reproductive isolation in Saccharomyces cerevisiae. Curr Biol 24(10):1153–1159

26. Li H, Durbin R (2009) Fast and accurate short read alignment with Burrows-Wheeler transform. Bioinformatics 25(14):1754–1760

27. Li H et al (2009) The sequence alignment/map format and SAMtools. Bioinformatics 25(16):2078–2079

28. Voth WP et al (2001) Yeast vectors for integration at the HO locus. Nucleic Acids Res 29(12):E59–9

29. Schacherer J et al (2009) Comprehensive polymorphism survey elucidates population structure of Saccharomyces cerevisiae. Nature 458(7236):342–345

30. Liti G et al (2009) Population genomics of domestic and wild yeasts. Nature 458(7236): 337–341

31. Muller LA et al (2011) Genome-wide association analysis of clinical vs. nonclinical origin provides insights into Saccharomyces cerevisiae pathogenesis. Mol Ecol 20(19):4085–4097

32. Skelly DA et al (2013) Integrative phenomics reveals insight into the structure of phenotypic diversity in budding yeast. Genome Res 23(9): 1496–1504

Chapter 20

Experimental Evolution and Resequencing Analysis of Yeast

Celia Payen and Maitreya J. Dunham

Abstract

Experimental evolution of microbes is a powerful tool to study adaptation to strong selection, the mechanism of evolution and the development of new traits. The development of high-throughput sequencing methods has given researchers a new ability to cheaply and easily identify mutations genome wide that are selected during the course of experimental evolution. Here we provide a protocol for conducting experimental evolution of yeast using chemostats, including fitness measurement and whole genome sequencing of evolved clones or populations collected during the experiment. Depending on the number of generations appropriate for the experiment, the number of samples tested and the sequencing platform, this protocol takes from 1 month to several months to be completed, with the possibility of processing several strains or mutants at once.

Key words Yeast, Chemostats, Fitness, Whole genome sequencing, Nextera, MiSeq

1 Introduction

Continuous cultures of microbes have been used since the development of chemostats in 1950 by both Monod [1] and Novick and Szilard [2]. As chemostats allow the culture to grow in a controlled environment at steady state, physiological changes such as transcription, protein and metabolite levels can be accurately observed and compared between strains and conditions [3, 4]. Chemostats also provide an ideal set-up to perform experimental evolution [5]. Microbial experimental evolution studies have demonstrated the rapid accumulation of genetic variation such as point mutations, copy number variations, and genomic rearrangements over time in response to laboratory environments [6–12]. Phenotypic changes in morphology, resistance to drugs, and fitness, for just a few examples, can be observed as early as a few generations [11, 13–15]. The recent advances in sequencing have contributed to the identification of many of the mutations contributing to adaptation. Whole genome sequencing of populations has also contributed to our

Frédéric Devaux (ed.), *Yeast Functional Genomics: Methods and Protocols*, Methods in Molecular Biology, vol. 1361,
DOI 10.1007/978-1-4939-3079-1_20, © Springer Science+Business Media New York 2016

understanding of the dynamics of populations undergoing adaptation by following the changes in frequency of mutations over time [10].

In this chapter, we describe an experimental system to perform long term evolution and competitive experiments to measure fitness of yeast using continuous cultures in chemostats. We also describe one method to perform whole-genome sequencing of evolved clones to detect de novo mutations associated with a fitness increase.

2   Materials and Equipment

2.1 Continuous Culture Using Chemostats

- Cultures tubes.
- Ethanol 95 %.
- "Ministat" chemostat setup and media as described in ref. [16].
- Kimwipes.
- Syringes: 1 ml.

2.2 Competition Against GFP Marked Strains

- 50 ml conical tubes.
- Ethanol 95 %.
- Kimwipes.
- Syringes: 10 ml.

2.3 FACS Analysis

- C6 Flow cytometer, CFlow Plus Software (BD Biosciences).
- Locking lid microcentrifuge tubes, 2 ml.

2.4 Tagmentation

- 0.5 ml PCR tubes with flat caps.
- Nextera TD buffer and TDE1 enzyme (From Illumina kit).
- Thermocycler.

2.5 Cleanup of Tagmented DNA

- 100 % ethanol.
- Eppendorf tubes.
- Nextera RSB buffer (Illumina).
- Zymo DNA clean & Concentrator-5 kit (Zymo Research).

2.6 PCR Amplification

- 0.5 ml PCR tubes.
- LightCyler 480 SYBR Green I Master (Roche Applied Science, dilute from 10,000× to 100× in DMSO).
- Nextera NPM and PPC (Illumina kit).
- Nextera P7 and P5 primers buffer (Nextera Index Kit).
- Thermocycler.

2.7 PCR Cleanup

- Agencourt AMPure XP magnetic beads (Beckman Coulter Genomics Inc).
- Eppendorf tubes.
- Ethanol 80 %.
- Magnetic plate: DynaMag-96 Side (Invitrogen/Lifetech Gibco BRL).
- Nextera RSB buffer (From Illumina kit).

2.8 Quantify and Pool Libraries

- 0.5 ml PCR tubes with flat caps.
- Eppendorf tubes.
- Qubit DNA quantification system.

2.9 Library QC

- 5× loading buffer :0.25 % bromophenol blue, 0.5 M DTT, 50 % glycerol, 10 % SDS.
- 6 % TBE gel.
- 50 bp DNA ladder with 6× loading dye.
- SYBR Gold 10,000× concentrate in DMSO.
- 10× TBE: 108 g tris base, 55 g boric acid, 40 ml 0.5 M EDTA (1 l at pH 8.0).
- XCell Sure Lock Mini-Cell CE Mark (Invitrogen Novex Minicell).

2.10 Library Denaturation

- 2 N sodium hydroxide (NaOH).
- Buffer EB :10 mM Tris–HCl pH 8.5.

2.11 Sequencing on the MiSeq

- Ethanol 70 %.
- Illumina MiSeq or HiSeq2000 sequencer.
- Milli-Q water.
- MiSeq reagent kit V2 PE 300 cycles (Illumina).

3 Methods

3.1 Continuous Culture Using Chemostats

In this protocol, we use the "ministat" chemostat system described in ref. [16] and diagrammed in Fig. 1a. However, the protocol can also easily be modified for use with other chemostat platforms (e.g., [17]).

1. Day 1: Inoculate a single fresh colony of each strain into a separate tube containing 2.5 ml of the media you will use for the chemostat, and let each culture grow to saturation overnight (Fig. 1b) (*see* **Note 1**). It is important to inoculate each culture with an independent colony in order to avoid shared mutations that might occur during the batch growth phase.

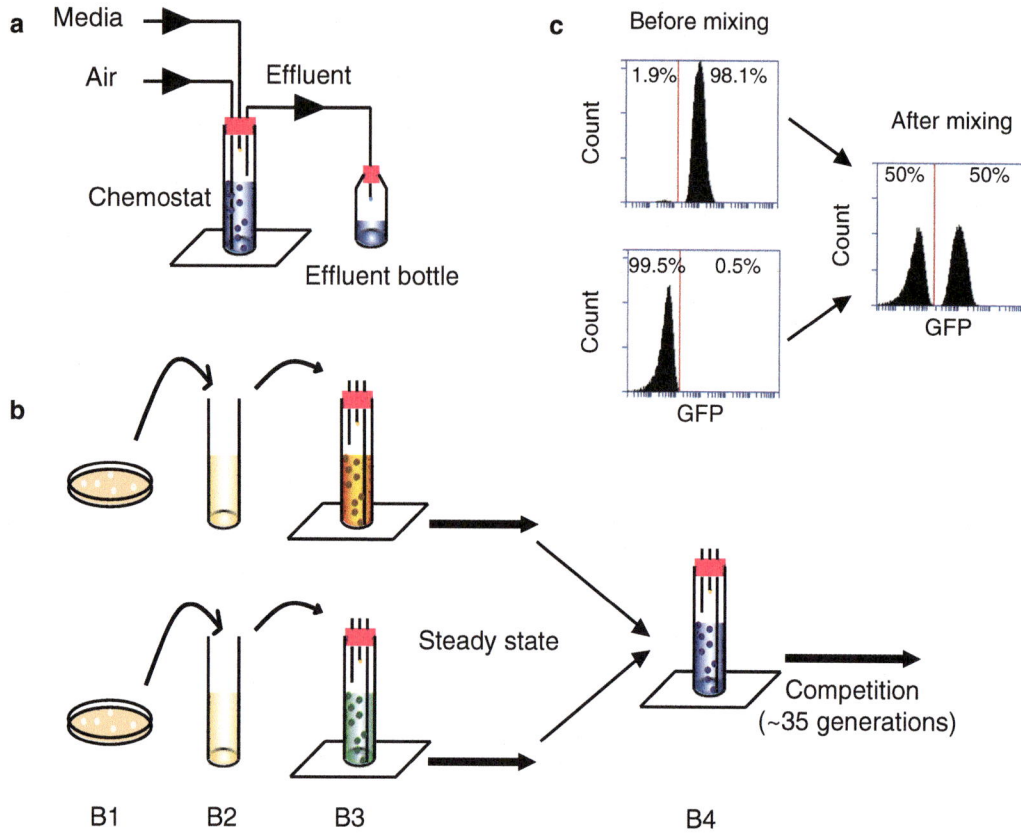

Fig. 1 FACS-based competitive assay. (**a**) Diagram of a ministat. The ministat chamber is a 50 ml glass tube stopped with a silicon cork (in *pink*). Air and media are delivered by two independent systems. Culture volume is determined by a third needle connected to an effluent bottle. Positive pressure ejects overflow into the effluent bottle. Samples can be collected passively by switching the effluent bottle with a collection tube. (**b**) Competitive assay in the chemostats. (**b1–b2**) Overnight cultures of both the clone and the reference strains from an isolated colony. (**b3**) Inoculation of the chemostat vessel with 0.1 ml of the overnight cultures. (**b4**) When the cultures reach steady-state, the two cultures can be mixed together. Samples will be collected twice daily for 35 generations. (**c**) FACS analysis of the samples. Prior to mixing, each vessel should contain either an unlabeled or a GFP-marked strain. After mixing, the ratio of both populations will be measured

2. Build your array of ministats as described [16] and prepare your media as appropriate for the selective conditions desired (Fig. 1a).

3. Turn the pump on to fill the ministats, and then turn it off when the media reaches the 20 ml mark.

4. Day 2: Sterilize the tops of the corks of the vessels with 95 % ethanol; wipe the cork using a Kimwipe.

5. Inoculate each chemostat vessel with 0.1 ml from one individual overnight culture using a syringe (Fig. 1b).

6. Thirty hours after inoculation, turn the media pump on to a dilution rate of 0.17 vol/h (*see* **Note 2**).

7. When the media starts to exit through the effluent line, turn off the air to adjust the culture volume to 20 ml by moving the sampling needle up or down (*see* **Note 3**). Turn the air on when done.

8. Once all the cultures reach 20 ml, empty the effluent bottles. Record the time and this will be your time 0.

9. Let the culture reach steady state (*see* **Note 4**).

Experimental evolution can operate for hundreds of generations. With a dilution rate of 0.17 vol/h, 200 generations will be reached in 35 days (*see* **Note 5**).

3.2 Competition Against GFP Marked Strains

The fitness of a strain of interest is generally measured against a matched reference strain to perform cross comparison. To discriminate between the two strains, the reference strain can be labeled with a fluorescent protein such as y*GFP* or *d-Tomato*, integrated at a neutral locus, such as *HO* [8] or *YEL014C* [17] (*see* **Note 6**).

1. Sterilize the tops of the chemostats with 95 % ethanol (*see* **Note 7**).

2. With a 10 ml syringe, collect 10 ml of steady state culture of the culture of interest. Do the same for the vessel containing the reference strain culture (*see* **Note 8**).

3. Unscrew the syringe from the needle and swap the syringes. Push each needle into the cork of the appropriate destination vessel, eject the culture into the vessels, remove the syringes, and dispose of them in an appropriate container.

4. Note the time of the mixing.

5. Wait 20 min for the volume to stabilize before collecting a sample.

6. Note the time and collect 1 ml passively by transferring the sampling corks into labeled 20 ml sterile sampling tubes.

7. While the tubes are filling, measure and record the effluent volume (Veff) that has collected in the effluent bottles.

8. Use the time elapsed and the Veff to calculate the dilution rate D ($D = $ Veff/(time elapsed $\times 20$)) and the number of generations elapsed Ge (Ge $= D \times 1.44 \times$ time elapsed) [4].

9. Wash effluent bottles and replace the sampling corks on the bottles.

10. Collect samples for analysis by cytometry for up to 35 generations to follow the relative abundance of the two strains over time. Two samples are generally performed on the day of the mixing, three the following day and one or two the third day to ensure that you will have enough points to fit a normal regression.

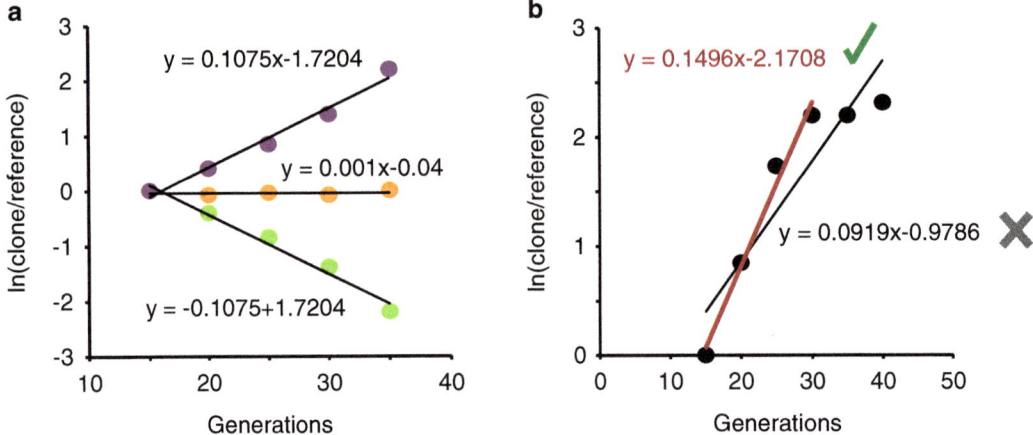

Fig. 2 Relative fitness. (**a**) The relative fitness of a strain is determined by linear regression of the natural log (ln) of the ratio of the strain over the control strain against the number of generations. A positive slope indicates an increased fitness of the clone compared to the control strain while a negative slope implies a decrease of fitness. A slope around 0 indicates a neutral or near-neutral fitness. (**b**) To accurately measure fitness, exclusion of the later time points might be performed

3.3 FACS Analysis

These instructions are specific for the C6 cytometer from BD Accuri, but could be modified for use with any other appropriate cytometer or FACS machine.

1. Perform a 1:4 dilution of your sample into water. You can either do the dilution in a 2 ml tube or in a 96-well plate.

2. Sonicate your sample to disperse mother and daughter cells and to break up cell clumps.

3. Count 50,000 cells at a medium speed on the cytometer.

4. Create a histogram plot to display the distribution of cell number (Count) and the GFP level (Channel FL1-A) (*see* **Note 9**).

5. Create vertical marker by clicking the cursor at the point along the *x*-axis to gate the histogram and separate the two subpopulations. BD CSampler Software will automatically display the percentage of events to the left and the right of the marker (Fig. 1c).

6. Run a cleaning cycle. When the cleaning cycle is finished, you can export your data for further analysis.

3.4 Analysis:
Selection Coefficient

1. Per generation selection coefficient is used as a measurement of fitness [4]. The selection coefficient is measured by linear regression of the natural log of the strain abundance ratio against the time in generations (Fig. 2a).

2. Before attempting to fit a linear model to calculate the selection coefficient, you should first determine if your data follow a linear model or if some points are out of the range. This happens generally for the last few time points as one of

the strains may have already reached fixation. Such points should be excluded from further analysis (Fig. 2b).

3.5 Whole Genome Sequencing

1. Whole genome sequencing of clones or populations is often used to determine the genetic changes associated with phenotypic changes such as fitness. Depending on the length of the reads, the capacity of the sequencer and the minimum coverage required to ensure quality results, you might want to sequence more than one library at a time. Using the dual index method you can multiplex up to 96 libraries together. 96 libraries can be processed in less than 3 h using the Nextera kit.

2. As of today, up to 4 libraries of individual clones can be sequenced on one lane of a MiSeq or 24 samples on one lane of a HiSeq for a 30× coverage minimum for each library with ~15 million reads of 150 bp. This level of coverage ensures that SNPs or CNVs can be accurately detected.

3.6 Tagmentation

Nextera technology uses in vitro transposition to create whole-genome libraries with a low input requirement (35 ng). This technique is a flexible and simple method capable of fragmenting and tagging DNA in a 5 min reaction, followed by a PCR [18] (Fig. 3a). Genomic DNA can be extracted from overnight cultures or frozen

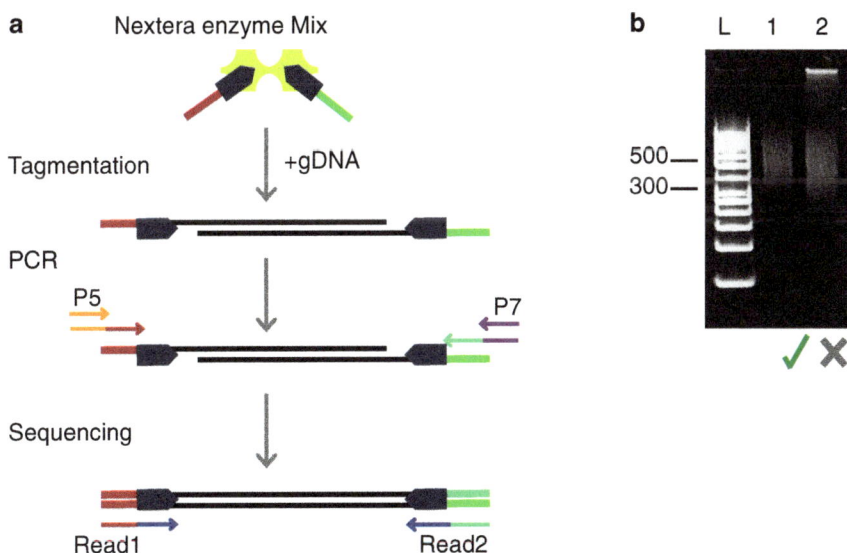

Fig. 3 Nextera library of yeast genomic DNA. (**a**) Generation of libraries by the Nextera method. Genomic DNA is fragmented and tagged with the Nextera Enzyme Mix, followed by a PCR reaction containing four primers to add compatible adaptors (*purple* and *orange*) to the sequencing library. The two sequencing primers correspond to the *red* and *blue arrows*. Figure modified from Adey et al. (2010). (**b**) Lane L: GeneRuler 50 bp DNA ladder. Lane 1: example of a high quality library. The size of the library ranges from 300 bp to less than 1000 bp. Lane 2: example of a low quality library. Large DNA fragments are present. Size selection should be performed on this library to ensure high quality sequencing results

pellets using the Smash-and-Grab method [19]. gDNA should be quantified using a fluorescence-based method such as Qubit.

1. Remove the 2× buffer (TD), the enzyme (TDE1), and genomic DNA from −20 °C and thaw on ice (*see* **Note 10**).

2. Ensure that the buffer and the enzyme are mixed by gently inverting the tubes five times, followed by a brief spin in a microcentrifuge.

3. In 0.5 ml PCR tubes mix 35 ng of genomic DNA, 25 μl of 2× buffer (TD), and 5 μl of enzyme TDE1 and bring to 50 μl total volume with ddH$_2$O (*see* **Note 11**).

4. Incubate in thermocycler at 55 °C for 8 min.

5. Hold at 4 °C.

3.7 Cleanup of Tagmented DNA

1. Thaw the RSB buffer at room temperature.

2. Add 180 μl of Zymo DNA binding buffer to each 50 μl tagmentation reaction.

3. Gently pipet up and down ten times.

4. Load the mixture into a Zymo-Spin Column and centrifuge at full speed (>10,000×*g*) for 30 s. Discard the flow through.

5. Wash twice with 300 μl of the DNA Wash Buffer.

6. Centrifuge at full speed for an additional 30 s to remove any remaining wash buffer.

7. Place column into a clean Eppendorf tube and pipet 25 μl of RSB onto each column.

8. Incubate at room temperature for 2 min.

9. Centrifuge at full speed for 30 s.

10. Eluted DNA can be stored at −20 °C.

3.8 PCR Amplification

1. Thaw the Nextera PCR Master Mix (NPM) and PCR Primer Cocktail (PPC) from the Nextera kit and the index primers (P7 (×12) and P5 (×5)) at room temperature. Plan primer pairs that are compatible for multiplexing and uniquely mark each sample.

2. Set up PCR in individual 200 μl PCR tubes, for easy removal from RT-PCR machine.

3. Mix 5 μl of the chosen index 2 (P7), 5 μl of the chosen index 1 (P5), 1 μl of SYBR green (final concentration $n = 0.25×$), 15 μl of the Nextera PCR master mix (NPM), 5 μl of PCR Primer Cocktail (PPC), and 20 μl of the tagmented DNA.

4. Perform PCR on RT-PCR machine using the following program: 72 °C for 3 min, 98 °C for 30 s, then 5–15 cycles of (98 °C for 10 s, 63 °C for 30 s, 72 °C for 30 s), followed by 3 min at 72 °C.

5. Remove the PCR tubes from the thermocycler when the absorbance starts to plateau and spin down using a microcentrifuge.

6. Tubes can be stored at 4 °C for a few days.

3.9 PCR Cleanup

1. Bring the beads to room temperature for 30 min before use and vortex the tube for 30 s.

2. Add 30 µl of the magnetic beads to each PCR reaction.

3. Mix well by gently pipetting up and down ten times.

4. Incubate at room temperature without shaking for 5 min.

5. Place the plate on a magnetic stand for 2 min or until the supernatant has cleared.

6. With the tube still on the stand, use a 20 µl tip to carefully remove the supernatant.

7. With the tubes still on the magnetic stand, wash the beads with freshly prepared 80 % ethanol, twice, as follows:

   (a) Add 200 µl of freshly prepared 80 % ethanol to each tube. Incubate on magnetic stand for 30 s until the supernatant appears clear.

   (b) Carefully remove and discard the supernatant.

8. With the tubes still on the magnetic stand, allow the beads to air-dry for 15 min.

9. Remove the tubes from the magnetic stand.

10. Add 32.5 µl of RSB to each tube, making sure to bring the beads into solution.

11. Mix well by pipetting up and down ten times, changing tips for each tube.

12. Incubate at room temperature for 2 min.

13. Place the tubes back on the magnetic stand for 2 min or until supernatant has cleared.

14. Label new tubes.

15. Taking care to not disturb the beads, while keeping the tubes on the magnetic stand, carefully transfer 30 µl of supernatant from each tube on the magnetic stand to a new labeled Eppendorf tube.

16. The libraries can be stored at 4 °C for a few days or at −20 °C indefinitely.

3.10 Quantify and Pool Libraries

We recommend measuring concentration using the Quant-iT dsDNA HS assay kit and a Qubit-iT fluorometer.

1. Set up your 0.5 ml tubes: you will need two tubes for the standards, and one tube per sample (*see* **Note 12**).

2. Dilute the Quant-iT reagent 1:200 in Quant-iT buffer.

3. Aliquot 190 μl of the diluted reagent in the two standard tubes and 197 μl in the tubes dedicated for the samples.

4. Add 10 μl of the standards in the standards tubes, and 3 μl of your libraries in the tubes dedicated for the samples.

5. Mix the solution by pipetting up and down and incubate the tubes for 2 min in the dark at room temperature.

6. Read tubes in Qubit fluorometer and note the concentration for each sample.

7. Save your individual libraries. Make sure they are clearly labeled.

8. If you plan on multiplexing your libraries, mix together equal ng from each library to make the pooled library.

3.11 Library QC

1. Set up 6 % acrylamide gel in the gel box. Add ~300 ml of 0.5× TBE.

2. Mix 5 μl of the library with 5 μl loading buffer.

3. Mix 5 μl ddH$_2$O with 5 μl loading buffer 5× and 0.5 μl 50 bp ladder.

4. Load the ladder and the libraries on the acrylamide gel.

5. Run at 160 V for ~45 min.

6. Incubate gel in ~50 ml of 0.5× TBE with 5 μl SYBR Gold on a shaker for 5 min.

7. Take a picture using a gel imaging system.

8. The Nextera method produces libraries with a broad range from 300 to 1000 bp. The ideal size of a Nextera library is around 500 bp (Fig. 3b) (*see* **Note 13**).

3.12 Sequencing of Your Library on a MiSeq: Sample Sheet

The sample sheet is the set of instructions that the machine uses to know how to sequence your sample.

1. Open the program called "Illumina Experiment Manager".

2. Select "create sample sheet".

3. Choose MiSeq, press "next".

4. Select "FASTQ Only" (*see* **Note 14**).

5. Fill in the field "Reagent Cartridge Barcode" using the barcode found on the side of the cartridge.

6. Fill in the information about the run—project name, experiment name, investigator name, and a description (*see* **Note 15**).

7. Fill out the information for your sample(s).

8. Remove the reagent cartridge from the freezer and let it thaw along with the tube of HT1 (hybridization buffer) (*see* **Note 16**).

3.13 Library Denaturation (Gloves Are Required for This Step)

1. Thaw your pooled library on ice along with your HT1 (*see* **Note 17**).

2. Prepare a fresh dilution of 0.2 N NaOH.

3. Dilute your library to 2 nM in EB buffer (*see* **Note 18**).

4. Mix 10 μl of your library at 2 nM with 10 μl of NaOH at 0.2 N.

5. Vortex briefly and centrifuge your tube for 1 min.

6. Add 980 μl of prechilled HT1 buffer to the tube containing the denatured DNA.

7. Dilute the denatured sample to 12 pM (*see* **Note 18**): 600 μl of the 20 pM denatured sample, 400 μl of the prechilled HT1 and 1.2 mM of NaOH.

8. Invert five times to mix the DNA solution.

9. Briefly centrifuge the DNA solution.

10. Place the library on ice until you are ready to load your library onto the MiSeq.

3.14 Load Sample onto Cartridge (Gloves Are Required for This Step)

1. Pierce the foil of the cartridge seal over "Load samples" with a clean pipette tip.

2. Pipette 600 μl of the sample onto reservoir (*see* **Note 19**).

3. Proceed directly to the MiSeq.

3.15 Start the Sequencing Run

1. Use plastic forceps and remove the flowcell by the base of the plastic cartridge from the storage buffer.

2. Rinse the flowcell with 2 ml Milli-Q water, making sure to remove the salt.

3. Using care dry the flowcell on a Kimwipe.

4. Clean the flowcell glass with a Kimwipe and 70 % EtOH to remove streaks and fingerprints.

5. Clean the flowcell stage in the flowcell compartment with a Kimwipe.

6. Place the flowcell onto the flowcell stage with the label facing upward.

7. Gently press down. You will hear a click.

8. Follow on-screen instructions (*see* **Note 20**).

9. Review the run parameters (Experiment name, analysis workflow, read length).

10. The MiSeq will perform a pre-run check.

11. The goal is to reach a cluster density of 750–1200k cluster/mm^2.

12. The duration of the sequencing depends on the number of cycles (i.e., read length): ~4 h for 2×36 bp run and up to ~39 h for a 2×250 bp cycle.

13. Always perform a post-run wash just after your run by following the instructions on the screen.

14. After the wash, leave the used flowcell, wash tray and wash bottle in the instrument until your next run.

3.16 Retrieve Your Data

Once the sequencing run is complete, store and back up your data before beginning analysis. The sequencing files (.fastq) will be in the Data/Intensities/BaseCalls/directory of your run folder. If the information about the multiplexing was present in the sample sheet, each file will correspond to an individual sample. The quality of the sequences can be assessed using quality control tools such as FastQC (http://www.bioinformatics.babraham.ac.uk/projects/fastqc/). The SNPs, INDELS, and CNVs calls can now be generated using various tools such as the SAMtools variants caller [20], mrCaNaVar [21], and SPLITREAD [22].

4 Notes

1. Overnight cultures can be started in YPD or another nutrient rich media and cells can be washed with water before inoculating, though it is preferable to use chemostat media for the overnight culture.

2. A dilution rate of 0.17 vol/h corresponds to between 5.75 and 6.5 rpm on the Watson and Marlow pump.

3. Before use of new chemostat vessels, fill the vessels with 20 ml of clean water and mark the glass to indicate the target volume.

4. The cultures are considered to have reached steady state when the cell count has been stable for 2–3 daily measurements (±5 %). Steady state is generally reached 3 days after the pump is turned on and before 25 generations, though some strains may take longer.

5. Evolution experiments can be ended when desired. However, experiments also can be terminated due to the appearance of clumping or wall growth or to contamination of the culture vessels by bacteria or fungi. The set-up as well as the samples must be frequently monitored and recorded.

6. The fitness of the reference strain versus a wild type unmarked strain needs to be tested before its usage to ensure no significant differences were introduced during strain construction.

7. It is good practice to save glycerol stocks and 2 ml of samples for DNA preparation before mixing in case you need to confirm strain genotype or other parameters such as plasmid copy number.

8. The percentage of mixing depends on the fitness estimated. If you do not know, a 50–50 % (vol/vol) mix is a good starting point. If you know your strain is going to be more fit than the reference you can lower its abundance to 20–80 % (vol/vol) to allow for a longer period of time over which informative samples can be collected. The volume of the ministats used in our laboratory is set at 20 ml, so a 50–50 mix will be 10 ml of each culture mixed together.

9. Using the initial pure cultures, set gates on the cytometer to discriminate the two populations (unmarked cells and cells expressing *GFP*). For each mixed sample, record the percentage of cells in each gate (Fig. 1c).

10. The working area and the pipettes should be cleaned with 10 % bleach to avoid cross-contamination.

11. This protocol has been adapted from the Nextera protocol (Illumina). Although 50 ng of genomic DNA is recommended by the official Nextera protocol, we recommend using 35 ng of DNA with 5 μl of the enzyme or 50 ng of DNA with 3 μl of the enzyme to ensure a good size selection.

12. The standards are stored at 4 °C and the buffer and the reagent at room temperature. Ensure all reagents are at room temperature before you begin. You will need 200 μl of working solution for each sample and the two standards.

13. Run each individual library and the pooled library on an acrylamide gel in order to confirm the size and quality of your libraries. The ideal range is around 500 bp. Larger or smaller libraries will not produce high quality reads.

14. This setting means that no on-machine analysis will be performed.

15. If you are not using the Nextera indices, you can at this stage select random sequence. The sample sheet can be edited later.

16. You can thaw the cartridge in a tub of water at room temperature. Be careful not to fill the tub over the line indicated on the side of the cartridge. Thawing in this way will take 1 h, after which you can then store the cartridge at 4 °C.

17. Do not start denaturing your libraries unless your cartridge is thawed.

18. Be sure to use large volumes to reduce pipetting errors. After the dilution you can also double check the concentration of the libraries using the Qubit-iT assay.

19. The Nextera sequencing primers are already loaded in the cartridge.

20. Do not leave the reagent chiller door open for extended periods of time.

Acknowledgments

Thanks to Emily Mitchell and Giang T. Ong for their protocols. This work was supported by grants R01 GM094306 and P41 GM103533 from the National Institute of General Medical Sciences from the National Institutes of Health, and National Science Foundation grant 1120425. MJD is a Rita Allen Foundation Scholar, and a Fellow in the Genetic Networks program at the Canadian Institute for Advanced Research.

References

1. Monod J (1950) La technique de culture continue, theorie et applications. Ann Inst Pasteur 79:390–410

2. Novick A, Szilard L (1950) Description of the chemostat. Science 112(2920):715–716

3. Skelly DA et al (2013) Integrative phenomics reveals insight into the structure of phenotypic diversity in budding yeast. Genome Res 23(9):1496–1504

4. Dykhuizen DE, Hartl DL (1983) Selection in chemostats. Microbiol Rev 47(2):150–168

5. Paquin C, Adams J (1983) Frequency of fixation of adaptive mutations is higher in evolving diploid than haploid yeast populations. Nature 302(5908):495–500

6. Dunham MJ et al (2002) Characteristic genome rearrangements in experimental evolution of Saccharomyces cerevisiae. Proc Natl Acad Sci U S A 99(25):16144–16149

7. Gresham D et al (2008) The repertoire and dynamics of evolutionary adaptations to controlled nutrient-limited environments in yeast. PLoS Genet 4(12), e1000303

8. Payen C et al (2014) The dynamics of diverse segmental amplifications in populations of Saccharomyces cerevisiae adapting to strong selection. G3 (Bethesda) 4(3):399–409

9. Gresham D et al (2010) Adaptation to diverse nitrogen-limited environments by deletion or extrachromosomal element formation of the GAP1 locus. Proc Natl Acad Sci U S A 107(43):18551–18556

10. Kvitek DJ, Sherlock G (2013) Whole genome, whole population sequencing reveals that loss of signaling networks is the major adaptive strategy in a constant environment. PLoS Genet 9(11), e1003972

11. Brown CJ, Todd KM, Rosenzweig RF (1998) Multiple duplications of yeast hexose transport genes in response to selection in a glucose-limited environment. Mol Biol Evol 15(8):931–942

12. Wenger JW et al (2011) Hunger artists: yeast adapted to carbon limitation show trade-offs under carbon sufficiency. PLoS Genet 7(8), e1002202

13. Hong J, Gresham D (2014) Molecular specificity, convergence and constraint shape adaptive evolution in nutrient-poor environments. PLoS Genet 10(1), e1004041

14. Adams J, Paquin C, Oeller PW, Lee LW (1985) Physiological characterization of adaptive clones in evolving populations of the yeast, Saccharomyces cerevisiae. Genetics 110(2):173–185

15. Zhang E, Ferenci T (1999) OmpF changes and the complexity of Escherichia coli adaptation to prolonged lactose limitation. FEMS Microbiol Lett 176(2):395–401

16. Miller AW, Befort C, Kerr EO, Dunham MJ (2013) Design and use of multiplexed chemostat arrays. J Vis Exp (72):e50262

17. Ziv N, Brandt NJ, Gresham D (2013) The use of chemostats in microbial systems biology. J Vis Exp (80):e50168

18. Adey A et al (2010) Rapid, low-input, low-bias construction of shotgun fragment libraries by high-density in vitro transposition. Genome Biol 11(12):R119

19. Hoffman CS, Winston F (1987) A ten-minute DNA preparation from yeast efficiently releases autonomous plasmids for transformation of Escherichia coli. Gene 57(2-3):267–272

20. Li H et al (2009) The sequence alignment/map format and SAMtools. Bioinformatics 25(16):2078–2079

21. Alkan C et al (2009) Personalized copy number and segmental duplication maps using next-generation sequencing. Nat Genet 41(10):1061–1067

22. Karakoc E et al (2011) Detection of structural variants and indels within exome data. Nat Methods 9(2):176–178

Chapter 21

Reconstruction and Analysis of the Evolution of Modular Transcriptional Regulatory Programs Using Arboretum

Sara A. Knaack, Dawn A. Thompson, and Sushmita Roy

Abstract

Comparative functional genomics aims to measure and compare genome-wide functional data such as transcriptomes, proteomes, and epigenomes across multiple species to study the conservation and divergence patterns of such quantitative measurements. However, computational methods to systematically compare these quantitative genomic profiles across multiple species are in their infancy. We developed Arboretum, a novel algorithm to identify modules of co-expressed genes and trace their evolutionary history across multiple species from a complex phylogeny. To interpret the results from Arboretum we developed several measures to examine the extent of conservation and divergence in modules and their relationship to species lifestyle, *cis*-regulatory elements, and gene duplication. We applied Arboretum to study the evolution of modular transcriptional regulatory programs controlling transcriptional response to different environmental stresses in the yeast Ascomycota phylogeny. We found that modules of similar patterns of expression captured the transcriptional responses to different stresses across species; however, the genes exhibiting these patterns were not the same. Divergence in module membership was associated with changes in lifestyle and specific clades and that gene duplication was a major factor contributing to the divergence of module membership.

Key words Arboretum, Gaussian mixture model (GMM), Phylogeny, Gene tree, Multi-clustering, Evolution, Transcriptional modules, Regulatory networks

1 Introduction

Comparative functional genomics aims to compare functional data such as transcript, protein, metabolite, transcription factor occupancy and histone modification levels across multiple species. With advances in genomics such studies are becoming routine in unicellular [1–5] and multicellular organisms [6–10]. However, computational approaches to compare these data across species are not well developed. For example, an important question in comparative studies of transcriptional regulation is to examine the extent of conservation and divergence of transcriptional modules, defined as sets of genes that are co-expressed across multiple conditions. Such modules represent genes that are often co-regulated and constitute

Frédéric Devaux (ed.), *Yeast Functional Genomics: Methods and Protocols*, Methods in Molecular Biology, vol. 1361,
DOI 10.1007/978-1-4939-3079-1_21, © Springer Science+Business Media New York 2016

entire or parts of a pathway. However, existing approaches to identify and compare modules across species have been limited to two [11] or three species [12, 13]. The challenge with comparing multispecies data from a complex phylogeny with large number of species is that often there is no one-to-one mapping of genes between species, because of gene duplication and losses. Such relationships are most naturally represented by a "gene tree" with leaf nodes representing genes in extant species, and internal nodes representing ancestral species. To address that challenge we developed Arboretum, a novel computational algorithm to systematically identify modules of co-expressed genes across species and study their evolution.

Two unique aspects of Arboretum are (1) the ability to reconstruct the hidden ancestral module membership, and (2) explicitly model gene trees that are not necessarily identical to the species tree because of gene duplication and loss events. The ability to handle gene duplication events, a major mechanism by which networks can rewire [14], allows us to work with large phylogenies with complex many-to-many relationships. The inferred ancestral module assignments can be used to trace the evolution of module memberships. In addition to the algorithm itself, we developed several metrics to examine evolutionary conservation and divergence of sets of genes across species.

We applied Arboretum to study patterns of conservation and divergence in yeast Ascomycota species under different stress conditions [15, 16]. We found that although the general patterns of expression were conserved across the different species, the genes exhibiting these changes were the same. In particular, we found several cases of genes that were orthologous in sequence but exhibited change in expression level. We found that gene duplication played a major role in this divergence of module membership. In this chapter, we describe the steps of applying Arboretum and downstream analyses to study the evolution of modular regulatory programs to new phylogenies.

2   Materials

Arboretum is a multi-species clustering approach and is based on a generative probabilistic model that simultaneously infers the modules across species, while exploiting the relatedness of species encoded in the tree topology. Arboretum is implemented in C++ and is publicly available as open source code at (http://pages.discovery.wisc.edu/~sroy/arboretum/). To compile and run Arboretum, the three requirements are a C++ compiler, adequate processor memory (<1GB) and disk space (a standard desktop machine is sufficient), and one third-party library called the GNU scientific library to be installed prior to compiling. The following paragraphs describe the key steps in installing and applying the

Arboretum approach to study evolutionary dynamics of transcriptional modules using an example dataset of five yeast species: *S. cerevisiae*, *C. glabrata*, *K. lactis*, *K. waltii*, and *C. albicans*. The expression data used here measure the transcriptional response of gene expression to a heat shock stress and were originally studied in the Arboretum publication [16].

3 Methods

3.1 Downloading and Installation of Arboretum

The box below shows the set of commands that are needed to have a running version of Arboretum. The steps are described for a Unix command line terminal. (1) Get the source code using the wget command. (2) Extract Arboretum code. (3) Change to the directory using *cd* and compile the code using make.

```
1. wget  http://pages.discovery.wisc.edu/~sroy/
   arboretum/gmm_crossspeciescluster.tgz
2. tar -xvfz gmm_crossspeciescluster.tgz
3. cd gmm_crossspeciescluster
4. make
5. cd./
```

3.2 Applying Arboretum to Cross-Species Data

The inputs to running Arboretum requires us to specify: (a) expression data and input module assignments, (b) species and gene trees, (c) location of results, (d) base species, (e) initialization of the transition probabilities. Example inputs are available from the website which can be obtained from the Arboretum website using:

```
wget      http://pages.discovery.wisc.edu/~sroy/
arboretum/example_inputs.tgz
   tar-xvfz example_inputs
```

Once all inputs and configuration files have been organized, an example usage of Arboretum is as below:

```
../gmm_crossspeciescluster/incAncClust
specorder_allclade.txt OGid_members.txt 5
species_prob_heat8spec.txt      cluster_conf.txt
rand result_dir learn Scer uniform 0.8
```

We now describe each of the arguments to the Arboretum program in detail below.

The first three arguments, specorder_allclade.txt, OGid_members.txt and species_prob_heat8spec.txt constitute the phylogenetic information, including the species and gene tree relationships, specified to Arboretum. Here the specorder_allclade.txt and OGid_members.txt files present the gene orthology information. The OGid_members.txt file specifies the sets of gene orthology relationships for the genes of each species. An example of the OGID file is available in example_inputs. Each OG is denoted by an ID, notated in the form OG#_#, which represents a gene tree with the first number in the ID, and the duplication level of the gene tree as the second number in the ID. In OGid_members.txt each line is associated with an orthogroup of genes, and begins with an OGID string, and is followed with the string of gene names from the respective species, each separated by a comma. The order of the genes from the species is in the order of the species as listed in specorder_allclade.txt. Where a specific species does not have a gene represented in a particular orthogroup "NONE" will appear instead of a gene name. Example lines from these files is as follows:

```
specorder_allclade.txt:
   Scer
   Cgla
   Kwal
   Klac
   Calb

OGid_members.txt:
OG16_1    YLL043W,CAGL0C03267g,Kwal55.20572,K
          LLA0E00550g,NONE
OG16_2    YLL043W,CAGL0C03267g,Kwal55.20572,K
          LLA0E00550g,NONE
OG16_3    YLL043W,CAGL0C03267g,Kwal33.15269,N
          ONE,NONE
OG16_4    YLL043W,CAGL0C03267g,Kwal33.15269,N
          ONE,NONE
OG16_5    YFL054C,CAGL0E03894g,Kwal55.20572,K
          LLA0E00550g,NONE
```

The species tree is used to define the phylogenetic relationships of the species and is specified by species_prob_heat8.txt file, which has the following format:

```
species_prob_heat8spec.txt:
```

#Child	LeftorRight	Parent
Scer	left	Anc4
Cgla	right	Anc4
Anc4	left	Anc9
Anc8	right	Anc9
Kwal	left	Anc8
Klac	right	Anc8
Anc9	left	Anc11
Calb	right	Anc11

The gene trees are expected to be directory named data/ TREES relative to the working directory of the issued Arboretum command. If a species and gene trees are not known, we recommend using resources where such data are available, e.g., in Paranoid [17] or Ensemble Compara (http://useast.ensembl. org/info/genome/compara/index.html) (*see* **Note 1**).

The next argument specifies the expression data and initial set of cluster assignments in the configuration file cluster_conf.txt, which is exemplified below:

```
cat cluster_conf.txt:
    Scer  heat8/Scer_clusterassign.txt  heat8/
Scer_expr.geneexp
    Calb  heat8/Calb_clusterassign.txt  heat8/
Calb_expr.geneexp
    Cgla  heat8/Cgla_clusterassign.txt  heat8/
Cgla_expr.geneexp
    Kwal  heat8/Kwal_clusterassign.txt  heat8/
Kwal_expr.geneexp
    Klac  heat8/Klac_clusterassign.txt  heat8/
Klac_expr.geneexp
```

The first column has the name of the species, the second the name of the input cluster assignment file for the species genes, and lastly the name of the expression data file for the expression measurements used for that species. The expression data themselves may need to be pre-processed as described in **Note 2**.

The input cluster assignments of individual species are needed to seed the Arboretum algorithm. These cluster assignments themselves are typically obtained by merging all the data across species into a single file and applying a k-means or a Gaussian mixture model algorithm [18] to cluster the merged data. In this case each row of the merged data represents the data values for those genes in a single ortholog set of genes across the represented species. The input cluster assignments for the ortholog set are used to assign initial cluster assignments to each of ortholog genes in each species. We emphasize that this is only an initial assignment for seeding Arboretum, and we do not advise interpreting such results as being more significant than that. This is because merging the data across species also requires us to deal with many-to-many mappings and in a flat merged clustering approach such complex mappings result in missing values (gene loss) or duplicated data (gene duplication) in the data rows.

The fifth argument is k, the number of modules. This is specified as user input. In **Note 3** we discuss possible ways to determine this argument. The sixth argument specifies how to initialize the initial set of module parameters and can be either "rand" or "none". This option will tell the program to either use the input clustering assignments of genes only to define cluster sizes (rand), or to use the input cluster assignments verbatim for initialization (none). For "rand" Arboretum randomly partitions the data using these size distributions. In **Note 4**, we discuss inspecting results for different random initializations of the Arboretum algorithm using this argument.

The seventh argument is the result_dir, a name for the output directory for the results.

The eighth argument defines the mode of the algorithm. This option will generally be set to "learn", which will run the clustering algorithm on the input expression data, or "generate" to invoke the generative model to produce data for simulation studies. The ninth argument is the name of a well-annotated base species (Scer in the example usage). This species is chosen to organize the output of Arboretum, especially the module assignments for the ancestral nodes.

The remaining two arguments relate more specifically to the functioning of the Arboretum multi-clustering algorithm. The tenth argument determines the mode of initializing the transition matrices for module membership of genes across species. These matrices will be of the size $k \times k$, where k is the number of modules specified in the analysis. There is one transition matrix for each non-root node in the tree and specifies the probability of a gene to maintain (diagonal entry) or switch its module membership (off-diagonal). There are two initialization methods for these matrices. If "uniform" is selected, the 11th argument is a real-valued number and used to initialize the diagonal elements of the transition matrices for all species. If "branch length" is used, the value of the 11th argument is interpreted as a file name with branch lengths and is used to initialize the transition matrices for each individual node. Typical settings for these two arguments are "uniform 0.8".

3.3 Arboretum Outputs and Their Interpretation

Once Arboretum has finished, outputs will be presented in the result_dir directory. These results include (1) inferred transition matrices, (2) inferred module assignments, for both ancestral and extant species.

There are several analyses that we perform to interpret the modules:

(1) Heatmaps to visualize patterns of expression in each module (Fig. 1a); (2) Global module conservation between extant species (Fig. 1b); (3) Patterns of divergence across a phylogeny; (4) Gene Ontology enrichment of modules; (5) Motifs of *cis-regulatory* elements; (6) Computation of various metrics to quantify the extent of conservation and divergence in modules. We present several of the most significant examples here.

Fig. 1 (continued) (**b**) Shown for each species pair are the 5 by 5 matrices of negative logarithms of Hypergeometric test *p*-value results. The diagonal elements, shown on a 0–40 color scale, in each matrix represent the comparison of the gene sets in different species with the same expression pattern (e.g., Module 1 of *S. cerevisiae* with Module 1 of *C. glabrata*). When the modules are highly conserved for gene content (or coherent) these matrices are highly diagonal. The off-diagonal entries of the matrix show the significance of overlap of modules of different expression patterns. These are generally lower than the diagonal entries and are shown on a scale of 0–5

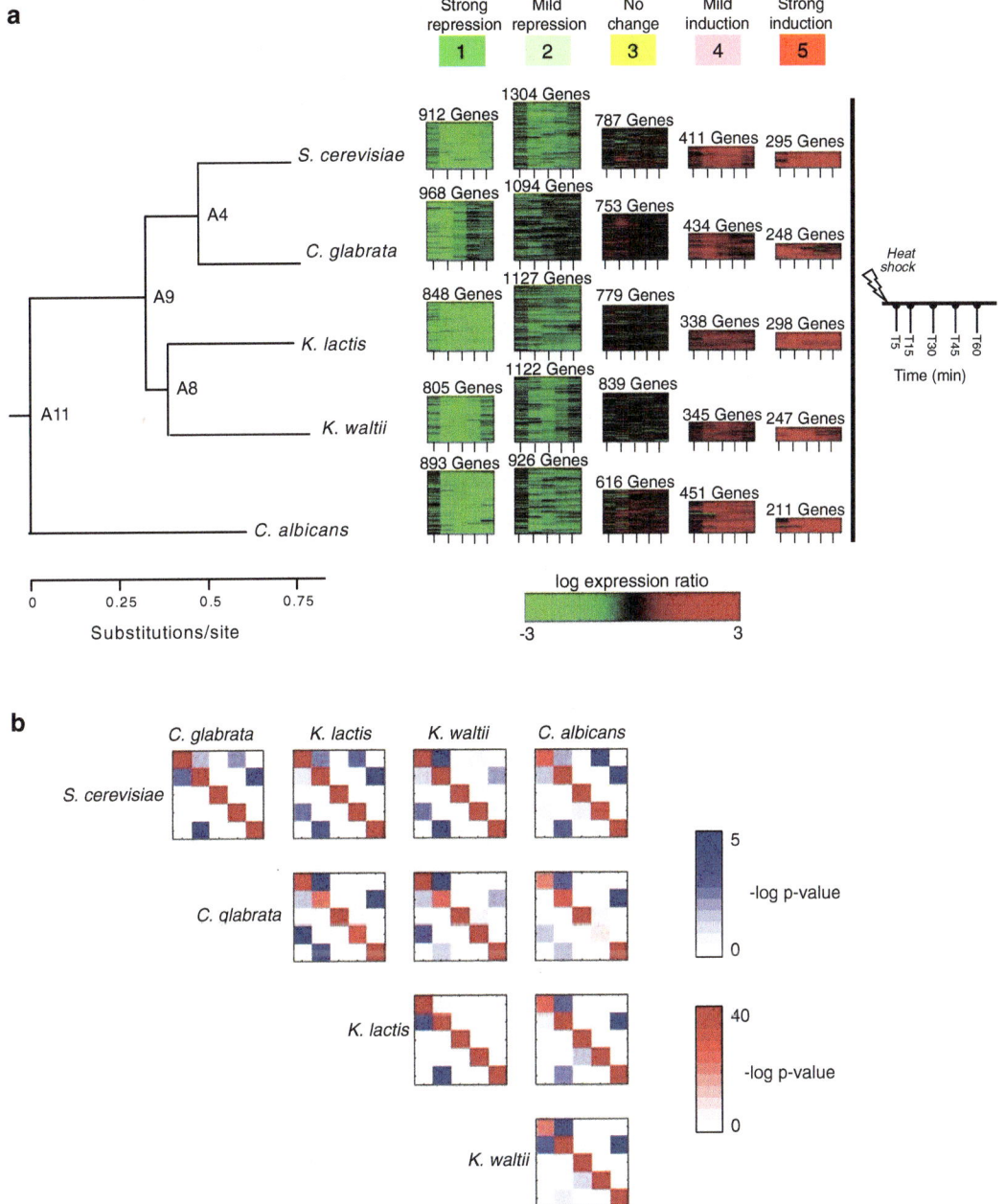

Fig. 1 Heatmap of expression modules and module divergence and conservation. (**a**) Shown are the expression module heatmaps for five modules in each of the five species inferred using Arboretum. On the *right hand side* are heatmap plots of the inferred co-expression modules. Here each *row* of heatmap plots are for the modules within each respective species. These modules, as labeled at the *top*, are associated with specific changes in expression, which are categorized and either repressive (*green*) or inductive (*red*). The center module (3) is associated with little change in expression in response the heat shock stimulus. The color scale of the individual heatmaps is also represented at the *bottom*. For each species module heatmap, the *horizontal axis* represents measurements at different time points after the heat shock stress was applied, represented at the *left* of this plot. The height of each heatmap represents the genes in each module and is reported on *top* of each module

Figure 1a presents the five modules inferred using Arboretum for the five-species example data set presented here. From these plots we see the pattern of expression of each module is highly similar for all five species. An important aspect of interpretation of plots like this is to assess the similarity of gene expression within each module, across each species, for example, are genes in module 1 of *S. cerevisiae* similar to genes in module 1 of *K. waltii*, noting that in both species, the expression patterns of the modules are quite conserved.

This brings us to a key computational measure developed for measuring the similarity of gene content between modules across species. For each pair of species we estimate the significance of gene orthology relationships between each pair of modules within those two species (Fig. 1b). These results are generated by using a Hyper-geometric test to assess the significance of overlap between all pairs of modules for these two species. For each pair of modules, one module from species A and another from species B, we obtain a *p*-value for the significance of the similarity of the module gene groups. We use the negative log of that *p*-value as a score for the similarity of those modules. Here in Fig. 1b, for each species pair among our species we have a 5 by 5 matrix of Hyper-geometric test *p*-value scores, represented here in heatmaps. The diagonal matrix elements, in red, illustrate that the set of modules are well conserved across species within each module. The off diagonal elements, in blue, show significant divergence relationships, where gene in one species have a given module membership, but their orthologs in another species are in a different module.

Figure 2 shows one such example of a set of ortholog gene groups, spanning all five species that diverge in their expression and module membership. These plots demonstrate a coherent change in expression for these ortholog gene groups from an induced state (red) in *S. cerevisiae* and *C. glabrata*, to a repressed state (green) in the three remaining species. This observation represents a phylogenetically coherent change in expression, and as such these sets of orthologous genes represent a interesting example of changes in gene regulation in evolution. Possible explanations for such changes include change in transcription factor binding in the different species, and DNA sequence features such as *cis*-regulatory elements, and other genomic markers and can subsequently be examined to substantiate such observations to provide insight to the mechanisms behind such changes in gene expression.

To assess biological meaning of our modules, we perform enrichment analysis of the modules. We use Gene ontology [19] terms to assess the significance of overlap of genes in each module with known annotated categories. For the base species, we usually have the GO annotation available. For the non-base species including the ancestral species we use the ontology terms mapped to the orthologs of the base species. We use the Hyper-geometric test

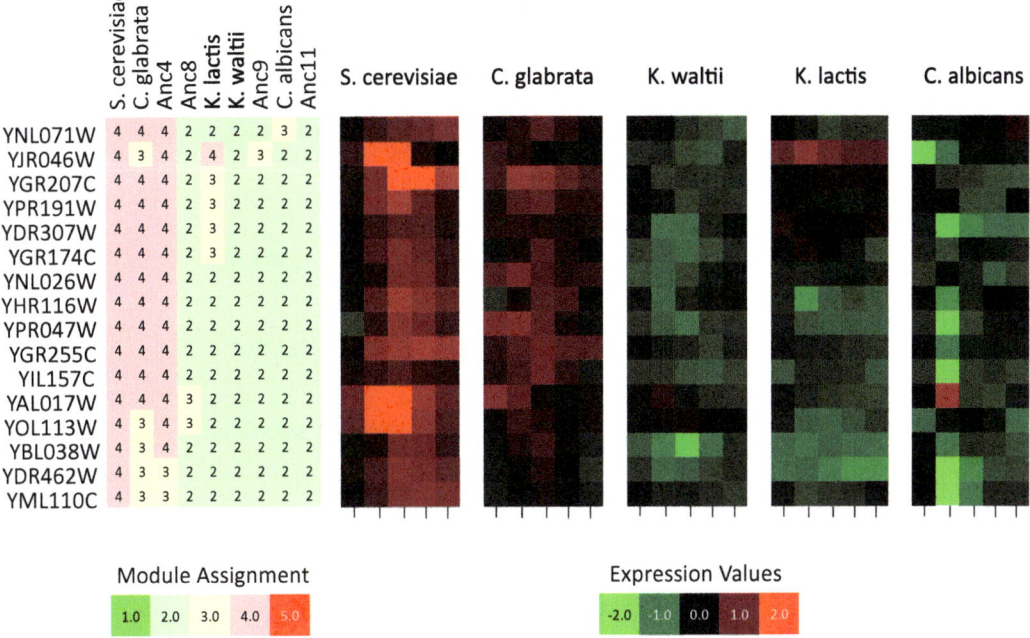

Fig. 2 Example of divergence of orthologous genes. Shown is a set of orthologous genes using *S. cerevisiae* systematic gene names that exhibit coordinated change in expression along the phylogeny. The *left-most panel* shows the Arboretum-inferred module membership for each genes in the five extant species, and the inferred membership for the ancestral species (Anc4, Anc8, and Anc11). Note that these memberships are generally for module 4 for *S. cerevisiae* and *C. galbrata*, and module 2 for the related ortholog genes in *K. waltii*, *K. lactis*, and *C. albicans*. The expression levels of these genes in each of the species are shown in the next set of five panels

with the Benjamini–Hochberg procedure for multiple hypothesis correction. Once GO enrichments for all species have been completed, we examine the patterns of enrichment across the species per module as well as per term. Figure 3 shows the GO enrichments for module 1 and it can be seen that there is significant conservation of processes as well. This presentation highlights cases where specific biological processes are associated with a specific module across all species (e.g., ncRNA processing). Such patterns demonstrate evolutionary conservation of biological function for these modules, which Fig. 1 shows to be highly conserved for co-expression and gene orthology. Figure 3 also shows examples of terms that are associated with a module in all species, but not necessarily the same module (e.g., chromatin organization), and this gives examples of processes that have likely diverged in their regulation between species. Similarly we assess the modules for motif enrichments. For each species, we used a motif collection generated by the Cladeoscape algorithm [20]. Once motifs have been identified across each module for each species, we visualize them in a similar way as the GO enrichment results.

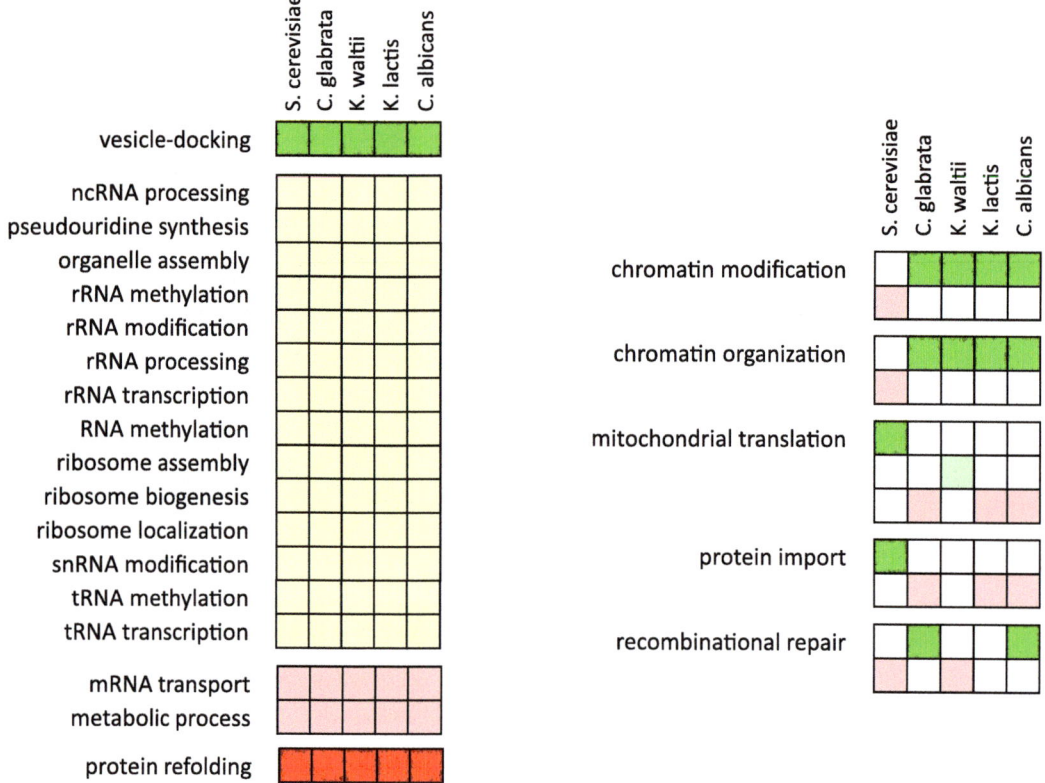

Fig. 3 Gene Ontology (GO) process enrichment of Arboretum modules. Shown is a summary of GO enrichments results from the inferred Arboretum modules in our five species example. Specifically, each module in each species is tested for association with the curated set of GO enrichment terms. The enrichment results shown here have been selected using two criteria: (1) a *q*-value threshold of <0.001 for each term and species-module association indicated, (2) the number of genes associated with each specific term within each individual species module, be at least a 0.2 fraction of the number of genes in that module. The terms on the *left* represent examples of these high confidence term-module associations that are conserved in the same module (denoted by the *color* of the box) across all five species. The terms on the *right* are cases where a given biological process shows significant association with at-most one module in each species, but a different module in each species. These are cases where orthologous genes associated with a given process have diverged in their module membership

In addition to assessing the orthology overlap significance, we have developed several measures to quantify extent of conservation and divergence between the modules. These include: (1) Ancestral module conservation index (AMCI), (2) Module expansion index (MEI), (3) Module contraction index (MCI), (4) Gene module divergence index (DI). All these indices are directly computed from the outputs of Arboretum. The AMCI metric uses the inferred transition matrix (one of the results of Arboretum), to assess the tendency of a species to maintain or switch its module assignment from its ancestor. This is a number between 0 and 1, and the closer it is to 1, the more likely is a species to maintain its ancestral module assignments, and the closer it is to 0 the more likely are modules going to diverge from its ancestral version.

The MCI and MEI metrics were designed to assess how modules diverge along a phylogenetic tree. The MCI metric measures the module contraction tendency, which happens when a module in a species loses genes compared to its ancestral version. The MEI metric measures the tendency of a module to gain genes in a child species compared to this ancestral version. Finally the gene module divergence index is a measure to assess the extent to which a gene switches its module assignments, and is defined as the number of switches divided by the total number of species the gene is present. Table 1 presents the calculated AMCI scores, the MCI scores for the most repressed and most induced expression modules (1 and 5, respectively), and the MEI scores for the same modules for this example analysis. Such results are given for each of the five extant species, and the (internal) ancestral species nodes (excluding the last common ancestor).

Table 1
Calculated metrics for examining Arboretum module dynamics

Species	AMCI	MCI in repressed module	MCI in induced module	MEI in repressed module	MEI in induced module
S. cerevisiae	0.77	0.14	0.10	0.18	0.22
C. glabrata	0.63	0.44	0.36	0.52	0.48
Anc4	0.74	0.22	0.11	0.31	0.35
K. waltii	0.68	0.31	0.15	0.43	0.52
K. lactis	0.71	0.30	0.12	0.36	0.58
Anc8	0.97	0.00	0.01	0.15	0.00
Anc9	0.77	0.15	0.28	0.16	0.05
C. albicans	0.64	0.33	0.24	0.42	0.47

This table highlights three measures: *AMCI* the ancestral module conservation index, *MCI* module conservation index, and *MEI* module expression index. Calculated AMCI values of each non-root species node are reported in the right most column of this table. The variation of this measure across the species of this example analysis is from 0.63 for *C. glabrata* to 0.93 for Anc8. MCI is the propensity for the same ancestral module to have shrunk. This measure is calculated for each individual species module, with results provided in columns 2 and 3 for the most repressed (module 1) and induced (module 5) modules respectively. MEI for a module measures the degree to which genes join that module from a different module in the immediate ancestor. Here the trend for the left most two data columns show the MEI results for the example analysis for the same two modules (1 and 5), which show widely varying degrees of expansion across species

4   Notes

1. Generating species and gene trees if not available: Often for a new phylogeny a species tree and gene trees may not be available. Such data can in general be obtained from public resources such as from ENSEMBL release 75 gene trees from ENSEMBL Compara resource. Our own gene trees were created using the SEMPHY [21] algorithm and the species tree were relearned for some subset of species by first doing a multiple sequence alignment of genes that are present in all species using MUSCLE [22] and learning trees using PAML [23].

2. Expression data pre-processing: The chief requirement for the input expression data is that the values can be modeled with Gaussian (normal) distributions. Typically the data used in Arboretum is gene expression data from a microarray or RNA-seq experiments. As microarray data is typically a comparative log ratio value of measurements from an experimental sample and a reference or background standard, such data is typically usable as is. In some cases, there might be missing values in the microarray. We recommend throwing away genes with >50 % missing data. For the remaining genes with <50 % missing, values can be interpolated using the mean calculated from the non-missing samples of that gene.

   For RNA-seq data, per-gene fragments per kilobase of transcript per million mapped reads (FPKM) expression values can be first Laplace-transformed, and then log-two-transformed. These expression data would finally then be row zero-meaned for Arboretum (rows corresponding to genes). To be explicit, the data for each species in the phylogeny being studied will have an independent input expression data file. When needed it is suggested to apply a row zero-mean transformation to data values across multiple measurements of a single species to facilitate the ability of the algorithm to model the variation in the data.

3. Determining the number of modules: An important question in applying Arboretum is to specify the number of modules, k. We do this in two ways: (a) use penalized maximum likelihood scores where the penalty is proportional to the number of parameters, which grows with the number of modules, (b) Use enrichment of GO and motifs to see if the clusters represent meaningful partitions. Manual inspection of the patterns is also helpful to determining the number of modules. Figure 4 shows the average penalized log likelihood plot for a range of module number settings, k, and for multiple randomized input cluster assignment. An un-penalized and penalized log likelihood score is generated by each instance of an Arboretum run. The un-penalized log likelihood is log of the inferred probability of

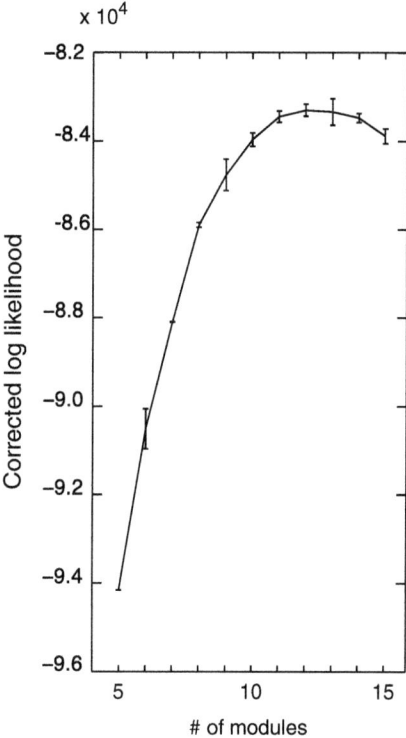

Fig. 4 Using penalized log likelihood scores to determine the upper bound number of modules in an Arboretum analysis. This figure shows the penalized likelihood scores determine from results for the same example data used here for $k=5$–15. Each data point in these plots represents the average scores for each setting of k, where the error bar represent the variation across 5 multiple random initializations of the input module assignments. The values for corrected log likelihood scores turn over at around $k=12$, which is indicated as the maximum number of modules to consider for this given input set of data

the algorithmic model to describe the final cluster assignments of a clustering run. The penalized log likelihood corrects for the complexity of the model for the number of genes and species, and modules involved. We see that the penalized log likelihood suggests that a $k>12$ is not going to be useful to describe the data. The attentive reader may wonder why the results in Figs. 1 and 2 are not for $k=12$ but $k=5$. We use this as an upper bound and use a combination of enriched terms and manual inspection to determine the final number of modules. The final determination for the optimal number of modules is subject to manual inspection of the expression patterns in the inferred modules, the ortholog coherence of modules across species, and finally the significance (including number and uniqueness) of GO enrichment terms for the results generated with varying values of k, hence $k=5$ was chosen in our example.

4. Stability of modules: Arboretum, like many clustering algorithms, is prone to local minima, which might result in a suboptimal solution. We recommend running Arboretum multiple times on the same data and inspecting the expression coherence of the modules to assess the reproducibility of the results and that they do not represent any aberrant patterns. This can be done by running Arboretum multiple times each time keeping the sixth argument set to "rand", or setting it to "none" but using a different initial module assignment set each time. In practice we have seen that Arboretum produces very stable results.

Acknowledgements

S.R. is supported in part by a NSF ABI CAREER award (DBI-1350677). S.K. is supported by an NLM training grant to the Computation and Informatics in Biology and Medicine Training Program (NLM5T15LM007359). This work was also supported by NIH grant 2R01CA119176-01 and a SPARC grant from the Broad Institute.

References

1. Jensen LJ, Jensen TS, de Lichtenberg U, Brunak S, Bork P (2006) Co-evolution of transcriptional and post-translational cell-cycle regulation. Nature 443:594–597

2. Gasch AP (2007) Comparative genomics of the environmental stress response in ascomycete fungi. Yeast (Chichester, England) 24:961–976

3. Wohlbach DJ, Thompson DAA, Gasch AP, Regev A (2009) From elements to modules: regulatory evolution in Ascomycota fungi. Curr Opin Genet Dev 19:571–578

4. Romero IG, Ruvinsky I, Gilad Y (2012) Comparative studies of gene expression and the evolution of gene regulation. Nat Rev Genet 13:505–516

5. Thompson DAA, Regev A (2009) Fungal regulatory evolution: cis and trans in the balance. FEBS Lett 583:3959–3965

6. Brawand D et al (2011) The evolution of gene expression levels in mammalian organs. Nature 478:343–348

7. Schmidt D et al (2010) Five-vertebrate ChIP-seq reveals the evolutionary dynamics of transcription factor binding. Science 328:1036–1040

8. Xiao S et al (2012) Comparative epigenomic annotation of regulatory DNA. Mol Cell 149:1381–1392

9. Barbosa-Morais NL et al (2012) The evolutionary landscape of alternative splicing in vertebrate species. Science 338:1587–1593

10. Merkin J, Russell C, Chen P, Burge CB (2012) Evolutionary dynamics of gene and isoform regulation in mammalian tissues. Science 338:1593–1599

11. Tanay A, Regev A, Shamir R (2005) Conservation and evolvability in regulatory networks: the evolution of ribosomal regulation in yeast. Proc Natl Acad Sci U S A 102:7203–7208

12. Waltman P et al (2010) Multi-species integrative biclustering. Genome Biol 11:R96+

13. Kuo D et al (2010) Evolutionary divergence in the fungal response to fluconazole revealed by soft clustering. Genome Biol 11:R77

14. Hittinger CT, Carroll SB (2007) Gene duplication and the adaptive evolution of a classic genetic switch. Nature 449:677–681

15. Thompson DA et al (2013) Evolutionary principles of modular gene regulation in yeasts. eLife 2, e00603. doi:10.7554/eLife.00603

16. Roy S et al (2013) Arboretum: reconstruction and analysis of the evolutionary history of condition-specific transcriptional modules. Genome Res 23(6):1039–1050. doi:10.1101/gr.146233.112

17. O'Brien KP, Remm M, Sonnhammer ELL (2005) Inparanoid: a comprehensive database of eukaryotic orthologs. Nucleic Acids Res 33:D476–D480

18. Hastie T, Tibshirani R, Friedman JH (2003) The elements of statistical learning. Springer, New York

19. Ashburner M et al (2000) Gene ontology: tool for the unification of biology. The Gene Ontology Consortium. Nat Genet 25:25–29

20. Habib N, Wapinski I, Margalit H, Regev A, Friedman N (2012) A functional selection model explains evolutionary robustness despite plasticity in regulatory networks. Mol Syst Biol 8:619

21. Wapinski I, Pfeffer A, Friedman N, Regev A (2007) Automatic genome-wide reconstruction of phylogenetic gene trees. Bioinformatics 23:i549–i558

22. Edgar RC (2004) MUSCLE: a multiple sequence alignment method with reduced time and space complexity. BMC Bioinformatics 5:113

23. Yang Z (2007) PAML 4: phylogenetic analysis by maximum likelihood. Mol Biol Evol 24:1586–1591

Chapter 22

Predicting Gene and Genomic Regulation in *Saccharomyces cerevisiae*, using the YEASTRACT Database: A Step-by-Step Guided Analysis

Miguel C. Teixeira, Pedro T. Monteiro, and Isabel Sá-Correia

Abstract

Transcriptional regulation is one of the key steps in the control of gene expression, with huge impact on the survival, adaptation, and fitness of all organisms. However, it is becoming increasingly clear that transcriptional regulation is far more complex than initially foreseen. In model organisms such as the yeast *Saccharomyces cerevisiae* evidence has been piling up showing that the expression of each gene can be controlled by several transcription factors, in the close dependency of the environmental conditions. Furthermore, transcription factors work in intricate networks, being themselves regulated at the transcriptional, post-transcriptional, and post-translational levels, working in cooperation or antagonism in the promoters of their target genes.

In this chapter, a step-by-step guide using the YEASTRACT database is provided, for the prediction and ranking of the transcription factors required for the regulation of the expression a single gene and of a genome-wide response. These analyses are illustrated with the regulation of the *PDR18* gene and of the transcriptome-wide changes induced upon exposure to the herbicide 2,4-Dichlorophenoxyacetic acid (2,4-D), respectively. The newest potentialities of this information system are explored, and the various results obtained in the dependency of the querying criteria are discussed in terms of the knowledge gathered on the biological responses considered as case studies.

Key words YEASTRACT, Transcriptional regulation prediction, Transcription regulatory networks, Transcriptomic data analysis, *Saccharomyces cerevisiae*

1 Introduction

Transcriptional regulation is a key first step in the control of gene expression. Consequently, the ability to understand, model and control transcriptional regulation is crucial for the understanding of all living processes, with implications in Health and Life Sciences and also in the application of living systems in Industrial and Environmental Biotechnology.

The YEASTRACT database (http://yeastract.com) was developed almost a decade ago [1] as a tool for the analysis and prediction

Frédéric Devaux (ed.), *Yeast Functional Genomics: Methods and Protocols*, Methods in Molecular Biology, vol. 1361,
DOI 10.1007/978-1-4939-3079-1_22, © Springer Science+Business Media New York 2016

of transcription regulation in the model eukaryote *Saccharomyces cerevisiae*. Given the huge amount of data generated for this model organism since its genome sequence was released [2], budding yeast has placed itself as possibly the best system to understand basic transcriptional regulation mechanisms in eukaryotes. Indeed, the YEASTRACT database has grown throughout the years to include more than 200,000 regulatory associations established between transcription factors (TF) and target genes (TG) in yeast [3–5]. This data can be used, through the queries made available in the database, to predict transcriptional regulation at the gene and genome-wide level. However, two facts have to be considered to allow a reliable prediction. First, the regulatory associations described in YEASTRACT are based on numerous experimental procedures, scattered throughout more than 1000 research papers. Some of the regulatory associations therein were identified through DNA-binding detection methods, and are thus considered direct regulatory associations, whereas some are based on expression evidence, often resulting from the comparison of the expression of target genes in the absence, presence or overexpression of a given transcription factor. In this case, the regulatory associations are considered indirect. Second, the action of a transcription factor upon a target gene is highly dependent on the environmental condition, and genetic background and so this fact should be taken into consideration when predicting the regulatory control of genomic expression.

In the newest release of the database [5], queries were included to overcome these limitations and thus reinforcing the confidence in the obtained results. Furthermore, new tools were provided to allow a finer ranking of the transcription factors predicted to regulate a given gene or genomic transcriptional change, including the recently developed TFRank algorithm [6], that prioritizes the relevant transcription factors by walking through the yeast regulatory network.

In this chapter, the use of the YEASTRACT database is described, step by step, and illustrated with two case studies. The first, is the analysis of the transcriptional control of the *PDR18* gene. *PDR18* encodes a multidrug efflux pump of the ATP-Binding Cassette superfamily, identified as conferring resistance to the herbicide 2,4-Dichlorophenoxyacetic acid (2,4-D), the metal ions such as Zn^{2+}, Mn^{2+}, Cu^{2+}, Cd^{2+} [7], and ethanol [8]. The biological role of Pdr18 was proposed to be the incorporation of ergosterol in the yeast plasma membrane, this physiological trait contributing to the MDR phenotype [7]. The second example is based on the transcriptome-wide response to 2,4-D dataset [9], which is used herein to show the potential of YEASTRACT to analyze genome-wide data. This second case study was selected given the relatively high degree of understanding of the mechanisms underlying the yeast response and adaptation to this herbicide [9–16], including some knowledge of the transcription factors involved in this adaptive response, such as the transcription factors Pdr1 and Pdr3 [13], the major regulators of multidrug resistance in yeast, and Msn2 and Msn4 [11], that control the environmental stress response.

2   Methods

2.1   Predicting the Regulation of Individual Genes (see Note 1)

1. Open the "Search for TFs (Transcription Factors)" query in the YEASTRACT website (http://yeastract.com).

2. Insert the gene name in the "Regulated Genes" box, in this case "*PDR18*" being used as an example.

3. Not changing any of the default parameters, press the "Search" button. The resulting table includes all the transcription factors that are demonstrated to directly or indirectly control Pdr18 and the references that provide evidence for each regulatory association.

4. Since this tool does not offer visualization options, alternatively open the "Rank by TF" query.

5. Insert "*PDR18*" in the "Target ORF/Genes" box and tick the "Check for all TFs" checkbox, to consider all transcription factors described in YEASTRACT.

6. Not changing any of the default parameters, press the "Search" button. The resulting table includes all the transcription factors that are demonstrated to directly or indirectly control Pdr18.

7. To visualize the network composed of the *PDR18* regulators, press the "Visualize Network" button, in the bottom of the results table. The resulting image is illustrated in Fig. 1a.

8. Since it is unlikely that 19 different transcription factors regulate a single gene simultaneously, the prediction of the actual *PDR18* regulators can be filtered by changing some of the default parameters in **step 3** (*see* **Note 2**). Using the default parameters in the "Rank by TF" query, insert "*PDR18*" in the "Target ORF/Genes" box and tick the "Check for all TFs" box. Filtering can now be done at three levels.

9. At the level of the experimental evidence underlying the definition of regulatory associations: In the "Regulations Filter" box select "Only DNA binding evidence". Press the "Search" button. The resulting table includes all the transcription factors that are demonstrated to directly control *PDR18* expression. Press the "Visualize Network" button, in the bottom of the results table. The obtained image is depicted in Fig. 1b.

10. At the level of the environmental conditions in which the regulatory associations were observed: In the "Filter Documented Regulations by environmental condition" box select "Unstressed log-phase growth (control)" group. Press the

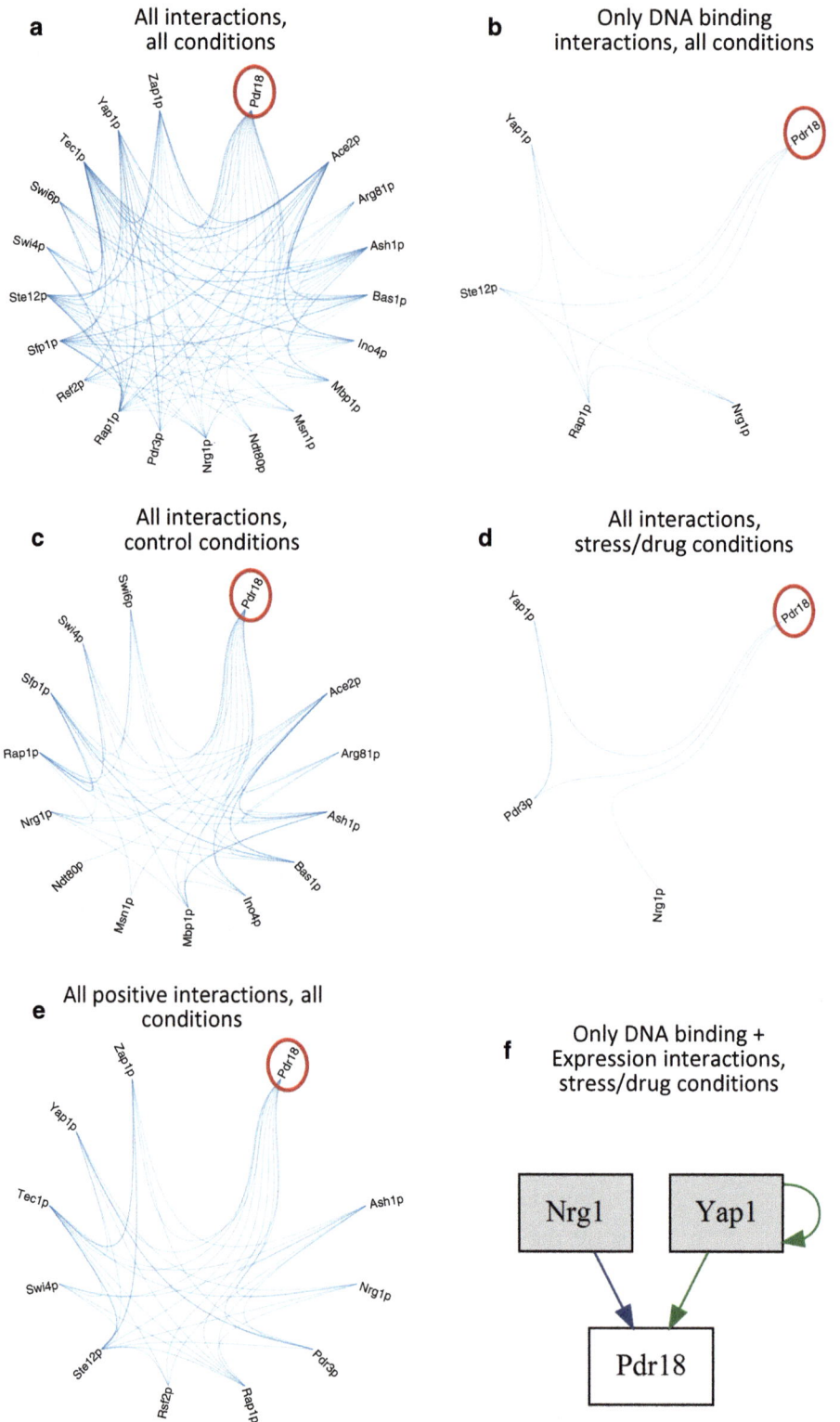

Fig. 1 Six different regulatory networks controlling the expression of the *PDR18* gene, highlighted in a *red circle*, obtained using the YEASTRACT "Rank by TF" tool. (**a**) Considering all known transcriptional associations registered in any experimental setup and any environmental condition. (**b**) Considering all known transcriptional associations registered in DNA-binding experiments, in any environmental condition. (**c**) Considering all

"Search" button. The resulting table includes all the transcription factors that are demonstrated to directly or indirectly control *PDR18* expression in yeast cells in the absence of stress. Press the "Visualize Network" button, in the bottom of the results table. The obtained image is displayed in Fig. 1c. In the "Filter Documented Regulations by environmental condition" box select "Stress" group. Press the "Search" button. The resulting table includes all the transcription factors that are demonstrated to directly or indirectly control *PDR18* expression in yeast cells under stress conditions. Press the "Visualize Network" button, in the bottom of the results table. The obtained image is depicted in Fig. 1d (*see* **Note 3**).

11. At the level of the mode of action of the transcription factor, which for each regulatory association may be working as an activator or repressor of transcription: In the "Regulations Filter" box, deselect the checkbox "TF acting as inhibitor". Press the "Search" button. The resulting table includes all the transcription factors that are demonstrated to directly or indirectly control *PDR18* expression in yeast cells, but only considering expression evidence cases in which the transcription factors are acting as activators of *PDR18* expression. Press the "Visualize Network" button, in the bottom of the results table, the obtained results are illustrated in Fig. 1e.

12. Being very strict in terms of the filtering approach, it is further possible to select only "TF acting as activator", for which there is "DNA-binding and expression evidence", detected in yeast cells under "Stress", induced by "Drugs/chemical stress exposure". Pressing the "Visualize Network" button, in the bottom of the results table, the obtained results are illustrated in Fig. 1f. Interestingly, this very small network of transcription factors was exactly the one identified to be required for the full activation of *PDR18* expression in yeast cells exposed to the herbicide 2,4-D [7] (*see* **Note 4**).

13. In a complementary approach, the user may search for the regulators of a given gene, based on the occurrence of TF binding sites in its promoter region (*see* **Note 5**). To do that, open the "Search for TFs" query, insert "*PDR18*" in the "Regulated Genes" box, and select "Potential" in the "Regulations filter" box. Tick the "Image" checkbox, and press the "Search" but-

---

Fig. 1 (continued) known transcriptional associations registered in any experimental setup in yeast cells which were not exposed to stress. (**d**) Considering all known transcriptional associations registered in any experimental setup in yeast cells exposed to stress in general or to stress induced by drugs or other chemical agents. (**e**) Considering all known transcriptional associations in which the TF acts as an activator of *PDR18* expression, registered in any experimental setup and any environmental condition. (**f**) Considering all known transcriptional associations registered in both expression and DNA-binding experiments in yeast cells exposed to stress in general or to stress induced by drugs or other chemical agents

ton. The result is provided with a figure illustrating the distribution of the TF binding sites in the promoter region of the gene under analysis (*see* **Note 6**), plus a Table including the exact TF binding sequences considered and the exact position and strand within the promoter region (Fig. 2).

2.2 Predicting the Regulators and Regulatory Network Underlying a Transcriptomic Response

2.2.1 Based on Documented Regulatory Interactions

1. Open the "Rank by TF" query in the YEASTRACT website (http://yeastract.com).

2. Insert the user's list of genes in the "Regulated Genes" box. As a case study, we use the list of 526 gene upregulated more than twofold in *S. cerevisiae* cells exposed to 2,4-D stress [9].

3. Not changing any of the default parameters, tick the "Check for all Tfs" checkbox and press the "Search" button. The resulting table includes all the *S. cerevisiae* transcription factors, since all of them are demonstrated to directly or indirectly control at least one of the 2,4-D-upregulated genes. In an attempt to highlight the most relevant transcription factors in the regu-

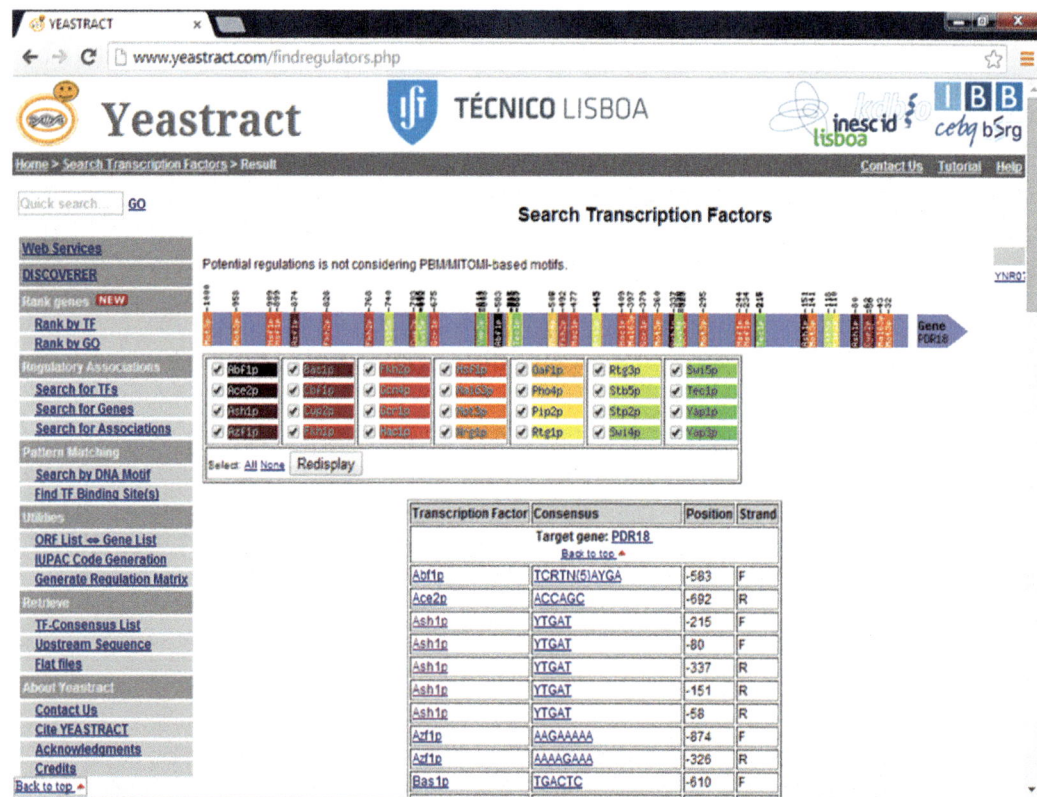

Fig. 2 Output of a Search for TFs query focused on the potential regulators of the *PDR18* gene. Graphic display of the transcription factor-binding sites found to occur in the promoter regions of the *PDR18* gene. The *table below* contains the list of all transcription factors identified as potential regulators of *PDR18*, followed by an indication of the exact TF binding site nucleotide sequence and position found in the *PDR18* promoter

lation in a given response, three ranking methods are offered by the YEASTRACT database.

4. Ranking based on the percentage (%) of regulated genes (*see* **Note 7**). The top nine TFs found to regulate the majority of the 2,4-D-induced genes, using these criteria, are depicted in Table 1, columns 1–6 (*see* **Note 8**).

5. Ranking based on the enrichment of the TF targets in the dataset provided by the user, when compared with the genome, as measured by a *p*-value (*see* **Note 9**). The top nine TFs found to regulate the majority of the 2,4-D-induced genes, using these criteria, are displayed in Table 1, columns 7–12 (*see* **Note 8**).

6. Ranking based in the TFRank method [6], which achieves the prioritization of regulators by computing a relevance measure reflecting their contribution within the whole genomic regulatory network (*see* **Note 10**). The top nine TFs found to regulate the majority of the 2,4-D-induced genes, using these criteria, are depicted in Table 1, columns 13–18 (*see* **Note 8**). For a visualization of the obtained network, select only the top TFs and press the "Visualize Network" button, in the bottom of the results table. The obtained results are illustrated in Fig. 3 (*see* **Note 11**).

7. Whatever ranking method is selected, the user may still chose to restrict the search to consider only the TF-target gene associations based on DNA-binding experiments, considered as direct regulatory associations. To do so, select "Only DNA binding evidence" in the "Regulations Filter" box and press the "Search" button. The top nine TFs found to regulate the majority of the 2,4-D-induced genes, using these criteria, are represented in Table 1, columns 3–6, 10–12, and 16–18 (*see* **Note 8**).

8. Whatever ranking method and the supporting experimental evidence is selected, the user may still choose to restrict the search to consider only interactions found to occur in specific environmental conditions. In the "Filter Documented Regulations by environmental condition" box select the "Stress" group. Press the "Search" button. The top nine TFs found to regulate the majority of the 2,4-D-induced genes, using these criteria, are represented in Table 1, columns 2, 5, 8, 11, 14, 17. If within the "Stress" group, the subgroup "Drug/Chemical stress exposure" is further selected (*see* **Note 12**), the resulting top nine TFs are depicted in Table 1, columns 3, 6, 9, 12, 15, 18 (*see* **Note 8**).

Table 1

Top nine TFs identified, using the YEASTRACT tools, as the key regulators of the transcriptome-wide responses to 2,4-D, in the dependency of the indicated parameters

Top ranking TFs according to % of regulated genes						Top ranking TFs according to target gene enrichment						Top ranking TFs according to the TFRank algorithm					
Considering all TF-TG associations			Considering only DNA-binding evidence			Considering all TF-TG associations			Considering only DNA-binding evidence			Considering all TF-TG associations			Considering only DNA-binding evidence		
All cond.	Stress	Drugs	All cond.	Stress	Drugs	All cond.	Stress	Drugs	All cond.	Stress	Drugs	All cond.	Stress	Drugs	All cond.	Stress	Drugs
Ace2	Msn2	Arr1	Spt23	Spt23	Yap1	Hsf1	Msn2	Arr1	Hot1	Spt23	Cad1	Ace2	Msn2	Arr1	Spt23	Spt23	Yap1
Sfp1	Msn4	Yap1	Ste12	Msn2	Pdr1	Msn2	Msn4	Pdr1	Spt23	Sko1	Yap5	Msn4	Msn4	Yap1	Yrm1	Msn2	Pdr1
Msn2	Yap1	Rpn4	Msn2	Cin5	Yap5	Msn4	Rlm1	Rpn4	Yap6	Yap6	Aro80	Sfp1	Cin5	Rpn4	Ste12	Cin5	Rpn4
Tec1	Cin5	Pdr1	Cin5	Yap1	Cad1	Sok2	Gis1	Pdr1	Sko1	Hot1	Met31	Tec1	Yap1	Pdr1	Cin5	Yap1	Pdr3
Ste12	Spt23	Pdr3	Yrm1	Sko1	Yap6	Hot1	Spt23	Pdr3	Adr1	Skn7	Yap1	Yap1	Pdr1	Pdr3	Abf1	Sko1	Yap5
Bas1	Pdr1	Mig3	Abf1	Yap6	Cin5	Ste12	Cin5	Spf1	Skn7	Yap6	Yap1	Cin5	Msn2	Arr1	Msn2	Yap6	Cin5
Sok2	Arr1	Aft1	Sok2	Skn7	Pdr3	Sfp1	Arr1	Crz1	Tos8	Smp1	Cin5	Msn2	Gcn4	Aft1	Rap1	Pdr1	Cad1
Gcn4	Rlm1	Met4	Rap1	Pdr1	Msn2	Phd1	Msn1	Cad1	Arr1	Cin5	Arr1	Sok2	Rpn4	Gis1	Adr1	Stb5	Swi4
Msn4	Rpn4	Gis1	Yap6	Stb5	Msn4	Mga2	Zap1	Yap1	Smp1	Cad1	Pdr3	Cst6	Rlm1	Met4	Yap1	Skn7	Msn2

In yellow, red, green, or blue are highlighted the transcription factors involved in general stress response, oxidative stress response, multidrug resistance, and amino acid limitation response, respectively. *TF* transcription factors, *TG* target genes, *All cond.* considering the transcriptional associations observed irrespectively of the environmental conditions, *Stress* considering the transcriptional associations observed in yeast cells under stress conditions, *Drugs* considering the transcriptional associations observed in yeast cells exposed to drugs or other chemical stress inducers

Transcriptional Regulatory Network

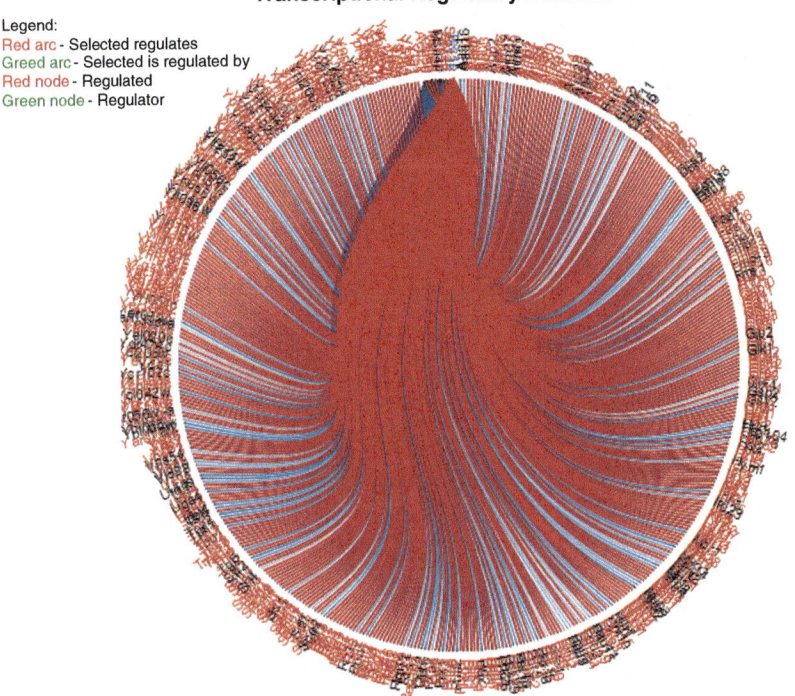

Legend:
Red arc - Selected regulates
Greed arc - Selected is regulated by
Red node - Regulated
Green node - Regulator

Fig. 3 Regulatory network comprising the top nine TFs regulating the 2,4-D responsive genes, according to the TFRank method, considering all regulatory associations and all environmental conditions. The *red arcs* represent the links between all of the 526 2,4-D-induced genes and the selected TFs. The *blue arcs* underneath represent all the connections between the 2,4-D-induced genes and the nine TFs considered

2.2.2 Based on the Identification of Overrepresented Sequences in the Promoter Regions of the Co-regulated Genes

1. Open the "DISCOVERER" section in the YEASTRACT website.

2. Select the "MUSA" algorithm [17].

3. Insert the user's list of genes in the "ORFs/Gene Name List" box. Again as a case study, we use the list of 526 gene upregulated more than twofold in *S. cerevisiae* cells exposed to 2,4-D stress [9].

4. Fill in the "Recipient's mail" and "Job description" boxes and, not changing any of the default parameters (*see* **Note 13**), press the "Run" button.

5. The top results are depicted in Fig. 4, which shows the nucleotide sequences found to be the most overrepresented in the promoter regions of the 2,4-D-induced genes. Results are provided as Position Weight Matrices (PWMs).

6. To check for the existence of TFs with a binding site similar to the identified overrepresented nucleotide sequences select one of them and press the "Match" button. The results for the third and fourth nucleotide sequences are depicted in Fig. 3,

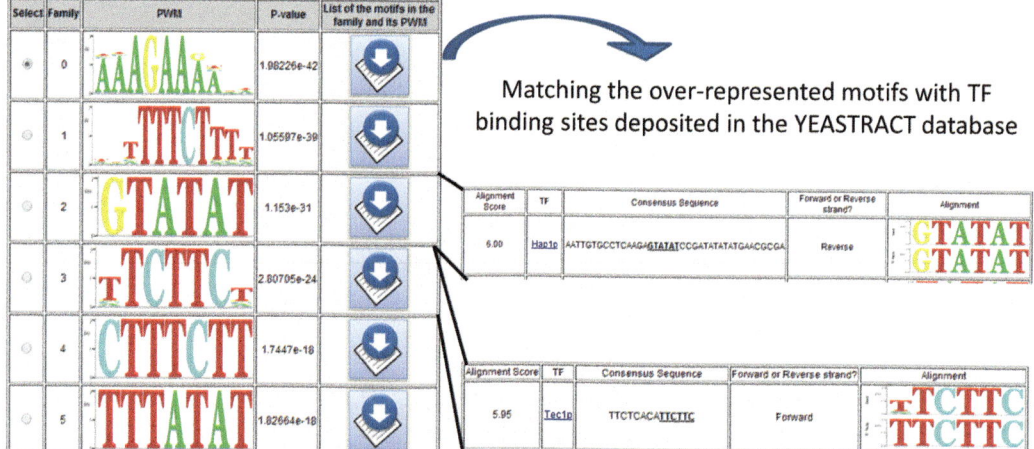

Fig. 4 Sample results showing the MUSA motif finder output in YEASTRACT-DISCOVERER presenting a PWM for each motif family found to be overrepresented in the promoter regions of the 2,4-D induced genes. The match output on the right was obtained by querying the database for TF binding sites that matched the selected PWM

showing perfect matches with the recognition sites of the transcription factors Hap1, involved in the complex regulation of gene expression in response to levels of heme and oxygen, and Tec1, a regulator of filamentation-related genes (*see* **Note 14**).

3 Notes

1. A similar approach can be used to predict which may be the target genes of a specific transcription factor, using the "Search for Genes" query in the YEASTRACT database.

2. The authors do not recommend any of the filtering approaches in detriment of the others. However, it appears reasonable to say that a conservative approach, were some filtering of the results is carried out, may spare the user of unnecessary experiments, which may result from testing all possible TFs.

3. The same result would be obtained if besides selecting the "Stress" group, the subgroup "Drug/Chemical stress exposure" is also selected.

4. Figure 1 highlights the fact that, depending on the filtering approach used, the prediction of the transcription factor network that underlies the regulation of a given gene may vary quite significantly. In the *PDR18* case the number of TFs can vary from 19 (Fig. 1a) to 2 (Fig. 1f). The prediction that the two TFs indicated in Fig. 1f, Nrg1 and Yap1, are the regulators of PDR18 appears to be the most reliable since it was verified to occur through expression analysis under 2,4-D stress [7], but also through DNA-binding experiments [18, 19].

However, the remaining TFs cannot be fully discarded, including Pdr3, found to control *PDR18* expression under 2,4-D stress [7] (Fig. 1d), or Rap1 [20] and Ste12 [21], found to interact with the *PDR18* promoter region (Fig. 1b).

5. Based on the occurrence of TF binding sites in the 6000 yeast promoters, more than 375,000 potential regulatory interactions are predicted, which appears to be much more than expected to really occur. Given the more uncertain nature of this predictive method, care should be taken when interpreting these results.

6. The promoter region is always considered to be the 1000 bp immediately preceding the START codon.

7. The result obtained using this ranking method is biased by the fact that some TFs are known to control the expression of hundreds of target genes, while others regulate only a few genes. These regulators with a low impact on the genomic regulation, but that may be crucial in specific transcriptional responses will never show up using this ranking method.

8. Table 1 highlights the fact that, depending on the used filtering approach, the prediction of the transcription factors that are the most relevant in the control of a transcriptional response may vary quite significantly. More specifically, the table indicates the top nine ranked TFs predicted to underlie the 2,4-D transcriptome-wide response in yeast. Interestingly, the TFs that were previously demonstrated to regulate the expression of specific genes in response to 2,4-D were uncovered in this analysis. These include the multidrug resistance TFs Pdr1 and Pdr3 [13] and the general stress responsive transcription factors Msn2 and Msn4 [11]. Identified TFs which were not previously linked to the 2,4-D response, but which make sense with the known effects of the exposure to this herbicide in yeast are also highlighted. For example, Yap1 and Skn7 are key regulators of the oxidative stress response in yeast and 2,4-D was demonstrated to act as a pro-oxidant molecule [15]. Also Gcn4, Met4, Aro80, and Met31 are involved in the response to amino acid limitation and 2,4-D has been shown to lead to a severe depletion of amino acids in yeast cells [14]. Many of the non-highlighted TFs regulate phenomena that had not been previously linked to 2,4-D induced stress pointing out to interesting paths to obtained a more in-depth understanding of the mechanisms of action of this herbicide.

9. TF score is given by a p-value denoting the overrepresentation of regulations of the given TF targeting genes in the list of interest relative to the regulations of that TF targeting genes in the whole YEASTRACT database. The *p*-value further denotes the probability that the TF regulates at least the number of genes found to be regulated in the list of interest if we were to

sample a set of genes of the same size as the list of interest from all the genes in the YEASTRACT database. This probability is modeled by a hypergeometric distribution and the p-value is finally subject to a Bonferroni correction for multiple testing.

10. Advantages of the TFRank algorithm include its ability to consider multiple levels of regulation and interactions between transcription factors in an integrated, rather than isolated-per-TF, network analysis perspective [6].

11. Figure 3 illustrates the degree of complexity that can be found while predicting the transcription regulatory network that underlies a transcriptomics data analysis, even when considering the top ranked transcription factors.

12. Given that 2,4-D is a chemical stress inducer, this set of selected environmental conditions appears to be the one that suits best the analysis of this case study. For other datasets, additional groups and subgroups are offered in the "Filter Documented Regulations by environmental condition" box.

13. For details on the meaning and use of MUSA parameters check the YEASTRACT-DISCOVERER Tutorial (http://yeastract.com/discoverer/tutorial.php).

14. Figure 4 highlights two TF binding sites, those recognized by Tec1 and Hap1, as being the most overrepresented in the promoter of the 2,4-D-induced genes. No previous implications of these TFs in the 2,4-D response or resistance mechanisms have, to our knowledge been highlighted. As Tec1 is a key regulator of invasive or pseudohyphal growth and Hap1 controls the response to heme and oxygen levels, it would be interesting to look further into the interaction between these two phenomena and the mechanisms of the toxicological action of, and adaptive reaction to, the herbicide 2,4-D.

Acknowledgements

Deep recognition is extended to all present and former YEASTRACT team members that over the years have contributed to the retrieval and curation of transcriptional regulatory data (Sandra Tenreiro, Nuno P Mira, Alexandra R Fernandes, Artur Lourenço, Sandra C dos Santos, Tânia R Cabrito, Catarina Costa, Joana Guerreiro, Margarida Palma, Marta Alenquer) and to the generation of computational tools (Ana T Freitas, Arlindo L Oliveira, Nuno Mendes, Dário Abdulrehman, Pooja Jain, Alexandre P Francisco, Sara C Madeira, Alexandra M Carvalho, Hélio Pais, Joana P Gonçalves, Ricardo S Aires, Sofia D'Orey) that make YEASTRACT such a powerful database. We are also grateful to colleagues and friends from the international Yeast and Systems Biology Communities for their encouragement and suggestions.

Funding for the development and maintenance of the YEASTRACT database was partially and indirectly obtained from Fundação para a Ciência e a Tecnologia (FCT) through PhD and postdoctoral scholarships to the members of the YEASTRACT teams. Funding received by iBB-Institute for Bioengineering and Biosciences(UID/BIO/04565/2013) from the Portuguese Foundation for Science and Technology (FCT) is acknolwedged.

References

1. Teixeira MC, Monteiro P, Jain P, Tenreiro S, Fernandes AR, Mira NP, Alenquer M, Freitas AT, Oliveira AL, Sá-Correia I (2006) The YEASTRACT database: a tool for the analysis of transcription regulatory associations in Saccharomyces cerevisiae. Nucleic Acids Res 34:D446–D451

2. Goffeau A, Barrell BG, Bussey H, Davis RW, Dujon B, Feldmann H, Galibert F, Hoheisel JD, Jacq C, Johnston M, Louis EJ, Mewes HW, Murakami Y, Philippsen P, Tettelin H, Oliver SG (1996) Life with 6000 genes. Science 274(546):63–67

3. Abdulrehman D, Monteiro PT, Teixeira MC, Mira NP, Lourenco AB, Dos Santos SC, Cabrito TR, Francisco AP, Madeira SC, Aires RS, Oliveira AL, Sa-Correia I, Freitas AT (2011) YEASTRACT: providing a programmatic access to curated transcriptional regulatory associations in Saccharomyces cerevisiae through a web services interface. Nucleic Acids Res 39:D136–D140

4. Monteiro PT, Mendes ND, Teixeira MC, D'Orey S, Tenreiro S, Mira NP, Pais H, Francisco AP, Carvalho AM, Lourenco AB, Sá-Correia I, Oliveira AL, Freitas AT (2008) YEASTRACT-DISCOVERER: new tools to improve the analysis of transcriptional regulatory associations in Saccharomyces cerevisiae. Nucleic Acids Res 36:D132–D136

5. Teixeira MC, Monteiro PT, Guerreiro JF, Goncalves JP, Mira NP, Dos Santos SC, Cabrito TR, Palma M, Costa C, Francisco AP, Madeira SC, Oliveira AL, Freitas AT, Sa-Correia I (2014) The YEASTRACT database: an upgraded information system for the analysis of gene and genomic transcription regulation in Saccharomyces cerevisiae. Nucleic Acids Res 42:D161–D166

6. Goncalves JP, Francisco AP, Mira NP, Teixeira MC, Sá-Correia I, Oliveira AL, Madeira SC (2011) TFRank: network-based prioritization of regulatory associations underlying transcriptional responses. Bioinformatics 27:3149–3157

7. Cabrito TR, Teixeira MC, Singh A, Prasad R, Sá-Correia I (2011) The yeast ABC transporter Pdr18 (ORF YNR070w) controls plasma membrane sterol composition, playing a role in multidrug resistance. Biochem J 440:195–202

8. Teixeira MC, Godinho CP, Cabrito TR, Mira NP, Sá-Correia I (2012) Increased expression of the yeast multidrug resistance ABC transporter Pdr18 leads to increased ethanol tolerance and ethanol production in high gravity alcoholic fermentation. Microb Cell Fact 11:98

9. Teixeira MC, Fernandes AR, Mira NP, Becker JD, Sá-Correia I (2006) Early transcriptional response of Saccharomyces cerevisiae to stress imposed by the herbicide 2,4-dichlorophenoxyacetic acid. FEMS Yeast Res 6:230–248

10. Fernandes AR, Durão PJ, Santos PM, Sá-Correia I (2003) Activation and significance of vacuolar H+-ATPase in Saccharomyces cerevisiae adaptation and resistance to the herbicide 2,4-dichlorophenoxyacetic acid. Biochem Biophys Res Commun 312:1317–1324

11. Simões T, Teixeira MC, Fernandes AR, Sá-Correia I (2003) Adaptation of Saccharomyces cerevisiae to the herbicide 2,4-dichlorophenoxyacetic acid, mediated by Msn2p- and Msn4p-regulated genes: important role of SPI1. Appl Environ Microbiol 69:4019–4028

12. Teixeira MC, Duque P, Sá-Correia I (2007) Environmental genomics: mechanistic insights into toxicity of and resistance to the herbicide 2,4-D. Trends Biotechnol 25:363–370

13. Teixeira MC, Sá-Correia I (2002) Saccharomyces cerevisiae resistance to chlorinated phenoxyacetic acid herbicides involves Pdr1p-mediated transcriptional activation of TPO1 and PDR5 genes. Biochem Biophys Res Commun 292:530–537

14. Teixeira MC, Santos PM, Fernandes AR, Sá-Correia I (2005) A proteome analysis of the yeast response to the herbicide

2,4-dichlorophenoxyacetic acid. Proteomics 5:1889–1901

15. Teixeira MC, Telo JP, Duarte NF, Sá-Correia I (2004) The herbicide 2,4-dichlorophenoxyacetic acid induces the generation of free-radicals and associated oxidative stress responses in yeast. Biochem Biophys Res Commun 324:1101–1107

16. Viegas CA, Cabral MG, Teixeira MC, Neumann G, Heipieper HJ, Sá-Correia I (2005) Yeast adaptation to 2,4-dichlorophenoxyacetic acid involves increased membrane fatty acid saturation degree and decreased OLE1 transcription. Biochem Biophys Res Commun 330: 271–278

17. Mendes ND, Casimiro AC, Santos PM, Sa-Correia I, Oliveira AL, Freitas AT (2006) MUSA: a parameter free algorithm for the identification of biologically significant motifs. Bioinformatics 22:2996–3002

18. Lee TI, Rinaldi NJ, Robert F, Odom DT, Bar-Joseph Z, Gerber GK, Hannett NM, Harbison CT, Thompson CM, Simon I, Zeitlinger J, Jennings EG, Murray HL, Gordon DB, Ren B, Wyrick JJ, Tagne JB, Volkert TL, Fraenkel E, Gifford DK, Young RA (2002) Transcriptional regulatory networks in Saccharomyces cerevisiae. Science 298:799–804

19. Salin H, Fardeau V, Piccini E, Lelandais G, Tanty V, Lemoine S, Jacq C, Devaux F (2008) Structure and properties of transcriptional networks driving selenite stress response in yeasts. BMC Genomics 9:333

20. Lieb JD, Liu X, Botstein D, Brown PO (2001) Promoter-specific binding of Rap1 revealed by genome-wide maps of protein-DNA association. Nat Genet 28:327–334

21. Zheng W, Zhao H, Mancera E, Steinmetz LM, Snyder M (2010) Genetic analysis of variation in transcription factor binding in yeast. Nature 464:1187–1191

INDEX

Frédéric Devaux (ed.), *Yeast Functional Genomics: Methods and Protocols*, Methods in Molecular Biology, vol. 1361,
DOI 10.1007/978-1-4939-3079-1, © Springer Science+Business Media New York 2016

Printed by Printforce, the Netherlands